"*Proving God* is an ambitious project that aims to bring the theistic science and mystical theology of Emanuel Swedenborg (1688-1772) into the 21st Century where it belongs. Edward F. Sylvia achieves this goal from necessity: In order to make sense of the astonishing implications of contemporary quantum physics, an intellectually defensible model for spiritual-natural causality is an essential next step. In Sylvia's judgment, this model has been waiting in the wings of science and religion for 250 years, for the science of ultimates to ask the proper questions.

That time is now, and *Proving God* matches seminal problems of matter and energy with Swedenborg's powerful ideas on being and becoming, with astonishing results. What emerges from this scientific treatment of divine metaphysics is a comprehensible world view in which spirit and matter are now finally and undeniably one. This is an important book."

> – Rev. Dr. Reuben P. Bell, Saco, Maine, USA
> President, Swedenborg Scientific Association
> Director, Samara Center for Practical Spirituality

"Sylvia follows the original route into metaphysics—straight through the heart of physics—and directly engages 'the elephant in the living room,' the utterly improbable symbiosis of complexity and coherence that is characteristic of both our universe and our minds. His argument that the universe and our minds are profoundly akin is carefully framed, and his conclusion that the coherence is an effect of love and the complexity an effect of wisdom is far removed from simplistic romanticism. In his able overview of Swedenborgian theology, he presents a complex and coherent view of spiritual reality that not only welcomes partnership with present-day physics but urges it to continue its pursuit of deeper understanding."

> – Rev. George F. Dole, Ph.D., Professor Emeritus,
> Swedenborg School of Religion
> Author of *A Book about Us: The Bible and Stages of our Lives*
> and *Freedom and Evil: A Pilgrim's Guide to Hell*

"Good work – original cutting-edge thinking around Swedenborg. This book is much-needed."

> – Rev. Carla Friedrich, San Diego, CA

"With breathtaking scope across the disciplines of natural science, Ed Sylvia brilliantly engages the metaphysics of Swedenborg's theosophy in a penetrating interdisciplinary contribution to the ongoing conundrums of 'who are we?', 'why are we here?' and 'where is it all going?'. This is a tour de force offering an engrossing conversation on every page."

> – Rev. Dr. James Lawrence, Berkeley, California
> Dean of the Swedenborgian House of Studies at
> Pacific School of Religion
> of the Graduate Theological Union

"This book presents a fresh new integration between the physics of the natural and the psychology of the spiritual. Grounded in the unique writings of Emanuel Swedenborg, this modern synthesis shows how the divine human Love in God is the spiritual substance that forms the scaffolding upon which the material world is built as an external covering. The quality of our struggle for happiness now and the ultimate success of our immortal life are determined by how we incorporate this Divine Human Love into our individual personality, either by loving goodness through rational principles or by inverting Love and pursuing selfishness through false principles. People interested in the rational resolution between science and religion will appreciate this lucid new account.

Sylvia presents a rational account of how Love is the essential or primary substance that gives rise to both the mental world and the physical universe. Divine Love substance is the mental or spiritual reality that establishes the existence and order of physical reality and nature. All objects are forms of love substance. Look into any physical object deeply enough and you will see the same mental reality of love substance determining the object's properties and utility, which act as the cause of its existence. To remove Love's purpose or teleology from something is to remove the object's etiology and existence. Without Love there is no existence of any kind. Love's substance exhibits itself outwardly as visible or measurable forms in infinite variety, and causes these forms to unite or combine into more complex clusters that become objects with specific properties and utility. All objects so formed by the purposes of God's Love, have the inherent propensity to unite in larger and more complex units or systems, and ultimately, in the universe as one integrated system.

Once you begin reading this fascinating book you'll want to come back to it again and again. It has a smooth topical flow that lifts the veil of ignorance concerning the most basic issues in life that people want to know. It is profound, with surprising and delightful insights on every page. This is a completely honest book. The content hangs together as a coherent and consistent whole. It succeeds in its claim to deliver a 'grand unified theory of everything' that 'can come only from revelation.' It presents in modern terms the unique revelation in Swedenborg's Writings. I intend to use *Proving God* as a text in my theistic psychology course."

> – Dr. Leon James, Professor of Psychology,
> University of Hawaii
> author of the Web-published *Theistic Psychology* Series
> His online books and articles are at
> http://e-swedenborg.com/tp/articles.htm

"This is one of the most important works to be published on Emanuel Swedenborg's Writings. You'd think Swedenborg was at a modern seminar on quantum physics. The author has Swedenborg talking the talk, using virtually the same vocabulary as the top scientists. And Swedenborg has the drop on everybody—the great physicists of our time are completely outclassed. Has Swedenborg come back? The author really takes the physicists to task and does it on their own terms, in their own language. Very convincing."

> – Stephen Koke,
> former editor, Blue Planet Press, Nevada City, CA

"Updating a once-famous engineer, scientist, philosopher, and theologian from the age of enlightenment, Sylvia revives a perspective of reality that uses the concept of a universal singularity to unify the duality controversy over mind versus matter. *Proving God* displays an amazing range of collected study relating Emanuel Swedenborg's theories and experiences in the 18th century to current sciences from archeology and anthropology to neurology and quantum mechanics. This awesome undertaking is an interdisciplinary study well worth the time and deep meditation required for grasping its profound concepts of universal reality."

> – Oliver R. Odhner & Rachel D. Odhner
> authors of *Mentics, A Mind Modeling Method*

PROVING GOD
Swedenborg's Remarkable Quest for the Quantum Fingerprints of Love
by Edward F. Sylvia, M. T. S.

ISBN: 978-0-9702527-1-5

STAIRCASE
PRESS

Staircase Press
P.O. Box 83 • Troy, IL 62294
www.staircasepress.com

Cataloging-in-Publication Data:

Sylvia, Edward F.
Proving God : Swedenborg's remarkable quest for the quantum
fingerprints of love / Edward F. Sylvia. – 1st ed. – Troy, Ill. :
Staircase Press, c2009.
p. ; cm.
ISBN: 978-0-9702527-1-5
Includes bibliographical references and index.
1. God–Love. 2. God–Proof. 3. Religion and science.
4. Swedenborg, Emanuel, 1688-1772. 5. Spiritual life–Christianity.
I. Title.
BT140 .S95 2009
231.6–dc22 2009

PROVING GOD

SWEDENBORG'S REMARKABLE QUEST FOR THE QUANTUM FINGERPRINTS OF LOVE

To Bernard,

Love is the ultimate science ! ♡

Edward F Sylvia

PROVING GOD

SWEDENBORG'S REMARKABLE QUEST FOR THE QUANTUM FINGERPRINTS OF LOVE

EDWARD F. SYLVIA, M.T.S.

WITH A FOREWORD BY IAN J. THOMPSON, PH.D

STAIRCASE PRESS
TROY, ILLINOIS

TABLE OF CONTENTS

Chapter One

IN THE BEGINNING (WE HAVE A PROBLEM)   1

Chapter Two

EMANUEL SWEDENBORG: WHO WAS HE?   11

TABLE OF CONTENTS

Chapter Eight

WELCOME TO THE SPIRITUAL WORLD 149

Chapter Nine

THE STARTLING SPIRITUAL HISTORY
OF THE HUMAN RACE 187

TABLE OF CONTENTS

Chapter Thirteen

HYPNOSIS: THE COSMIC MANIPULATION OF LOVE

Chapter Fourteen

LOVE: THE ULTIMATE SCIENCE

Foreword

It is well known that there are many severe problems yet unsolved in the foundations of physics, not least the question of whether and how to unify the dynamic geometries of general relativity with the superpositions of quantum mechanics. There are even more difficult problems when it comes to understanding minds and how they can be related to the physical world. Most scientists these days want to accept some kind of "non-reductive physicalism," but there are still persistent debates about whether such a view is even internally consistent. And there is always the question of how God can possibly be understood, and how anything Divine can be related to the physical world. Can we say anything scientific, for example, about how God could influence the evolution of life on earth? Most scientists and philosophers want rather to accept some kind of "dual magisteria," whereby science and religion are allowed to peaceably coexist within their own realms, and as long as they are not allowed to disturb each other.

These commonly held views are all based on the desire to leave science alone; to let it proceed autonomously and not to disturb it. However, the views are all based on ignorance of connections. They all reflect the fact that we do not yet have any scientific knowledge that connects general relativity with quantum mechanics, or connects minds with the physical world, or connects anything Divine with the universe. They are all therefore susceptible to revision if we do have some good theory about any of these connections. Many today say that there are no connections, but that again is from ignorance. If someone does propose a theory for these connections, then that proposal should be worked out as best as possible, as it may be a chance for solving our severe problems.

Developing such a connecting theory is what Ed Sylvia is trying to do in this book, based on some neglected ideas found in the works of Emanuel Swedenborg. Swedenborg, a Swede who lived from 1688 to 1772, claimed to have received extensive instruction in philosophical, spiritual and theological knowledge after his "inner sight was opened" in his 50s. Before that stage, Swedenborg had demonstrated a very independent and penetrating scientific mind, and published a Principia to explain his theory of how physical objects may be constructed by the rapid spiral motions of microscopic points.

This is not the place to discuss the entire veracity of Swedenborg's writings, but his ideas do certainly appear to be relevant to all our contemporary problems as listed above. This book starts by using Swedenborg's early physics ideas to see a more modern account of how a "pregeometric" realm might be constructed. Ed then works to link that account with Swedenborg's later ideas about how a spiritual realm might exist, and how such a realm might function in relation to the physical world. In a most interesting manner, Swedenborg and Sylvia see the spiritual world as continuously existing "alongside" the physical, and continually generating the physical world to sustain it in apparently stable forms. This, they argue, gives the appearance of physicalism, as the world functions "as if" from its own powers; but the powers are themselves derived from some other (spiritual) cause. And it would go some way to explain the apparent autonomy of the physical world.

Of course, anyone can make such claims: the proof is in the details. And there are certainly many details known today about the world that could not have been known in the 18th century. It is therefore a challenge to present Swedenborg's ideas again in relation to what we now know about physics, biology and neurology. Sylvia certainly rises to that challenge.

Ian J. Thompson, Ph.D.

Lawrence Livermore National Laboratory, Livermore, California
and Department of Physics, University of Surrey, Guildford, United Kingdom.

Aug 28, 2009.

Introduction

Unifying science and religion is a high-risk venture. Landmines and dangers are everywhere on both sides of the issue. Yet, the history of human exploration is full of individuals who have risked even death to find what they are seeking. The passion of the human mind and spirit is such that visionary people will always feel it is worth making the attempt to explore the unknown.

For that very reason, there is growing interest among scientists, theologians, and laypeople to explore another uncharted region and resolve whether science and religion can both answer the same questions about reality and have real points of interaction. I like to think of myself as a part of this exciting and mentally stimulating movement. This book is my contribution to this discussion.

Both religion and science make truth claims about ultimate reality. Science deals with *facts* and religion deals with *values*. Because of this, some people feel that science and religion address different issues and should be kept apart.

But, can these two powerful endeavors ultimately satisfy the human psyche by keeping them apart? Einstein said in 1941 that, "science without religion is lame, and religion without science is blind." Religion is weak on the *how* of creation, and science is weak on the *why*. In other words, science shuns teleology or purposefulness in the universe as a legitimate category of explanation. In place of a purposeful creation, scientists embrace the concepts that fundamental reality consists of *irreducible* chance and that everything must be describable exclusively in physical terms and physical quantities.

Many scientists also believe that metaphysical principles cannot be a part of real science because such principles and philosophies make claims that are not testable. Ironically, physicists who have jumped on the bandwagon of string theory and a multi-dimensional universe have embraced concepts that also cannot be tested. Checkmate.

If God created the world, then God created the laws of nature as well as the tenets of virtuous living. But theology offers us no further rational help here. It offers only faith and expects belief. Does God create one set of laws for nature and another set of laws for the human heart? Or are God's laws wholly self-consistent? (Inconsistency implies imperfection.) If the ubiquitous law that *everything in the universe proceeds by the most economical means* flows out from the action of the Creator, then there must be a top-down *causal link* between God's nature and the laws of nature.

This book attempts to show that the laws of nature emerged out from God's spiritual principles and values. That is, the laws of nature and its forces are actually spiritual laws and forces extended into spacetime constraints. While this is daunting and challenging enough, it is not the only challenge of this book!

Many other tricky problems are associated with attempting to write a book like this. Each of these problems is one more landmine ready to explode when stepped on. In spite of this, I have decided to step everywhere and not purposely avoid any dangers. The first big landmine is best expressed by the quote:

> "I cannot give you the formula for success, but I can give you the formula for failure: which is: try to please everybody."
> – *Herbert B. Swope*

I did not write this book to please anyone. People have different and strong opinions about things. Theologians argue with theologians, scientists argue with scientists, and theologians and scientists argue with each other, often bitterly. In science, we have competing theories, even within the realm of quantum physics. In religion, we have competing theologies, even within the realm of a single "ism." For instance, did God create the world and let it run on its own (Deism) or is God continually active in the world (Theism) and interested in our personal happiness? If the latter is true, which interpretation of quantum mechanics do I use (assuming one is correct) for demonstrating how God acts in the world?

So, in my attempt to unify science with religion, I must answer the question: *which scientific model do I use and which interpretation of theological doctrine do I use?* Two wrongs do not make a right, and my attempt will surely lead to an enormous backlash, since most of my readers will have their oxen gored no matter what choices I make.

In our post-modern world, it is taboo even to suggest in any way that one religion or worldview is "superior" to another (and I would do this if I picked one). But there is a big difference between respecting everyone's deepest beliefs and suggesting that these belief systems can be improved upon; few people are experts concerning their own faith systems anyway. Does any theology excel over others in addressing scientific issues? Does any theology even adequately address such issues as the virgin birth, miracles, the resurrection, the Second Coming, and the nature of heaven from a scientific perspective? (I have already tipped my hand that I will try to unify science with Christian theology.)

Even if I enjoyed special enlightenment and chose the best interpretation from science to describe reality and the best interpretation from theology, the problem still exists that science and theology use wholly different languages. The differences must be addressed and bridged. And, unless I plan to sell this book only to a handful of intellectuals, I also need to reach the understanding of normal but serious-thinking laypeople while still challenging their minds.

Another problem is that God will stand in the way of my ultimate success. Why? I believe God does not want to be proven in any way that would threaten a person's freedom of thought and discernment. Otherwise God would use coercion and constantly interfere with all our daily activities. Besides, what constitutes proof? For instance, if experiments reveal that prayer and worship have a positive effect on one's health (and they do), is this proof of a Divine Architect? One might just as easily explain that religious faith is an evolutionary strategy of selfish genes to calm the human mind from stressful thoughts about the inevitable fate of one's death and enable us to live longer and have more chances at reproduction. So even if such an experiment in faith were repeatable, it would still be open to interpretation.

I have also put myself in the uncomfortable position of going against the experts. Therefore, I run the risk that this work will be summarily dismissed. However, since none of the experts has all the answers, I have invited myself to the table.

> "A leader must have the courage to act against the expert's advice."
> – James Callaghan

My calling is to go against the advice of the experts, to shake things up and stir up the dust. I come to the table with the wish to stimulate healthy discussion. I have not shied away from making choices, and you will find my choices to be quite unexpected; in many cases they will be quite new to you.

I have chosen to use the scientific *and* theological ideas of Emanuel Swedenborg, an eighteenth-century scientist, philosopher, mystic and theologian. Using the ideas of a little known eighteenth-century thinker to straddle complex twenty-first century issues may seem like intellectual suicide. But I have studied this extraordinary man for more than 35 years, and I am confident that he has provided the world with scientific ideas that have yet to be grasped (like quantum gravity) and a theology that is most suited to interface with the discoveries of modern science. My undertaking will live or die on that choice.

Who is he? Emanuel Swedenborg (1688-1772) is one of the most overlooked thinkers in human intellectual history. His theology, while Christian, is radically inclusive and teaches that all those who sincerely live according to their religious beliefs and conscience and strive to do good from spiritual principles are welcomed into heaven. He states:

> "All people who live good lives, no matter what their religion, have a place in heaven."

This universal idea of the essence of religion to seek goodness in one's life was shared by Einstein, who said:

> "True religion is real living; living with all one's soul, with all one's goodness and righteousness."

Swedenborg's Christian theology was so universal that Buddhist scholar T.S. Suzuki wrote a book about him, comparing his ideas to Buddhism and calling him the "Buddha of the North." Swedenborg demonstrated that similar universal principles could be found at the heart of all the world's religions.

His most remarkable idea is that God's Holy Word was more than a historical account of the human predicament. It was a scientific and multi-dimensional document. The Holy Word, which encompasses God's wisdom, not only teaches us how to live, but also contains deeper levels of meaning that offer insights into the true nature of God and the scientific principles, laws, and symmetries that emerge from this Divine nature and Divine order.

God and science are one.

All true knowledge is connected because it leads to Love and Wisdom. Knowledge that does not lead us to wisdom is incomplete and disconnected from the bio-friendly laws of the universe. This idea of the ultimate interconnectedness of knowledge is not simply New Age drivel or philosophical naiveté. Real Science seeks knowledge for the goodness and benefit of society. How else is human achievement to be a blessing? How else can human society reach true greatness? Again, Einstein:

> "All religions, arts and sciences are branches of the same tree. All these aspirations are directed towards ennobling man's life, lifting it from the sphere of mere physical existence and leading the individual towards freedom."

Also:

> "Intelligence makes clear to us the interrelationship of
> means and ends. But mere thinking cannot give us a sense
> of the ultimate and fundamental ends. To make clear these
> fundamental ends and valuations and to set them fast in
> the emotional life of the individual, seems to me precisely
> the most important function which religion has to form in
> the social life of man."
>
> *Famous Einstein quotes –*
> *www.some-guy.com/quotes/einstein.html*

Swedenborg underscores Einstein's sentiment that knowledge must lead us beyond head-intelligence and move toward the heart:

> " To understand and to be wise are two altogether distinct things,
> for we may understand and still not be wise; but one leads us
> to the other, namely, science to the cognition of truth *(veri)*
> and truth *(veritas)* to the cognition of good, and it is the good
> which is sought for. But in order that we may be wise, it is
> necessary, not only that we should know and thus understand
> what truth and good are, but that we should also be affected
> with the love of them."
>
> *Worship and Love of God, Part 3,* [footnote b]

Love is an emotion, and only recently has neuroscience begun to look at the importance of emotion within human cognitive function and consciousness. All human thought links itself to some emotion, appetite, desire, intention, volition, or derivative of love, and emotion is now recognized as a vital part of human reason. In other words, the neural networks are subservient to affection, which modifies the activity that animates, focuses our attention, and shapes our very thoughts and memory.

Swedenborg anticipated these "modern" ideas about the brain more than 250 years ago, even taking these ideas into deeper structures within the neuron. He believed that passion, emotion, intention, and love modified the neural structures of the brain, and the resulting modifications represented the analogs, ratios and equations that produce human thought. Thoughts are the outer forms of our intentions. Said another way, emotions and affections are the inner life of our thoughts, and from these thoughts come our speech. No information, idea, or subject can connect itself to our personal lives without some affection. Our worldview is an internalization of our loves.

The importance of emotion in all this is that it links neuroscience to personal-level experience and contributes an important connection between hard science, the human heart and a heavenly God of Love.

In spite of all the problems that come with writing a book like this, there is a way out of the challenge of pleasing all my readers. *Everyone responds to Love.* This book is about Love! Therefore, no matter what beliefs you hold, you are invited to experience a most pleasant surprise—that Love is the ultimate reality. I am not a betting man, but I wager that, quietly, you will root me on!

Edward F. Sylvia, M.T.S.

Acknowledgments

At the top of the list, I thank my wife Susan. She not only provided the spiritual support for this challenging seven-year project but made extreme emotional and financial sacrifices which included my going out of state to earn a Masters degree in Theology. As a professional graphic artist she also helped bring this project into the world by designing the cover and layout of the entire book.

Next, I would like to thank the Swedenborg Foundation, the Swedenborg Society of London, and the Swedenborg Scientific Association for making Swedenborg's valuable scientific and theological writings available to the public. Without the publishing efforts of these organizations this project would have been impossible.

I benefited greatly from personally taking classes with physicist/theologian Robert John Russell at the Graduate Theological Union (GTU) and joining his dynamic organization, Center for Theology and the Natural Sciences (CTNS) in Berkeley, CA., which exists to promote serious dialogue between science and theology.

I am greatly indebted to theoretical nuclear physicist Ian J. Thompson for reviewing parts of this project and offering valuable suggestions. Other physicists with whom I had valuable exchanges include Nobel Laureate physicist Brian Josephson, quantum physicist Amit Goswami, and materials physicist William A. Tiller. They are pioneers in developing paradigms to compete with current scientific models. Dr. Tiller went an extra step and wrote a letter of recommendation so that I could attend seminary at the Pacific School of Religion (PSR).

I would also like to thank Rev. Dr. James F. Lawrence, Dean of the Swedenborg House of Studies at the Pacific School of Religion for supporting my desire to be a writer of theological topics as opposed to taking the path towards ordination.

Many thanks to the spiritual friends who have offered their suggestions to this challenging project. Lastly, I am eternally grateful to the Lord God in Heaven for the gift of life and the favorable providence to make this book a reality.

PROVING GOD

Chapter One

IN THE BEGINNING (WE HAVE A PROBLEM)

How old is the universe? How did it come to be? These questions have meaning only if it *had* an actual beginning. Most scientists today accept the idea that the universe did indeed begin. But the more they inquire into the origin, the less useful their knowledge of physical things becomes. Let me give you a simple and short overview of how science came to this awkward predicament.

In 1929 astronomer Edwin Hubble discovered that the universe was not static but expanding. Physicists assume that if we reversed this process of expansion and turned time backwards, we would find out how old the universe is. Based on current calculations, the universe has been estimated as being about 14 billion years old, give or take a billion years.

If we faithfully follow this line of reasoning to its logical end it shows us that everything originated from an infinitely compressed state with a zero radius. This state of infinite compression is called a *"singularity."* Einstein's general theory of relativity (which equates gravity to spacetime geometry) also predicted the existence of the singularity as spacetime crumpled into an extreme state of infinite curvature and gravity.

The Big Bang theory postulates that the universe was created when this singularity expanded violently outward in an instant before anything else existed. This was not just a dispersal of energy and matter into time and space. It was the creation of time, space, energy, matter and law itself, seemingly spontaneously.

An Inflationary Model of the Big Bang theory was later developed to explain why everything seems *isotropic*, or equally spread out as it expands. This model holds that for a period of time the expansion of the universe took place at an inflationary rate, actually exceeding the speed of light. In other words, space expansion was the one thing faster than light. The existence of cosmic microwave background radiation is taken as the strongest evidence for supporting a Big Bang beginning in that it seems to show the thermal after-effects of an explosion.

However, it was later discovered that the expansion of the universe is actually *accelerating*. The expansion can no longer be explained simply as the momentum from an original cosmic event. Scientists have recently added dark matter and dark energy to the equation in attempts to explain why the expanding universe keeps accelerating. Dark matter helps gravity hold the universe together while dark energy pushes the universe apart. Dark energy is winning. If this is not the answer scientists will have to change their assumptions about physics. Some believe that this expansion may actually be a characteristic of spacetime itself.

What does religious faith have to say about this? There is an ongoing debate between science and religion as to the meaning of the birth of the universe. Was the dispersal of energy and momentum a natural event in which some unknown force overcame the immense power of gravity in the Big Bang singularity? Or was it the willful act of a Creator-God? Was the creation of the universe a mindless, violent explosive act or a conscious act of infinite Love?

Does the answer lie in the singularity? If it does, then scientists and theologians have more common ground than they might want to admit.

British Astrophysicist Paul Davies describes the essential singularity of Big Bang Cosmology as "the nearest thing that science has found to a supernatural agent,"[1] for it can only be described as an edge or boundary of the physical universe, where spacetime warps into apparent nothingness. In other words, the essential singularity of the universe must somehow contain everything in a non-physical state. Think about that. How does something with the mass of an entire universe become less physical as it diminishes in size?

Even a layperson knows this doesn't make sense. We have a problem here. Actually, there are several problems with this scenario.

One problem with brutally crushing everything into a "point" with an infinitely decreasing radius is that scientists are no longer able to identify a causal principle that could exist in such a hyper-natural condition. Such a causal principle would have to explain agency and those special initial conditions that end up giving us a unique, coherent, finely tuned universe, one that permits the lawful emergence of ordered structure and complexity— from the organization of stars into galaxies to the origin of sentient life.

The main reason why the ultimate principle of creation is so hard to grasp from the point of physical science is that in the singular state of hyper-density, gravity would be so powerful that even space and time cease to exist. All known physical laws would break down. No one seems to know how matter and energy would behave under these conditions. As physicist-theologian Robert John Russell explains it, "Scientists do not know how to eternalize matter."[2] The concept of a singularity puts materialistic philosophy on shaky ground because without time and space the cosmic beginning *could not have had a physical cause*. There was nothing physical to cause anything to happen. The mystery of the initial conditions of the universe, when not even space or time exists, lies beyond physics.

▶ The Two Pillars of Modern Physics Do Not Support a Unified Universe

There is yet another, more embarrassing problem in trying to figure out how time, space and matter began without a physical cause. As we move back towards the beginning of time, the known laws of physics, as we understand them, come into conflict. In general relativity theory, spacetime is squeezed into a singularity whose radius equals zero and disappears into a void as mentioned above. But in quantum physics spacetime fluctuates, coming in and out of existence from *anywhere* (called quantum foam). How can the probabilistic froth of quantum mechanics coexist with the radical pinpointing of forces taking place within the infinitely shrinking radius of a singularity? While these two pillars of modern physics both lead us to a pre-space void, they do so in quite different ways. This incompatibility makes scientists uneasy.

Cosmological physicist Stephen Hawking has suggested that there may have been an infinite number of singularities, all with quantum mechanical properties. But this does not tell us how the laws of physics (or anything else) originated. If the universe emerged out of a void, how do the foundations of physics originate from a non-spatial and non-temporal nothingness (*creatio ex nihilo*)? In fact, we cannot use the known laws of nature to explain anything before 10^{-43} seconds into the creation event (the earliest moment that scientists believe ideas of time and space have any meaning).

It seems that a new paradigm is needed. New principles need to be identified, leading to a deeper theory that will unify all knowledge. And scientists such as Roger Penrose, Brian Josephson, Ian J. Thompson, Michael Heller, William Tiller and Lee Smolin, are embracing such a possibility.

One thing is clear. Whether they are talking about relativity theory or quantum mechanics, physicists have backed themselves into a corner of an infinite "something" that resides in a void.

A singularity consists of *infinite* curvature and a quantum state consists of *infinite* coexisting possibilities. This presents another tricky problem for theoretical physicists because they feel that infinities point to flaws rather than to an essential part of physics and unity. Contemporary science, which works with quantities, is unable to make sense of the Infinite. The result is that physicists hate infinities showing up in their equations. On the other hand, Big Bang cosmology and quantum physics offer today's theologians an idea they can fully embrace—that the universe emerged out of some infinite dynamic. A God of infinite Love, power and wisdom certainly fits that bill.

Theologians can embrace other aspects of modern physics as well. The Big Bang theory supports the notion that there was an actual creation event—if there was no beginning and the universe always was, then you do not need a God to create it. Big Bang theory and quantum physics also support the idea expressed in Genesis that creation emerged out of a pre-space void or vacuum. Even more suggestive, quantum mechanics is also providing us with some evidence that *consciousness* plays a fundamental role in the foundations of physics and the nature of the physical universe. If so, then we have a scientific construct that is potentially compatible with the theological premise that the universe is the purposeful workmanship of a God of *infinite Love* and *wisdom*.

Science needs to be founded on a non-physical principle of agency. "Mind" and consciousness are good candidates and seem to be the only dynamical processes in the universe capable of agency outside the restrictions of time and space (mind cannot be said to be located in space).

Is science to be founded on mind and consciousness? Furthermore, should consciousness, if it is fundamental to the universe and its laws, be given theological considerations? Could the laws of physics have originated out from the mind of God and Divine order?

Putting one's faith (or other leanings) aside, there would be only one possibility open to us for intellectually and rationally determining whether the laws of nature originated in a Holy and infinitely creative consciousness. Could there be some kind of relation and *correspondence principle* between God and nature? This top-down symmetry would exist if the universe were created in God's image. Such a possibility would require that the universe be fine-tuned so that the laws of physics and all dynamical, causal processes are physical analogs of God's essential nature.

The theme of this book is that physical laws and forces are actually spiritual laws and forces constrained by what exists in time and space. In fact, the metrics of spacetime structure do emerge out of the pre-space metrics of *Love* and *Wisdom*. If God's mind functions from Divine Love and Wisdom, what is there about these two *psychical* qualities that has anything to do with physics? Mental process and physical process can actually be described by similar properties, patterns and sequences (dynamics). I am not talking about poetic metaphors. Rather, I propose there is a universal science that gives the metaphors between physical and psychical processes real ontological status.

To do this, we first need a new understanding of what *substance* is. Then, we need a new understanding of what *Love* is (beyond its usual romantic connotation).

▶ Fundamental Substance is Propensity and Endeavor

Science tells us that within the quantum vacuum of so-called empty space, fundamental reality consists of infinite or pure *propensities*. Theoretical physicist Fred Allen Wolf describes these pre-space states as fundamental "tendencies to exist."[3] These propensities pop out of nowhere and somehow strive towards embodiment as forms of stabilized matter. This endeavor continues through nature's ongoing compulsion for complexity and self-organization.

Substance is an *essence from which forms emerge*. In quantum theory, propensity and endeavor are fundamental *substance*. Propensity or disposition is substance that is in continual effort to take on an embodied form and become geometrically circumstanced. Substance can therefore be non-material or pre-material. Quantum physics offers us a powerful hint that the key to discovering the true nature of agency in the universe may well lie in a non-physical first principle.

Over 250 years ago, scientist-theologian Emanuel Swedenborg made a similar claim: the essential substance of a "thing" was its dispositional properties. Today, this idea is expressed in philosophical circles as *dispositional essentialism*.* Swedenborg took the idea of a non-physical primal substance into theological territory by stating that propensity and disposition were derivatives of spiritual Love. To Swedenborg, Divine Love was primal substance (*substantia prima*) and Divine Wisdom was substantial form (the measure, or metric, of that Love).

> ✳ *Theoretical nuclear physicist Ian Thompson has done a masterful job of applying sophisticated kinds of dispositions to the Schroedinger equation.* See "Derivative Dispositions and Multiple Generative Levels" at http://ianthompson.org/philosophy_papers.htm. But Swedenborg would have issues with the anti-geometrical nature of quantum theory's probabalistic features and its failure to meet the requirements of the correspondence principle.

Love (volition) and wisdom (discernment) are two aspects of mind that have their analogs in the dynamical relationships between substance and form or between *essence* and *existence*. Our own thoughts are forms generated from our inclinations, appetites, intentions and affections. In other words, our Loves find form in our thoughts. These thought-forms reveal the quality and metrics of the particular Love or volition. Our Loves and thoughts find physical embodiment in our actions, which are the final measurement outcomes of the original pre-space volition in our psyche. So the metrics of one level have reference to the metrics of the other (correlation).

In this scheme, Love, which is pure propensity, acts as the formative substance and causal agent in the whole "top-down" process. As an actual substance, Love provides us with a theological response to the modern scientific notions of quantum entanglement and non-separability, because everything in the universe emerges from one dynamical formative substance.

When Love is understood as primal substance, it will lead us to a better understanding of the laws of nature and physical processes. In fact, Love can lead us to an exact science. This may seem improbable, but by the end of this book you will see how Love, as causal agent, requires both the principle of least action and the fine-tuned constants of law in order to manifest as a physical actuality. Love is the origin of these laws because Love can only create through *self-representation*. As we will see throughout this book, Love and its dispositional properties organize force, action and form into coherent unified wholes. Love is the key to the mystery of self-organization in the universe, from the thermodynamics of gravitating systems (like galaxies) to the ordering of experience in the human psyche (our world views).

Creation through self-representation means an ordered sequence by which a propensity or determined Love finds form corresponding to its original "end." *Love adapts form to its disposition*. This top-down causal sequence through progressive self-representation allows for "purpose" to supervene in nature.

The essence of Love is to unify; Love cannot manifest itself unless it perfects cosmic unity (holonomy). This is why everything in the universe is interconnected, interdependent and interrelated. Existence is relationship. And, since Love is *living conscious force* (a propensity with real goals), it is also the means by which intelligent processes work in nature and why biological complexity, over time, emerges out of Love's eternal endeavor towards the perfection of distinct things through increased unity. The biosphere and its ecology are lawful outcomes as the non-material force of Love evolves increasingly complex *forms of utility* in order to perfect cosmic unity and relationship (and more perfectly reflect God's qualities). Evolution is a spiritual process.

According to Swedenborg, the causal power of nature (agency) comes from the uninterrupted flow of Divine Love and Wisdom into the various forms of nature (which are recipient forms and inert as of themselves), making them similarly dynamical by *attaching usefulness to structure*. In other words, nature's processes all follow (co-respond to) the patterning principles and dynamics of Divine propensity and Divine consciousness that seeks actual and useful goals. This striving towards useful ends is a property of *goodness*.

▶ Seeking First Causal Principles

In a created world, there is no greater parsimony (Occam's razor) or constancy of law than something that is the outward expression of its first causes. The best way to give scientific credibility to the theological assumption that God created the world would be to show that the laws of nature are consistent with God's character. Can God's character—Divine Love and Wisdom—really tell us something about the way time, space and matter came into existence? Can God's essential nature also tell us something about the "arrow of time" and where everything is heading? This is the precise Divine Scheme that Swedenborg discovered.

Causal processes, based on dispositional essentialism and the inflow of active information from a non-physical and Divine Source, open the door for science to be founded on wider and more dynamic principles that transcend the physical world (and free science from scientism). This will lead to a newer and more comprehensive model of cosmology that brings ethical and moral values into the equation. This is significant because one of the biggest obstacles in unifying science with religion is that the former deals with impersonal facts and the latter with life values.

Nobel Laureate physicist Richard Feynman was once asked by a friend's mother why physics was important. He told her that physics was not important, only Love was. Swedenborg would have found great irony in Feynman's statement. He would have told the friend's mother that Love was the ultimate substance of reality, the law-giving agent in creation, and therefore, infinitely more substantial than matter. Love *is* physics.

Love is the ultimate formative substance and reason why all action, process, and structure in the universe, is integrated, ordered and has orientation. Science and religion are united because an inquiry into Love can lead us to an exact science with explanatory and predictive powers! In Swedenborg's cosmological model of the universe, *Love* startles the frontal lobes. Love is not only important for our inner well being; it is also important for the laws of the universe.

The best way to show this is to explore with Swedenborg what spacetime is and how it emerges from Divine Love. The Swedish sage had to solve the cosmological problem of the singularity, and the challenge of conjoining the non-temporal and the temporal. That is, finding a nexus between the Infinite and the finite. In theological terms, this is equivalent to reconciling God's Divine transcendence and Divine immanence.

Before you prepare yourself to make this steep, uphill climb, you will need to become familiar with Swedenborg's amazing life. Hopefully, this will help to neutralize some of your healthy skepticism that an 18th century thinker is the right man for the job.

SUMMARY

- ▶ The universe had a beginning.
- ▶ First causes are pre-space, non-physical principles.
- ▶ The mystery of creation lies beyond physics.
- ▶ The singularity needs rethinking.
- ▶ Quantum and relativity theory both have flaws.
- ▶ Science cannot escape the Infinite.
- ▶ We need to understand the Infinite.
- ▶ Love is ultimate substance.

PREDICTION:

In this century there will be a new movement by scholars
to reevaluate the full extent of
Swedenborg's scientific and theological achievements.

Apocalypsis
Revelata
in Qua
Deteguntur Arcana
Quæ ipei
Tradita sunt
H. P. HANSENS C.

Chapter Two

EMANUEL SWEDENBORG: WHO WAS HE?

"No single individual in the world's history ever encompassed in himself so great a variety of useful knowledge."
– Robert L. Ripley, 1934
Believe It Or Not

Germany, 1763.

Immanuel Kant sat at his writing table. He reached for his writing quill but hesitated a moment before putting it to paper. Kant was obsessed over stories reaching him concerning Emanuel Swedenborg. Swedenborg was well-known to the European scientific community, and his scientific writings had received favorable reviews in Germany. This work had certainly gained the attention of Kant.

Kant was aware that Swedenborg had developed and published a work on the Nebular Theory of Creation twenty-one years before Kant's own great work of 1755, *Allgemeine Naturgeschichte und Theorie des Himmels* (General Natural History and Theory of the Heavens). The nebular theories of both men proposed that stars and planets evolved from nebular material and that the universe was filled with groupings of stars, today called galaxies. Kant even admitted to his closest acquaintances that there were similarities between his work and Swedenborg's (Kant, however, would gain the title and recognition of being the father of modern cosmology).

On this particular day, Kant was thinking about Swedenborg, not from any professional envy but from great perplexity. For several years word had been spreading all over Europe that Swedenborg possessed a second sight and could communicate with spirits in the other world—both in heaven and in hell.

A woman, Charlotte von Knobloch, had recently written a letter to Kant, asking him if he could make any sense out of these strange stories. Kant, sitting at his writing table, now felt compelled to offer her his opinion.

One of the "strange" stories that were circulating around Europe was that the Queen of Sweden had summoned Swedenborg before her court to prove to her that he had conversations with those in the other world. Queen Lovisa Ulrica was the sister of Frederick the Great of Prussia. She was a highly intelligent woman and not someone you would trifle with. Nonetheless, Swedenborg agreed to appear before her, King Adolphus Frederic, various ambassadors, counts and senators to answer all her questions.

The queen challenged Swedenborg to give her secret information from her recently deceased brother that only she and he could possibly know. Swedenborg returned less than three weeks later and whispered something into her ear, whereupon she stepped back and fainted.

As luck would have it, Kant actually met a Danish officer who had attended a dinner in Copenhagen where a letter was read from the Mechlenburg ambassador. The ambassador claimed to have been present at the royal residence when Swedenborg made this fantastic disclosure to the queen.[1]

Kant concluded that the event must have indeed taken place, since any representative of the queen would have committed a grievous crime for spreading false rumors about the queen. But Kant, the ever-cautious scientist, asked a good friend who was planning to go to Sweden to see what he could uncover.

Kant received two letters from his friend. The first letter reported that some of the most trustworthy people in Stockholm accepted as truth the strange stories surrounding Emanuel Swedenborg. The second letter informed Kant that his friend had actually succeeded in meeting Swedenborg at his home. He found Swedenborg to be "an intelligent, pleasing, open-hearted man." Swedenborg assured him, without reserve, that God had given him the miraculous gift of being able to communicate with departed souls at will.[2]

Kant looked down at his table and put quill to paper. He now felt he could reply to Charlotte von Knobloch. He wrote that he was not one inclined towards fantasy, nor had he ever been accused of fabrication. But after examining stories around Swedenborg, he wrote, he could only conclude that they were true. Following this, Kant ordered and read the entire 8-volume Latin edition of Swedenborg's *Arcana Coelestia* (Secrets of Heaven).

Then, three years later something had changed. In *Dreams of a Spirit-Seer*, Kant depicts Swedenborg as crazy—"the arch fanatic of all fanatics."[3] What happened?

Dr. Gregory Johnson and Dr. Dan Synnestvedt both believe that Kant was worried about his reputation and no longer wanted to be associated with Swedenborg in any way. However, they have found evidence that, privately, Kant actually valued Swedenborg's theological thinking.

Unfortunately, the "crazy" label has stuck and Swedenborg's contributions to the world remain largely overlooked in the history of human ideas. One of the purposes of this book is to rectify that.

⚶ ⚶ ⚶ ⚶ ⚶ ⚶ ⚶

Emanuel Swedberg (his last name would later change to Swedenborg when his family was ennobled) was born on Sunday, January 29, 1688 in Stockholm, Sweden. Jesper Swedberg, a Lutheran minister, named his son Emanuel, because the name means "God with us" and would serve to constantly remind the boy throughout his life of God's special union and relationship with humankind. This reminding factor would turn out to be prophetic.

Jesper strongly believed that angels were real and closely involved with the lives of individuals on earth. He claimed even to have heard angels singing after his ordination. This faith was passed on to his son Emanuel—in a big way.

As a young child, Swedenborg was rumored to have had invisible friends and playmates. While it is natural for children to play "pretend" and make up games in which they imagine to be playing with others, these rumors implied that he actually saw things adults could not see.

Swedenborg himself claimed that between the ages of four and ten, he often meditated deeply about God and spiritual topics. From the ages of six to twelve he would often engage visiting clergymen in discussions on the true essentials of faith. These are not the kinds of things a normal pre-pubescent lad usually talks about. Even more unexpectedly, he held a strong pietistic position that was at odds with orthodox Lutheran tradition, which promoted salvation by faith alone rather than good works. The young child held that Love to the neighbor was faith in action, and that only those who practiced such Love and charity were capable of receiving true faith from God.

Emanuel often said things that astonished his parents, who had not taught him the fine points about pietistic theology. How did he come upon such sophisticated theological ideas? Did he learn these things from his invisible friends, or did he learn them from angels, as his parents believed?

▶ A Four-Year Old Yogi?

Swedenborg himself never confirmed stories of having invisible friends. However, he did admit during his later life that between the ages of four and ten, he was engaged in "absorbing speculations" and observed profound changes in his breathing during this intense mental activity. While he reflected deeply during prayer and when reading the Holy Bible, he observed that his normal breathing became quiescent (tacit) and a more interior respiration became detectable. He believed that this deeper form of respiration was the *breath of his spirit.*

There has been some debate among scholars as to whether Swedenborg had stumbled onto actual yogi breathing techniques. The evidence from Swedenborg's own words is clear that changes in his breathing came from deep thought rather than the other way around. There is general agreement that he did experiment with breathing. In one such experiment he states that he synched up his *inner* breathing with his heartbeat.

Swedenborg would later discover that the Lord Himself was preparing his breathing and mind for a higher calling. This Divine discharge was more than intellectual. The Lord was kindling his inmost life and heart with the desire to love and be open to special Divine influence and revelation. The significance of this is that his absorbing speculations proceeded from increased receptivity to the dynamics of spiritual love.

At the age of eight, a tragic and serendipitous event occurred, one that ultimately provided the future direction for his unique powers of deep mental penetration. Both his mother and older brother died within a short period of time of each other. That same year, Emanuel's cousin and tutor, Johan Moraeus, gave him a book on the human anatomy. The death of loved ones and the book about the marvelous, organic workings of the human body raised questions in the young boy's mind about the existence of the soul and eternal life. Did we have souls? If not, then Nature went to a lot of trouble just to make worm food.

Emanuel grew up in a world of mystery. Not only did his father believe in the reality of angels, but Jesper possessed impressive hypnotic healing powers, and from time to time he would cure patients of hysteria and mental ailments through strong suggestion and reading the Bible for long periods of time to the sufferers.

▶ A Confluence of Worldviews

Emanuel grew up under another powerful influence. This was the Enlightenment that so inspired the college town in which he lived. In 1703, when his father became a Bishop and moved to Brunsbo, Emanuel stayed with his older sister Anna and her husband, Erik Benzelius, a university librarian, who would begin to steer the course of his life. During the time Emanuel attended the University of Uppsala; out of sight of his father, he chose to study science and mathematics. His father, who believed that one only needed the Bible to find truth, had misgivings about science. Jesper even blamed Descartes, who died in Sweden, for the new freedom of philosophy and scientific investigation taking hold in the educational institutions of his country.

Fraternity records show that Emanuel participated in a variety of student debates and showed an interest in discussing such antithetical topics as "Divine Providence" and "Natural Law." However, we get no real inkling that the young student was concerned with bringing biblical truth and scientific truth together until he went on a class field trip to explore Sweden's tallest peak, Mount Kinnekulle. One of the main curiosities about this mountain was the wagonloads of seashells and marine fossils explorers could gather on it. On this particular field trip, Emanuel discovered the fossil remains of an ancient aquatic reptile. Several years later, his cousin Moraeus (the same one who gave him his first anatomy book) discovered the bones of a whale 150 miles inland. Emanuel believed all this to be scientific evidence for the biblical flood (he would later change his mind).

After Emanuel graduated, Jesper offered to pay for his son to travel to Europe and round out his education. His voyage to England was delayed for a year because Sweden's war with Denmark made sea travel too dangerous. In the meantime, Swedenborg busied himself by improving his organ playing at church and learning the art of bookbinding, which would be valuable in later life. He was as good with his hands as he was with his mind.

Tired of waiting for the war to end, in 1710 the rambunctious young man managed to talk a ship's captain into taking him to England. He escaped death no less than four times. At sea he survived an attack from privateers and a broadside from an English warship that mistook his ship for that of the privateers. He got stuck on a sand bar. And when he finally reached London, he was almost hanged.

When Swedenborg left Sweden, a plague was sweeping through the countryside. When his ship reached London it was quarantined for six weeks. Having London within view and unable to resist the excitement of setting foot in the city, he hailed a small passing ship whose passengers were Swedish and living in England. Within minutes of setting foot in London he was arrested for not having an official bill of health. It was only through Swedenborg's high-level connections in Sweden and the efforts of the Swedish ambassador that he narrowly escaped the gallows. Having a proper bill of health would again become a critical issue 35 years later.

▶ A Brain like a Sponge

During his five-year tour through Europe, starting with London, Swedenborg brushed up on the English language in order to read their books and perfect his skills in advanced algebra and higher geometry, as well as to introduce himself into a circle of astronomers that included John Flamsteed and Edmund Halley (of Halley's comet fame). Swedenborg would eventually become proficient in eight languages.

The precocious Swede made contact with other distinguished members of London's Royal Society such as the zoologist Hans Sloane and geologist John Woodward, who was the first to put forth the theories that fossil remains represented life from earlier epochs and that the earth's evolutionary history could be deduced by its soil and mineral strata.

The 22-year-old Swedenborg also purchased and sent back to Sweden badly needed scientific instruments, including a microscope and a 24-foot telescope. He even found time to enter a contest for finding a system by which ships can determine the terrestrial longitude. After some mental exhaustion, he took a short break to read the classics of English literature and write poetry. In fact, as a young man, he was considered a talented poet in his own right. And he wrote his poetry in Latin, which is not easy.

Swedenborg wrote to his brother-in-law that he constantly changed landlords in order to learn different "useful trades." He picked up the arts of engraving, instrument making, clock making and cabinet making. He also witnessed the consecration of St. Paul's Cathedral. Most significantly, he continued to take great interest in the debates that the collision of science and theology was stimulating just when he arrived in London.

In Holland he checked out the great observatory. He studied it closely so he could bring the technology back to Sweden. In Holland, he also learned the art of lens grinding.

In Paris he met the mathematician Paul Varignon, and the astronomer Philippe de la Hire. He sought out Liebniz (who had correspondence with his brother-in-law Benzelius), but the famous German philosopher was in Vienna at the time.

Before returning home, he realized that he had nothing to show for his five years away from Sweden. He knew his father, always suspicious of science and his son's exaggerated ideas, would certainly judge him for the practical use he made of his European journey. Jesper would want to see some proof of his achievements. Emanuel therefore quickly designed 14

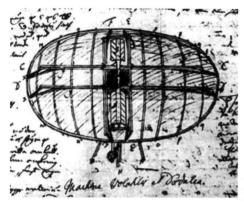

Swedenborg's overhead-view sketch for a "flying carriage."

mechanical inventions, including a submarine, a pneumatic machine gun, a water-clock, and a "flying carriage" that is now on display in the Smithsonian Institution in Washington, D.C. His flying machine was the first heavier-than-air, fixed-wing aircraft ever designed. And if that was not enough, he even devised a method of psychoanalysis, almost two centuries before Freud.

Back home, he was hired by Christopher Polhem (Sweden's Archimedes) to work on engineering projects. One huge project required Swedenborg to prepare the technical drawings and calculations for the building of a canal that would link the center of Sweden to the Atlantic seaboard. Another project involved helping King Charles XII move seven warships over 15 miles of rugged Norwegian countryside to a fjord where they could catch Danish boats completely by surprise and lay siege to the fortress of Fredrikshald.

Swedenborg was eventually appointed by the king to the position of Assessor to Sweden's Board of Mines. This gave him access to laboratory facilities where he could conduct mineral and chemical experiments and increase his knowledge of the natural world. He also applied his mechanical genius to inventions that would benefit Sweden's mining industry.

By 1719 he published Sweden's first scientific journal, *Daedalus Hyperboreus*, and completed a book on vibrations, *On Tremulation, or the Anatomy of our Finest Nature*, which was a bold move from physics to the operation of mechanisms in the human anatomy. He wrote various other treatises such as *The Motion and Position of the Earth and the Planets, On Swedish Blast Furnaces*, and *New Ways of Detecting Mineral Veins*. Meanwhile, his family had been elevated to the peerage by Queen Ulrika Eleonora, after the death of Charles XII, and given the name "Swedenborg." In 1721 the new Baron Swedenborg published his *Principles of Chemistry*. A year later, he completed a work entitled *Miscellaneous Observations*, which inquired into various geological topics, including the properties of matter.

During this period of study he revisited his idea that Sweden's unique geological features and fossil deposits were evidence for a universal flood that once covered the earth. But this time, he studied Sweden's geological formations with new eyes (having been influenced by John Woodward) and concluded that he was observing the results of processes that had occurred over long periods of time, and that the geology, therefore, was not scientific evidence of the biblical flood. If, according to the Bible, the deluge lasted about a year but the scientific evidence suggested otherwise, something had to give to square God's Word with scientific discovery.

He also could not imagine why God would have given humankind such an inquisitive and ample brain, then turn around and demand only *faith*. God did not promote ignorance. Swedenborg saw philosophy and science as the bridegroom of religion.

He published nothing else for another dozen years.

▶ Pondering the Infinite

During those years Swedenborg may well have been contemplating the great intellectual challenge of how God both created and acted in the world. To reconcile God's Divine transcendence with Divine immanence required nothing less than providing a causal nexus between the Infinite and the finite. As I mentioned in the first chapter, scientists tend to shy away from anything dealing with the Infinite. That Swedenborg had been absorbed in such matters is clear when he finally returned to publishing.

In 1734 he completed two ambitious books, the *Principia*, which addressed the creation of the world from the Infinite (and even proposed that energy and matter are interchangeable more than 150 years earlier than Einstein), and *The Infinite And Final Cause Of Creation*, addressing philosophically the connection and causal link between the Infinite and the finite world.

We will see in the next chapter that he theorized the existence of a zero-dimensional point of pure action that acted as a medium between God's infinite activity and the finite world.

This supernatural entity has many similarities to the *essential singularity* in modern cosmology but offers new insights as to the special conditions of the universe before time and space appear.

Swedenborg believed that he found patterns in the universe that showed all dynamical processes proceeding according to the same rules. These rules had a theological origin, Swedenborg asserted, because the Infinite created the finite through self-representation. Nature displayed God's character, because Divine Love itself had to be mirrored in all things of a universe that embodied an ultimate and interconnected purpose. This was the origin of the unfathomable fruitfulness we see manifested in the physical world.

Swedenborg's science was clearly becoming based in Divine consciousness and moving toward expressing the theological doctrine of Love. His theory that God creates through self-representation eventually evolved into a *science of correspondences* (a principle entertained by quantum physics). He then focused his scientific attention back on the human anatomy, calling it "the temple of all sciences." This endeavor embraced such diverse branches of knowledge as geometry, mechanics, physics, chemistry, optics, acoustics, pneumatics, and (because of the human brain) even logic and psychology.

He vowed to "trace out the nature of the human soul" and to seek out the soul through the hidden operations of the physical, organic body.

In the late 1730s he traveled to the surgical schools of France and Italy to observe and experience first-hand the dissection of human cadavers. (Churches of less progressive countries, like his own Sweden, were horrified by this research, because they believed these bodies would need to be resurrected at the time of the Second Coming.) Swedenborg interviewed many of the leading anatomists of the day and read their most current work.

▶ New Scientific Tools

His next series of writings covered blood chemistry, the vital organs, the cerebrum, and psychology. In his writings on the brain, he is the first to articulate a neuron theory. Since current scientific ideas were inadequate for helping him rationally proceed from the material world to the soul he was forced to create new ways, or doctrines, to help him complete this task. He called these new methods of investigation the "Doctrine of Forms, the Doctrine of Order and Degrees: also the Doctrine of Series and Society: the Doctrine of Influx: the Doctrine of correspondences and Representation: lastly, the Doctrine of Modification."[4]

Swedenborg believed all knowledge was connected. His true genius was in identifying principles that could be applied to all things. His new methods of investigation were designed so that the deeper he probed into things, the more he could step back and get the big picture. His success ultimately required discovering principles that could be shared by both theology and science.

Swedenborg recognized that for knowledge to be holistically connected, it had to reveal a grand *organic* scheme. So rather than postulating a universe that was created from nothing and evolved into complex life-forms from dead molecules through a string of highly unlikely events, Swedenborg believed that the physical universe emerged from a living principle (vitalism). He identified this unifying and living principle as Love.

Swedenborg was convinced that in a world created by a God of Love, as is taught in all religions, the principle of Love had to be fundamental to even the physical laws of the universe (which are certainly bio-friendly) and offer insight into how time, space, and matter came into existence. Swedenborg used his new methods of investigation to show that these physical laws had their origins in spiritual laws.

Dealing with such profound concepts, it was only a matter of time until Swedenborg would begin experiencing profound changes in his life. He was shocked to discover that when he removed his ego and pride in his own intelligence from his research, he was left flat. In spite of his noble investigations into the ultimate reality, inwardly he found himself still motivated by fame, prestige, and personal honor. From his religious upbringing, he began to suspect that his self-love was an obstacle to discovering truth. Accordingly, he decided to put his life in God's hands.

The more he subjected himself to harsh introspection, the more his experience of inner conflict intensified. To purge and purify himself he began a regimen of meditation and impassioned prayer. Throughout all this, his mental state moved between depression and ecstasy. He alternated between feelings of unworthiness to God and bliss (especially after overcoming a particular inner temptation).

One of the outcomes of this struggle was that he reported losing all sexual desire. This may not have been pathological; it may well have been that his sexual and creative energies were being redirected inwardly, towards the formation of a new inner being.

He also started recording and interpreting the symbolic meanings of his dreams. A year earlier, in 1742, he had written in his book *Rational Psychology*, that dreams were the activity of an innate intelligence operating beyond the conscious, habitual mind. He was intellectually curious about the symbolic language of dreams. But when he began to interpret his own dreams, he saw them as shots fired across his bow, revealing his deepest inner conflicts and struggles.

In one of his earliest dreams he observes a man who is boiling in water. He realized that he was the boiling man and interpreted this to mean that he was in "hot water." Later in his journal he described a dream with hideous specters and a beast that attacked him but not a small child who was also in the dream. He also recorded a dream in which he was hanging on the ledge of a cliff, overlooking a deep abyss below. In another dream, he found himself lying in bed with a beautiful woman. As he turned to her, she told him that she was pure, but that he stank!

Other dreams seemed to offer insights for his scientific investigations. In one dream he was given special clues about the functions of the thymus and renal glands. The dream communicated to him that the general purpose of these organs in the body was to separate the serum from the blood, filter and purify it, then return the fresh serum back to the blood. But the dreams that he paid most attention to were those that began to dominate his sleep time; these communicated to him that he was to play a powerful role in God's Divine scheme.

Interesting phenomena also occurred while he was awake. As early as 1738 he recorded seeing things that were not there. While working on *The Five Senses*, he described seeing a fly land on the page of his manuscript. It surprised him because it was not a real fly! He also experienced mysterious lights and flames (*photism*). He saw these lights as signs from heaven that his studies were moving in the right direction.

Swedenborg had been trying to tap into his higher levels of mind since his early meditations and changes in his breathing as a child; now a whole new world seemed to be opening up to him. He not only saw strange lights, he began to hear voices. At last, he began to sense that he was in the presence of others, at first vaguely, until he determined that these were people whom he knew who had already passed away.

▶ Gaining Access to a Pre-Space Research Facility

Swedenborg was about to enter a new phase in his life that would move his scientific investigations from the outer, physical world, into a pre-space inner world. He would soon embark on one of the most remarkable journeys ever recorded in human history. It is during this next phase that his name "Emanuel" attained providential significance.

In Swedenborg's own words, during the years from 1710 to 1744, he was led by the Lord into the natural sciences and prepared for a more important purpose. One evening in the spring of 1744 he experienced a Divine challenge.

He was in Holland, one of the countries (besides England) he felt comfortable with publishing his progressive material. He had traveled from The Hague to Delft in a profound state of meditation. The day before, which was Easter, he had partaken of the Lord's Supper. He now found himself faced with the final challenge to abandon his scientific projects and totally give up his own life for the Lord. He believed he was ready to make the commitment.

But that night he read in the Bible about how God's miraculous powers were imposed on Pharaoh and Egypt through the mediation of Moses. Doubts about the verity of the story cropped up in his mind. For a moment he found himself fluctuating between his strong faith and his scientific cynicism. Did these incredible things really happen? He caught himself and blamed this short-lived inner detour on the "Tempter."

Swedenborg then looked at the flames in the fireplace of his room and gained a new insight. He blamed his own intelligence and its fallacies derived from the five senses for his doubts about the miracles of the Bible. "This is why," he concluded, "the angels and God revealed themselves to shepherds and not to a philosopher." [5]

He then went to bed. It was 10 PM.

A half-hour later, he was awakened by a roaring noise, one he described as the sound of many winds rushing together. His body began trembling from head to foot. He also sensed the presence of something "indescribably holy" in his room.

His body now shook so violently that he was thrown out of bed and landed on his face on the floor. Frightened, Swedenborg quickly clasped his hands and began praying profusely. Suddenly a hand came out from nowhere and pressed strongly against his hands. Before Swedenborg could react, he found himself lying in the Lord's bosom and looking at him—face to face!

Swedenborg stated that the Lord's countenance was of a holy demeanor, that could not be described by words. He then described the ensuing, brief conversation. The Lord spoke first, saying, "Do you have a bill of health?"

Swedenborg answered, "Lord, thou knowest it better than I."

"Well, then do it!" was the Lord's command.

Swedenborg woke up still trembling the next morning. He did not know what to make of the encounter. Was it real? Was it a dream? Swedenborg contemplated his conversation with the Lord. He understood the question, "Do you have a bill of health?" to mean, was he fit for a higher calling? These words reminded him of his arrival in London when he began his scientific journey without a bill of health. In view of his oath, the day earlier, to give his life to the Lord, he understood the Lord's reply, "then do it," to mean that he was now being challenged to take up the task.

A full year later, on an April evening in 1745 while eating at an English Inn, he was again confronted by the Lord, and everything came to a head. On this momentous occasion he learned the extraordinary nature of his next vocation. (See the extended discussion of this event in Chapter Eight.)

Through all this, Swedenborg's mental faculties remained sufficiently crisp enough to continue his prodigious anatomical writings. One such work included an amazing treatise on the human brain and presents the first articulation of a neuron theory (a nerve cell with its processes). His anatomical material was so massive that much of it remained in draft form. In spite of the profound life changes Swedenborg was experiencing, he had even started a new work, *The Worship and Love of God.*

This new writing project was a departure from his previous scientific works. Its style blended science and symbolic poetry. It was a remarkable cooperative effort between the right and left hemispheres of his brain. Significantly, he believed he was writing more for angels than his peers. He bemoaned the fact that most scientists of his age reasoned only from the five senses. "They see everything, but understand nothing." In one of his dreams he had even been assured that this was a holy work.

However, he never finished this new book. In the spring of 1745, he abruptly abandoned all his scientific works. He made this abrupt decision after his encounter with the Lord at a London Inn, mentioned above, in which he was given his important new writing task. Four years later, during the summer of 1749, the first volume of a series of strange books began to appear in bookstores. The books were entitled *Arcana Coelestia* (Secrets of Heaven), and to the shock of European readers, described things "seen and heard" in the world of spirits by an anonymous author. It would not be until 1759 that Swedenborg would be identified as the author.

At the age of fifty-seven, Swedenborg gave up his exploration of the natural world and, through Divine intervention, began a three-decade long exploration of the spiritual world. His mission was to share with his terrestrial human brethren new doctrines and secrets of Scripture. In the spiritual world he witnessed the Second Coming, and the descent of the Holy City, the New Jerusalem. He discovered that the Second Coming would not be a *physical* event, but the infusion of new information for men and women of the earth. This new information would eventually cause a big shake up. The earthquakes and great battle depicted in Revelation would take place on psycho-topological turf—the inner landscape of hearts and minds.

It is through this theological paradigm shift that the Lord will make all things "anew," once the opposition to this new dispensation, symbolized by the Great Red Dragon and his army, has been overcome. The Second Coming on earth happens one person at a time, as individuals accept these new teachings.

▶ Why Swedenborg?

Suspend any disbelief you might have about such a wild claim. For the sake of discussion, assume Swedenborg received virtually unlimited power to explore the "other world" from the

Lord. Why Swedenborg? Why a scientist? Why not a member of the clergy, a poet, a military leader or a government official?

This was the Age of Enlightenment. The Christian worldview was under attack. A new freedom of thought was taking hold. The world no longer had to be mysterious; instead, everything in the visible world was measurable and under laws. Empirical investigation was trumping faith.

If a spiritual world really did exist, then who better than a scientist could explore it and report back to earth? Swedenborg's scientific approach would preclude a purely emotional or dogmatic account of what was "seen and heard" there. Since Swedenborg had also devised a method of psychoanalysis and written a significant work on psychology, who would be better prepared to "interview" spirits and angels? A scientist would perform experiments and make observations about the other world, describing such things as the properties of the special substances found there, the flora and fauna, and the unique dynamics of its time and space. Swedenborg himself states that the Lord directed his life of scientific search in a special providential way:

> "I could see at last that the tenor of Divine Providence has ruled the acts of my life from my very youth, that I may at last come to this end; so that, by means of the knowledges of natural things I might be able to understand the things which lie deeply concealed in the Word of God, and thus serve as an instrument for laying them bare."[6]

From the observations of the Spiritual World (including Hell) which he made over three decades, Swedenborg was able to provide a systematic theology that identified the universal sense of unity and wholeness in the Divine scheme. This new theology was carefully developed throughout his publications, which eventually numbered about thirty books (enough to fit an entire bookshelf).

At the beginning of his new vocation as a theologian he had chosen to publish his astonishing books anonymously. But after about ten years of secretly publishing, events that took place in front of numerous witnesses displayed his augmented mental powers in a shocking way. This blew his cover.

▶ Extra, Extra Sensory Perception

On Saturday, July 19, 1759, Swedenborg was a dinner guest at the home of a prominent merchant, William Castle, who lived in Gothenburg. Gothenburg is a port city on the west coast of Sweden, some three hundred miles from Stockholm, on the east coast of Sweden.

The dinner began at about four o'clock. About two hours later, Swedenborg abruptly left the gathering for the privacy of the garden area. He returned looking pale and agitated. To the horror of the guests, he announced that a terrible fire was, at that very moment, quickly spreading through South Stockholm.

Still restless, Swedenborg went back and forth between the garden and dining room, keeping the guests informed about the status of the fire. At one point he reported that a friend's house was already in ashes and that the fire was headed for his own house. He had lost valuable manuscripts in the past due to fire, and this one threatened to destroy his most recent writings.

Then, around eight o'clock that evening, he rushed back into the room from the garden and with a look of relief, told the frightened guests, "Thank God! The fire is extinguished, at the third door from my house!"

That evening, the account of Swedenborg's story was spread all over Gothenburg, even reaching the provincial governor. The next morning, a Sunday, Swedenborg was summoned to the governor's house to answer questions concerning the incident. Swedenborg gave the governor a full account of the disaster, with precise details of how it started, the length of time the fire lasted and how it was put out.

On Monday evening, a messenger finally arrived in Gothenburg with news of the Stockholm fire. The next day, a royal courier arrived at the governor's mansion and verified Swedenborg's entire account of the event. This would be one of the several stories that eventually reached Immanuel Kant in Germany, and obliged him to write his letter to the female acquaintance who had sought his professional opinion of these matters.

After the Stockholm fire, it was not long until someone put two and two together and figured out that it was Swedenborg who had authored the strange books that were appearing throughout Europe. Swedenborg soon admitted that he was indeed the author—which raised even more eyebrows, since he was such a respected scientist and a member of the Swedish House of Nobles.

Swedenborg instantly became a great curiosity and celebrity, and soon he was at the top of everyone's list of "dinner party guests." While Swedenborg kept mostly to himself and his writing, he did make himself available out of a sense of civic duty. He answered the many questions the anxious public bombarded him with.

And he pulled no punches.

During one gathering of royalty, Swedenborg told Count Klas Ekeblad that his grandfather, de la Gardie, had met and married Empress Elizabeth of Russia, daughter of Peter the Great—in heaven! Imagine sitting among notable friends and guests and being told that your grandfather was no longer the husband of your grandmother.

There were other awkward moments at these events as well, especially when someone inquired about the status of an acquaintance or family member and was told that the individual in question was in hell.

At another dinner party in Amsterdam (July 17, 1762), Swedenborg's facial expression suddenly changed in the middle of a conversation. He seemed as if he were no longer present. When he finally "came to," he was asked what had happened. At first he refused to reply. But after being hounded, he finally said, "Now, at this very hour, Emperor Peter lll of Russia has died in prison." Swedenborg then told the astounded party guests a gruesome detail—the Russian Emperor had been strangled by a conspirator, and that he, Swedenborg, personally witnessed the horrible event. Not long after this, the newspapers confirmed everything that Swedenborg had said about the murder.

During Swedenborg's many travels overseas, ship captains became very fond of him. They believed that perfect sailing weather miracously followed him on every ship he boarded. One story from a Captain Dixon is particularly interesting.

Swedenborg boarded Dixon's ship to travel from London back to Sweden. The Swedish traveler told the captain that he had more provisions on board than was required for such a trip because, he assured the captain, they would be in Sweden in only eight days.

The ship left port on September 1, 1766, and arrived in Stockholm on September 8! Swedenborg's presence, according to Captain Dixon, provided the ship with more than a favorable wind—the wind became more intense and turned into a "perfectly controlled storm" that carried the ship all the way to Stockholm in record time. Captain Dixon reportedly said that in all his life he had never seen such a favorable wind, a wind that followed him at every turn he made.

Swedenborg not only became the talk of Europe but apparently created a "buzz" in the spiritual world as well. Swedenborg shared his account of the funeral of his old mentor, the famous engineer, Christopher Polhem. Swedenborg claimed that he was accompanied at Polhem's funeral by none other than Polhem's own spirit in the other world. Through Swedenborg's eyes Polhem was able to look back into the physical world and witness his own burial. He asked Swedenborg why they were burying him when he was still alive. And why did the minister, who was presiding over the funeral service, say he would rise at the Last Judgment, when he had been fully resuscitated for several days now? (The funeral was held on Monday, August 31, 1751, three days after his death.)

Swedenborg also claimed to have shared his physical eyesight with his deceased aunt Brita Behm (from his mother's side), and that she, too, had been able to witness her own funeral. According to Swedenborg this caused quite a commotion in the other world, where other human spirits there called it "the miracle of all miracles."

▶ Was Swedenborg Schizophrenic?

In view of fantastic stories like this, Swedenborg's sanity has to be addressed. Was he schizophrenic? Did he go mad? Men of great genius are not immune from schizophrenia (as creative geniuses like John Nash, Edgar Allen Poe, Franz Kafka and Vincent van Gogh illustrate). Those who strive to diminish the validity of Swedenborg's claims have more recently offered additional diagnoses (such as temporal lobe seizures) from the strictly materialsitic viewpoint that all mental phenomena leading to transcendental experience are pathological states.

But here lies the big difference. Individuals suffering from schizophrenia or other abnormalities generally cannot function without medication or profound care. Swedenborg did not enjoy the advances of modern medicine, nor was he ever under a doctor's care during this time, yet he became increasingly active in society and government. He was an active member of the Swedish Parliament all through this unique period, working on fiscal reform, general legislation, and programs addressing Sweden's problem with alcoholism. So he was never dysfunctional.

Sufferers of acute schizophrenia or other mental illness often become helpless during their episodes. When Swedenborg was experiencing his "visions," he was still able to compose himself and muster the intellectual forces to observe these events with the objective methodology of a scientist. He made a rational study of the phenomena occurring in his inner world (which was later turned into a complete and systematic development of a novel theology spanning thirty volumes). He maintained his coolness and objectivity even in the "charged atmosphere" of coming upon old acquaintances in hell.

Swedenborg admits that he kept his supra-experiences hidden from others. His assumed pathology or temporal lobe seizures are deduced merely from the unique and startling information of his writings—not his behavior around others. It seems unlikely that such pathology and mental illness could possibly be hidden from friends and associates.

No one who was closely associated with Swedenborg, or even in his presence, ever recorded him suffering from mental illness or bouts of madness.* The oddest thing witnesses reported is that he had a peculiar way of drinking wine. At dinner parties he would fill his wine glass half way with sugar before he permitted the server to pour the wine.

There are two ways to look at this odd routine. Swedenborg's vast theological works were either the result of a "sugar high," or his intense mental activity simply required more glucose (brain fuel) than the average person's.

John Wesley, the founder of the Methodist Church, had heard of Swedenborg's theology. Since both were in England, Wesley was interested in meeting him. This interest was particularly sparked because Wesley and Swedenborg were both Pietists, that is, they both

* **Those interested in learning more** about scholarly arguments against the claims of Swedenborg's supposed mental illness, please see the Jan-June 1998 issue of The New Philoosophy devoted to "The Madness Hypothesis." You can find it online here:

http://www.newphilosophyonline.org/ journal/index.php?page=archive#JJ1998 challenged the fundamental concept of the Reformed Churches that *faith alone saves*. Swedenborg had claimed that *Love, charity*, and *good works*, were faith put into action and therefore the very life of faith. True faith was not belief, but action. Similarly, Wesley favored a gospel of action over faith.

As the story goes, Wesley was with a close friend in England when a servant brought in a letter addressed to him. Wesley opened the letter, and his friend watched Wesley's facial expression change.

Concerned, the friend asked Wesley about the letter. Wesley informed him that it was from Swedenborg, and that Swedenborg had accepted an invitation to meet with him. This did not seem unusual to Wesley's friend until Wesley admitted that he was indeed interested in meeting with Swedenborg, but had not yet shared that desire with any living soul.

Wesley wrote back to Swedenborg with a possible date and place for such a meeting. Some time went by. Wesley's friend was again present when Swedenborg's reply was delivered. This time when Wesley opened it, the friend noticed an even greater shock on his face. When the friend inquired about the letter, Wesley told him that Swedenborg would be unable to meet him on that particular date. The reason Swedenborg gave was that he would be departed from the earth.

Swedenborg had learned from the spiritual world the exact day and time of his death. At five o'clock, on the afternoon of March 29, 1772, he passed into that world. The two men never had a chance to meet.

On April 7, 1907, Swedenborg's remains were brought back to Sweden from London aboard a Swedish naval cruiser. But it was in England, and not Sweden, that Swedenborg's theological writings would take hold and form the basis of a new Christian denomination.

It is time now to return to the challenge of this book—the unification of science with theology. We will need Swedenborg's unique discoveries in science and theology to accomplish this. He offered original ideas for addressing creation and the emergence of bio-complexity (as a physical analog and expression of God's Love and Wisdom). His thesis for unity is that physical laws and forces have their origin in pre-spatial spiritual laws and forces. It is to this subject that we now turn.

SUMMARY

▶ Swedenborg grew up in a religious and scientific environment.

▶ His scientific studies involved chemistry, hydrostatics, mechanics, magnetics, mathematics, pneumatics, geology, anatomy, astronomy, hydraulics, optics, metallurgy, acoustics, physiology, psychology, cosmogony, cosmology and dynamics.

▶ He was the first to theorize a nebular hypothesis for the formation of stars, the first to propose the idea of numerous galaxies in space, the first to address the relationship between energy and matter and the first to introduce a neuron theory of the brain.

▶ He invented new scientific doctrines that went far beyond the models and formulations of his era (and may offer important insights into the challenges of today's "New Physics".)

▶ He gave up his scientific pursuits to serve God.

▶ He explored the spiritual world and conversed with its citizens for almost thirty years.

▶ He witnessed the Second Coming.

▶ He wrote numerous volumes on theology, which he claimed to be a new dispensation from the Lord God.

PREDICTION

Swedenborg will one day find his rightful place within the history of human thought and ideas.

TIME, SPACE AND MATTER FROM THE INFINITE

"It is my opinion that our present picture of physical reality, particularly in relation to the nature of time, is due for a grand shake-up..."
— Roger Penrose
The Emperor's New Mind

In Chapter One we ended on the cosmological problem of the singularity. If a singularity represents something where space and time cease to exist it should be obvious to both the atheist and the faithful alike that it could not have a physical cause. There are no physical principles at hand to help us here. Put bluntly, *physics cannot tackle the problem of creation*.

This suggests that the origin of spacetime structure and order, as well as the laws of the universe, have their source in some invisible, non-physical and timeless realm. A singularity represents a limit or boundary of the physical universe. It is not a physical thing, yet it seems infinitely pregnant with the potential to create all physical things. Science can deal with this only if it is prepared to make metaphysical assumptions concerning ultimate reality. (This goes against the typical attitudes of a natural scientist or of *scientism*.)

In *God & The New Physics*, Paul Davies informs us that some cosmologists believe a "singularity is the interface between the natural and the supernatural,"[1] or a door where influences pass from one world to another. He also states that while there is no unanimous agreement concerning singularities, the issues surrounding singularities are at "the frontier of modern theoretical physics" and give "a new slant to the debate about God and the existence of the universe."[2]

In the late 1720s to the mid 1730s Emanuel Swedenborg was pondering these very same issues while he was working out a complete theory of the physical universe. He too, believed that the *physical* universe had its start from a non-physical singularity, which he called a *first natural point*. And, like some of today's cosmologists, he also imagined it to be a door between two worlds—a dimensionless medium between the Infinite and the finite—one that existed prior to the creation of the world. Theoretically, this pre-geometric condition allows for all possible curvatures of spacetime and principles of action. Interestingly, the idea that spacetime is not fundamental but is derived from "pregeometry" is credited to the 20th Century physicist John Wheeler. Yet, Swedenborg is clear that his singularity was generated by, and had its existence, from God, as a boundary between the physical world we can measure and an even more fundamental non-physical, spiritual realm, whose standard of measurement (metrics) is psychical. Physical reality is not the ultimate reality. (This has theological implications for addressing the disputed interpretations of quantum physics.)

Rather than working backwards from outer physical things (the process that produced the modern Big Bang theory) Swedenborg described his singularity as a door by which God's *Love* (spiritual substance) and *Wisdom* (spiritual form) is converted to the physical realm we can measure. In other words, unlike the theoretical foundations of the Big Bang singularity, matter is not externalized or exploded outwards from a super-crushed state. Physical laws and forces have their origins in spiritual laws and forces. The singularity is the medium by which one transitions into the other.

This is a bold intellectual move. I suspect many scientists would see it as intellectual suicide. For it suggests that God is perfectly self-consistent and has created all things in the universe from the self-projection of Divine Love into measurement outcomes. God can only create through self-representation. This perfectly self-referable activity is the basis of God's Infinite consciousness and Infinite Personality and the means by which the Creator can be in perpetual relationship with creation. The universe is fine-tuned from a *theological patterning* principle, which has a moral component (so the Divine purpose is fulfilled by the human race evolving towards conscious spiritual Love). However, even scientists who are inclined to believe in God would prefer it if the Creator made special laws just for the physical universe and another set of laws for moral character. But Swedenborg maintained that one had its equivalent in the other (correspondence principle).

Roger Penrose's assertion (at the beginning of this chapter) that our present notion of reality needs a shake-up is actually an understatement. We are about to see that time has its origin in Wisdom (states of truth) and space has its origins in Love (states of affection). But first we have to straddle the concept of the Infinite.

► The Reality of the Infinite

While science is the study of finite things, some physicists, mathematicians and philosophers do embrace the concept that infinity is also real. Liebnitz believed that the finite was a limitation of the infinite. Mathematician Rudy Rucker, in *Infinity And The Mind*, states, "it is unlikely that the Calculus could ever have developed so rapidly if mathematicians had not been willing to think in terms of actual infinities."[3] He adds that as mathematicians sought "to get a precise *description* of the continuum or real line, it became evident that infinities in the foundations of mathematics could only be avoided at the cost of great artificiality."[4] Nineteenth century mathematician Georg Cantor came up with a system for describing a hierarchy of types and different orders of infinity called transfinite numbers. I will spare you his mathematics, but he made this challenge to those who favored the nonexistence of infinities:

> "The fear of infinity is a form of myopia that destroys the possibility of seeing the actual infinite, even though it in its highest form has created and sustains us, and in its secondary transfinite forms occurs all around us and even inhabits our minds."[5]

In *Science and Creation*, theologian/physicist John Polkinghorne agrees with Rucker that reality is richer than that which can be expressed by the finite mind and quotes his statement that, "Reality is, on the deepest level, essentially infinite."[6] Physicist David Bohm embraced the idea of thinking about reality in terms of an "infinity of generative and implicate orders."[7] Astrophysicist Paul Davies in *The Mind of God*, states humorously, "Whether it is an infinite tower of turtles, an infinity of parallel worlds, an infinite set of mathematical propositions, or an infinite Creator, physical existence surely cannot be rooted in anything finite."[8]

How do we grasp the idea of a singularity having its existence from the Infinite? Swedenborg admitted that a primal singular state, subsisting and existing from the Infinite, could not be fully grasped geometrically. Instead, he believed it could be comprehended rationally.

He reasoned that for something finite to exist, something prior and less *finite* first had to be *limited*. Quantum physics actually supports this idea of finite events emerging from a more dynamical and expanded (non-local) prior reality. But Swedenborg has a superior approach to helping us visualize this transition whereby forms lawfully put on new geometrical constraints in their descent from non-physical substances to terrestrial matter (discussed in Chapter Five). Therefore, nothing finite can exist alone or from itself. Finite entities must spring from things other than themselves and greater than themselves. Swedenborg takes this contingency argument all the way to the Infinite. Physical things are not simply contingent upon something else, but every finite thing can be generated only when something under less geometrical constraint or greater freedom takes on additional limits, a process going all the way back to the Infinite itself taking on limits. Theology calls this Divine self-limiting process *kenosis*. (Only the infinite does not have a cause. It is uncreated because nothing is greater than the infinite.)

So, as one probes deeper into reality, things get more complex and subtler (one is moving closer to the Infinite). This flies in the face of traditional science thinking that things get simpler as you look at smaller bits of the universe (reductionism). It also suggests an answer to the question, where does complexity come from?

Swedenborg's singularity is not static. Rather, it is a supernatural, zero-dimensional, perpetual dynamo that carries the force of God's universal patterns. Its ineffable flux contains an idea representative of all dynamical, holistic process. Swedenborg describes this idea of Divine intelligent flux (active information) focused at a "point" as a *mathematical intuition of ends*. This ensures that all processes in nature proceed by the same precise steps and rules. This is why all created things represent a microcosm (small universe) of the macrocosm (large universe). It is a theological version of a Theory of Everything (TOE). The singularity can carry this "density of information" because it is a form of determined and purposeful activity adapted to God's Divine disposition. It is the mathematical precision by which all coherent organization emerges from the principle of Love (the schematic for this mathematical precision is the topic of the final chapter). In his introduction to Swedenborg's great philosophical work, *The Infinite and Final Cause of Creation*, Lewis F. Hite points out that Swedenborg did not simply see "the Infinite as a mere postulate of rationality, but a moral and religious principle of controlling influence."[9] God is an infinitely conscious Being with an eternal goal. This goal is the drive-belt of the singularity.

To further understand the origin of the singularity and its non-physical nature, we have to understand how subtle dynamical process and *complexity* have their origins outside of time and space. For science to make real and bold new advances it will have to grasp how action can take place in the absence of a physical world.

▶ God's Grand Unified Field

In Swedenborg's theological writings, made from his personal observations of the spiritual world (which was addressed in Chapter Two), he describes how God's Love and Wisdom project out like a sphere or unified field. God first puts limitations (constraints) on His infinity by sending out *spiritual* substances from Himself. These primal and non-material substances are various *finite* qualities of Love, that is, they become distinct qualities and recipient forms of some aspect of God's perpetual influence.

This creative activity is not to be understood in terms of physical motion. Rather than a movement from one location to another, these successive progressions are *changes of state*—from one quality of Love to another. These changing forms of Love represent God's character by displaying Divine utility through infinite distinction and infinite unity. However, these qualities of serviceability actuate nothing "real" because they lack fixedness and permanence. They are potentials seeking to be manifested in physical time and space and they are manifested when they take measurable, physical form.

God must convert the psychical measurements of Love (substance) and Truth (form) into physical measurement. This geometrization of spiritual goodness requires a special medium or pre-geometrical entity.

The Divine Will focuses its non-physical activity (its determined Love) at a "point." This kenotic (Divine self-limiting) step of pinpointing spiritual activity creates an essential singularity. It is an entity with placement *although not yet in space or time.* So brace yourself. What follows will be difficult to grasp.

Swedenborg suggests that we think of this micro-unit of pure and total action as an internal effort (*conatus*) towards outward motion. It may also be called *internal state.*[10] The singularity is a disposition or perpetual attempt to generate force, kinetic energy and velocity. From its internal endeavor, the singularity can convert itself into external or local motion.

Even when we understand Swedenborg's notion that propensity or endeavor is both real substance and the causal principle behind motion, we still have to account for the mass of the moving singularity. Disposition, force and its kinetic energy remain a potential unless they can accelerate a mass. But how can mass be understood in a non-physical entity? Swedenborg treats the trajectory of a singularity as a *virtual* event. How can a virtual process have mass?

Nuclear physicist Ian Thompson points out that in Quantum Field Theory (QFT), which makes use of virtual events, the kinetic part of these processes is not a "bare mass" but a "dressed mass."[11] The concept of dressed mass allows us to think about mass even in non-physical ways.

Simply put, dressed mass is the "clothing" which gives form to a disposition. This clothing can even represent *metaphysical mass.* As noted above, the psychical is fundamental. Love is substance and Truth (information) is form. In the case of a singularity, its dressed mass can be described as a density of information (God's patterning principle and Divine cosmic code focused at a point).

In Chapter One (page 6), we saw that with a little introspection, we can understand that we clothe our dispositions and intentions in the information of our thoughts and ideas. Our intentions (derivatives of Love) acquire form and structure through the process of thinking. Because thoughts clothe or dress-up our volitions and reveal their quality, they also give them their dressed "mass."

We instinctively understand this metaphysical principle of density when we recognize that some ideas are more complex and carry more "weight" than others. This is the mass that is involved in the kinetic part of mental action. Our intention and thought goes somewhere, that is, it has a *trajectory* (but not in physical space). When a thought "moves" towards some goal there is acceleration of metaphysical mass, according to the power of the desire. What this all means is that the measurement of physical momentum, in terms of *mass*, *distance* (length), and *time*, has its origins and analog in the dynamic magnitudes and non-physical processes of heart and mind. In psychical momentum, mass (the quantification of Love) involves *values*, distance is described by *affinities*, and time by the progression of *thought* (which is a measure of our understanding).

Again, in Swedenborg's cosmological model, physical laws and forces emerge from spiritual laws and forces, that is, from the mind of God. The dressed mass of a cosmic singularity represents the "density" of the Divine *telos*, or the ethical core of God's Infinite *living* nature. This is the dynamic behind the bio-friendliness of the universe. A singularity, therefore, is a cosmic seed containing a complete idea of the universe!

Are you still with me? If I've lost you, hang in there. These are most difficult ideas to grasp, especially the ones that follow. For most of us, these are completely new ideas. But our current picture of reality is in need of a grand shake-up.

The dressed mass of a singularity, even when it has been put into a kinetic energy mode (trajectory), still does not lead to an actual physical outcome. Swedenborg's singularity acts very much like a virtual particle in Quantum Field Theory—with the one big important difference: it can operate both IN SPACE and NOT IN SPACE!

How can that be? How can something involve a real displacement of location and still not be operating in physical space and time? The answer is crucial for understanding how spacetime and matter first made their appearance.

▶ A Conceptual Universe

Physicist Lee Smolin, in *The Trouble With Physics*,[12] makes the challenge that "We need a theory about *what makes up space*, a background-independent theory." It is assumed by some physicists that a background-independent theory is the key to unifying Einstein's general relativity theory with quantum physics, and yielding the correct theory of quantum gravity. What is a background-independent theory of spacetime? It is a theory of creation that does not start with a fixed spacetime background or anything moving in such a space. Space is derived from a deeper principle.

Other scientists, like David Bohm and more recently, Michael Heller, believe that finding deeper principles requires challenging the Cartesian description of geometrical space. Treating geometric space as a manifold or Cartesian grid (coordinate system) is too rigid and inadequate for uncovering new and hidden orders of reality. Bohm states, "Notions of wholeness, non-locality, and indivisibility within the quantum order are at odds with the Cartesian order."[13] Heller believes that a more flexible method for defining space that allows scientists to make further generalizations can be based on algebraic concepts and functions.[14] Here the geometry of space is structured by a function (a mathematical trajectory). But something dynamic must generate this structure or we will be leaving out the real physics from the fancy math.

Swedenborg offers such a theory of *structured* or *differential spaces* based on dynamic functions. Over two centuries ago he understood that spacetime was not fundamental. He did not view space as a background upon which events take place, but it derived and resulted from the EVENT. In other words, a dynamical function or kinetic event makes its own kind of space—they are consubstantial (the same thing). As we will see later, this gave Swedenborg the flexibility of describing spaces in a way that was radically more dynamical than even Einstein's ideas. A "space" could be described by a physical action as well as a mental or spiritual action. Therefore, spacetime is more than a continuum. It also contains discrete levels of structure, which manifest different qualities (that get closer to God's nature as we go inwards).

Swedenborg's model of the causal structure of space is hierarchical. In this model, gravity not only can be described as discrete units (quanta) but also *qualitatively* distinct units of flux that operate in different "kinds" of spaces and under different geometrical principles. Later in the book we will explore how gravity has its analog in mental and spiritual dynamics. The possibility of non-physical gravity is quite beyond what most of today's physicists are expecting.

Swedenborg's background-independent theory hits all the hot buttons that Smolin identifies in his book as successful approaches to quantum gravity—that spacetime is *emergent*, consists of *discrete* structures, and involves *top-down* causation.[15]

The reason for the intelligibility of the universe is that spacetime is derived from a real concept and rational plan. The universe obeys mathematical rules. While this is taken for granted, no one has provided an explanation for this. Physicist Eugene Wigner wondered why the human mind, which invented mathematics, expresses the same conceptual plan of nature. Physical scientists shy away from the idea that the intelligent order of the universe comes from intelligence itself.

The most important idea to grasp is that there is no spacetime until it has coherent structure, orientation and lawfulness. Spacetime becomes manifest through the kinematic display of a rational concept.

Now we can return to the extraordinary premise mentioned above—that a singularity could have a trajectory in either a pre-space condition or in real space. The difference has to do with the *concept of order*.

One thing moving alone in the universe (like a singularity) has no meaning. You wouldn't even know if it was moving. Without a notion of speed and distance, space has no meaning, either. You would need a singularity moving relative to other singularities to discern whether it is really moving. This would also allow you to determine if the singularities were moving in an *orderly* way, creating orientation in spatial relationships.

Swedenborg's theory of creation starts not with one essential singularity, like the Big Bang, but with an infinite number of them filling the prenatal universe (quantum vacuum). Since one singularity rules out a quantum mechanical beginning, Stephen Hawking has also surmised that the universe could contain an infinite number of singularities. While Hawking was ultimately frustrated by singularities as a "disaster for science," especially in terms of how they might contribute to the lawfulness of the universe, Swedenborg was spared from this frustration. He had a different assumption about physics and lawfulness. His singularities emerged from a pregeometric spiritual realm of infinite Divine order as predicates of Love (with a propensity to form relationships). Swedenborg literally applied the spiritually-based principle of Love to the lawfulness of the universe and the generation of spacetime structure.

Swedenborg's prenatal (structureless) universe is very similar to the quantum vacuum, which consists of pure potentialities (called a probabilistic froth). But there are notable differences. Unlike the wave mechanics of the Schrodinger equation in quantum physics, which describes *no trajectories* (a mixed state of being in many places simultaneously), his idea of possibilities in the micro-world are more like physicist Richard Feynman's idea of quantum events as consisting of *all possible trajectories* (called path integrals). Moreover, since Swedenborg's singularities are derived so immediately from the Infinite and spiritual world they surpass every degree of velocity; in fact, these velocities are indeterminate and can express all possible angles and curvatures.

Because of this, it is impossible to track the trajectory of a singularity. They can only produce a measurable outcome through cooperation. In other words, coherent spacetime structure emerges through some incarnation of Love.

▶ The Emergence of Space and Time

We now have a prenatal universe consisting of the indeterminate activity of numerous singularities. Space and time still do not yet actually exist within this pregnant void of virtual activity, this probabilistic froth. There is as yet no coherent structure from which space and time can be described or have any meaning. For even when singularities are pressed forward from their internal effort and forced into external motion, their indeterminate trajectories remain pregeometric, that is, their actions cannot yet create coherent *units of time* or a physically *structured space*.

Existence is relationship. Spacetime structure does not emerge until singularities form coherent and stable relationships. How they form these relationships is from the dynamical agency of Love. Love is the theological means by which spacetime structure and law emerged to form a rational and intelligible universe.

All spacetime structure must be *generated*. Since the virtual events of singularities are the only things at hand in our prenatal universe, spacetime structure can only arise from the *cooperation* of these virtual events. The activities of individual singularities are virtual events because alone they actualize nothing. When virtual events cohere into an organized region of activity, actual space and time are the outcomes, giving us the birth of the physical universe. The challenge here to our normal way of thinking is that a physical space does not exist until after virtual events come together in a subordinated and coordinated manner.

So how do virtual events actually congeal from the spiritual dynamics of Love? And how does their cooperation create physical time and space from non-time and non-space?

Motion organizes and creates order. It is by motion that all things tend to their equilibrium and find their "place" in the universe. This is called gravitational *order*. Since motion is more fundamental than rest (it has its origins in God's perpetual flux focused into a singularity), an equilibrium end state or stabilized scale can only be obtained when singularities conspire towards some unifying geometrical situation that can support *continuous action*. This is why gravitating systems have a tendency to grow structure spontaneously. Motion is what adapts figure or structure to disposition (and why spacetime structure involves causality). In the early universe, structure must adapt itself to God's eternal disposition, that is, find form in perpetual motion. Furthermore, this unrelenting motion must carry the dynamics and patterns of God's ultimate conceptual scheme.

According to Swedenborg, God's ultimate scheme involved creating a heaven from the human race. This involves both an outward movement, from God's non-physical realm to the *creation* of a physical world, and an inward movement, back towards God, through the *evolution* of intelligence and consciousness of the non-physical human heart (*will*) and mind (*understanding*).

Since Swedenborg's singularities or "points" of virtual activity have their origins from the Infinite, they each carry the force of this Divine disposition for outward and inward movement. They each share a perpetual endeavor to move from a center to peripheries and from peripheries to centers (a dynamical reciprocation that unifies motion and is, therefore, a physical analog or image of Love—a cosmic heartbeat). So singularities, having similar dispositions, will exercise the same effort in their motions. It follows that by continuing their motion they will eventually be brought into a situation conformable to the motion and figure of each.

This is the dynamics behind instantaneous self-organization. The combined and unanimous activity of flux and reflux among the singularities is the means by which virtual events transition into actualized events. From being mere possibilities without any discernible relativity (depicted in Figure 3.1), they have lawfully conformed to each other's action and begin to coexist as a coherent unity, producing a geometrical form with a *real* physical center and a *real* physical periphery (Figure 3.2). This is why Swedenborg maintained that for something new to exist, previous things must coexist. Coexistence is from the spiritual principle of Love, the essence of which is to unite.

Since each of the points or virtual particles is sustained by a perpetual influence of the Infinite (Divine order with zero entropy), these virtual events by geometrical necessity conspire towards a unanimous form from their continuous actions, which cannot be sustained or conceived except in a spiral figure. For nothing can spirate continuously unless it spirates reciprocally and assumes the figure of a vortex.

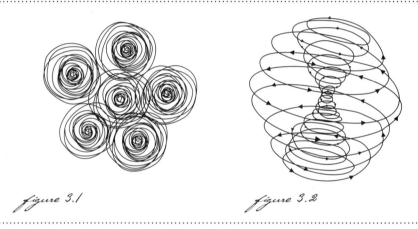

figure 3.1   figure 3.2

Again, Figure 3.1 is my simple depiction of virtual processes (singularities) operating independently of each other and without orientation in a pre-space mode of potentialities. But in Figure 3.2 we now have a discrete unit (quantum) of spacetime composed of an elementary "loop" of spiraling energy that describes a structure with a real center and a real periphery (from which poles, radii, an equator and an ecliptic emerge).

So, the difference between virtual processes (potentials) moving in a pre-space void and things "really" moving from place to place is that real motion needs properties that give rise to intermediate distance, such as an up and down and a horizon, elements that create real, place-to-place directions. Swedenborg maintained that "place" and "distance" can only be conceived when action is oriented to some center or in respect to other entities. This is why Swedenborg envisioned the first unit of spacetime structure as a spiral vortex in which virtual particles describe a real center, axis, radii and peripheries from an action principle. This is how spacetime can emerge through the operation of dispositions and can be described by functions rather than differential manifolds.

This process of virtual particles to congeal is an analog of love because the difference between virtual processes operating in space or not in space is *collective behavior, cooperation* and *periodicity*. This provides us with a good first example of how God creates through self-representation. God's Unified Field of activity (Divine Love) has generated a field-entity on a discrete new micro-level. In doing so, God's sphere of Love has transitioned to its physical analog (mirror image) in a dynamical structure fully dependent on mutual cooperation and a reciprocation of motion. This cosmic dance follows the Creator's Divine order and perpetual influence. This is why motion and organization is a more native state of the universe than rest (inertia) and entropy (disorder).

Similar micro field-entities are generated everywhere in the emerging universe from congealing singularities. These are the building blocks of spacetime. In this cosmological model, spacetime *emerges*, its structure is *discrete*, and it all has involved top-down *causality* (from the Creator). As I mentioned above, these are the features of spacetime that some of today's pioneering physicists are seeking in the hopes of developing a correct theory of quantum gravity (discussed in next chapter).

Since action precedes spacetime, Figure 3.2 is background-independent. It is a *structured* or *differential* space as opposed to a "manifold" space described by coordinate systems based on local grids. This gives us a unit of spacetime structure based purely on functions that describe the kinetic mode of the internal dispositions and forces of the singularities. Figure 3.2 is Swedenborg's depiction of the actual ontological status of primal spacetime structure as geometrically constrained, coordinated process. It is a distinct unit of spacetime. We need to understand space structure in this way for science to make further advances.

Spacetime order arises from dynamical and periodic process taking the shape of a vortex. Why? This follows from the fact that in a dynamic universe, where everything can be defined by process and change, it is only through the reciprocal flux and reflux of vortical patterns that the concept of *place, order, relation* and *orientation* can emerge among the singularities.

Motion traces out both distance and moments. However, as stated earlier, motion in the universe has no meaning if only one thing is moving. Motion can be discerned only relative to something else. But even relative motion has no meaning without orientation. Add to this the fact that time has no meaning if it is not *periodic* (time is measured by the cyclic motions of the earth and other heavenly bodies). This necessitates that kinetic energy must loop itself into a coherent, continuous, repeating cycle, like that of a spiral vortex, which forms a stabilized scale. From this we can see that Swedenborg's kinetic model is in agreement with Max Planck that action comes in wholes (holonomy).

The vortical structure (field-entity) in Figure 3.2 gives the virtual motion of the singularities true relativity, with orientation and order, and all this is put within a repetitive cycle or unit of action (quantum). This dynamic process gives *periodicity to time* and *curvature to space*. Think of it as a *space-clock*, that is, an activity that "temporalizes" space and "spatializes" time. In this way, time and space are related in a fundamental way to God's holistic nature.

▶ **The Emergence of Matter**

Collectively, the indeterminate velocities of the points or singularities moving reciprocally in this vortical looping pattern create the appearance of a solid surface and stable topology. As these streaming virtual particles come together, they also create a new order of *dressed* mass. Not only do we get a unit of spacetime, we now also have an actual, finite particle of primitive matter and the beginnings of physical or bare mass! There is no need for the modern theory of a Higgs field to create the mass of particles by slowing them down as they pass through it because Swedenborg's organic approach relies on aggregations of particles (coexistence) to place new limits and constraints on velocities, and gradually introduce the principle of inertia into the universe.

Thanks to the zero-dimensional point, or singularity, *the beginning of nature (space, time and matter) and the beginning of geometry (topology) are the same event.* This is why the universe can be so successfully expressed by mathematical concepts and is so intelligible. As mentioned earlier, science has yet to find an explanation for why mathematics fits the universe like a tailored suit.

Swedenborg brings all these ideas into clearer relief within a geometrical formalism that unifies interaction (coordinated relative motion) and describes primal units of both space and matter. In the early universe he makes no distinction between spacetime structure and matter. They are consubstantial (the same stuff). The notion that a primal unit of spacetime structure is also a primal unit of matter certainly seems like an odd idea, even an anti-intuitive one. However, contemporary physicists believe that time is closely linked to space and that matter and energy are deeply connected to space and time. In *The Fabric of the Cosmos*, for example, physicist Brian Greene observes that in a theory where only relative relations matter, "the distinction between spacetime and more tangible material entities would largely evaporate, as they would both emerge from appropriate aggregates of more basic ingredients in a theory that's fundamentally relational, spaceless, and timeless."[16]

From Swedenborg's idea that singularities or points are states of infinite propensity and endeavor with zero entropy, we can conclude that the spacetime fabric it weaves consists of motion or kinetic energy that has adapted form from its original spiritual disposition and influence (God's Infinite Love). The fabric of reality therefore consists of irreducible utility rather than irreducible chance.

This turns modern physics on its head. Since all created things are *recipient* forms that exist and subsist from God, the native state of the universe must be profound motion and order, not rest or aimless energy, as traditional physics supposes. Moreover, this Theo-dynamical perpetual activity rules out any possible destruction of the expanding universe from cooling off, or falling into disorder. Swedenborg's description of how spacetime and matter are first generated challenges Newton's first law of motion that bodies tend to remain in either a state of rest or of movement in a straight line. Space is curved because curvature is a universal principle of action and economy (the principle of action to remove resistance).

▶ God's Character displayed in the Creation Event

Figure 3.2 (on page 44) also illustrates that God's Divine Love (goodness) and Divine Wisdom (truth) find their physical analog represented in the dynamical process of creating time, space and matter. In his theological works, Swedenborg describes God as all form (so complexity emerges from complexity). Specifically, he described God's Divine form as consisting of things both infinitely distinctive and infinitely united.

Distinctiveness and *unity* come from two properties of *mind*—the intellect and the will. Divine Truth, God's intellectual mode, distinguishes one thing from another (making all things discernible), and Divine Love, God's volitional mode, unites all things (making all discernible things into an interconnected whole). Complexity in the universe emerges and perfects itself at the same time as distinctiveness generates greater unity (thus becoming more God-like). Complexity and self-organization are also analogs of Love because all unity is for the sake of *utility*. Nothing enters into the created universe unless it perfects unity through some distinct usefulness. And nothing can be utilized until process takes on stable form. So physical reality mirrors spiritual reality through utility, which are forms representative of God's goodness.

Divine activity cannot flow into dissimilarity or discord. This would obstruct top-down causation. All dynamical process follows the spiritual creative dynamic by which Love conjoins itself to Truth (a cosmic marriage), so that, *differentiation perfects unity*. This principle of relational holism is behind all the emergent features of the physical universe, including biocomplexity in evolution. Whether we speak of the formation of starry systems and galaxies or biological systems, dynamical structure obtains greater complexity as increasingly distinct things increasingly perfect unity. This dynamical process of unified differentiation can easily be shown in the simple action or process of something moving in a circular loop.

Motion is the displacement of position (Figure 3.3). It is through this displacement and change that motion distinguishes itself and becomes discernable. When this progressive action is unified in the form of a circular orbit or loop, the *change creates constancy* and stability (Figure 3.4). This process is analogous to self-conscious action and foresight (Divine Providence). When something moves in a circle (returning to where it starts), its action must be perfectly self-referable at every *infinitesimal* point along

figure 3.3

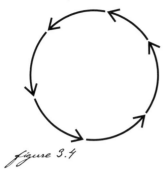

figure 3.4

its trajectory. Otherwise it will fly off in a tangent. This self-referable action includes both *perfect change of direction* and *perfect timing.* It is the power to unite the past with the future at every "present" moment. A circular action provides us with a simple geometrical principle for how God can behold the past, present, and future simultaneously within physical dynamical process.

This circular display of Divine consciousness in process produces a unified whole. This is why the Infinite can be self-represented in, take geometrical form in, and occupy a finite moment in time. God's eternal character and infinite nature is present at every finite moment and place of a dynamical process—as long as it is oriented to an eternal goal. The eternal goal is represented at every infinitesimal "point" of a cyclical process.

For instance, if we express a *transcendental* circle (Figure 3.5) which describes God's creative act as moving outwards from a non-physical spiritual realm to the formation of inert matter, then returning back to God and heaven through the evolution of human intelligence and consciousness—by means of spiritual values—we can see how the eternal goal is *simultaneously* present in each part of the sequence. The goal of creating and perfecting a heaven from the human race is eternal because it is never-ending.

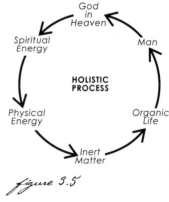

figure 3.5

This explains Divine transcendence and Divine immanence—an Infinite God who can relate to, and act in, a temporal and finite world. The new idea that Swedenborg puts on the table is that the nexus between the Infinite and the finite is *itself infinite*, because God's infinite activity directs, and resides in, all finite events from an eternal goal. (This purposefully precise activity brings up the nasty problem of

Theodicy—the issue of evil in the world—which I will address in a separate chapter.)

In geometry, a circle consists of an infinite change in angle around a circumference with a finite radius, which makes this form a *bounded infinity*. Circular rotation and angular momentum are ubiquitous in nature (unlike straight-line motion) and are attributes of matter, whether in the quantum microworld or the macro-orbits of planets.

But some forms of flux and reflux surpass the perfection of the circle, and are, therefore, more suitable as *recipient* forms of God's Infinite action, which suggests that finite forms can receive the Infinite to a greater or lesser degree of perfection. A spiral form, as in Figure 3.2, is *perpetually circular*. Since a circle represents infinite change of angle and a spiral is infinitely circular, in the spiral, change of direction and timing takes on an even higher degree of perfection by means of what Swedenborg called a new *analysis of infinity*. This "ascent of form" allows for greater complexity and modification (changes of state) to enter into the dynamics of flux and reflux. We will see in Chapter Five how Swedenborg's unique idea of submitting geometry to new analyses of infinity allows process and form to lawfully remove itself from the constraints of time and space and gain new freedom of action, far beyond that which modern physics describes.

Recall that physical trajectories involve changes of location while spiritual or non-physical trajectories involve *changes of state*. As form and dynamical process move in the direction of more fundamental things, they must therefore be capable geometrically of spontaneously running through increased changes of state and modification. This allows for more responsiveness to Divine activity.

Because of this increased capacity for changing states through the systematic removal of geometrical constraints, as process moves towards a more fundamental level, things become more energetic, more life-like and ultimately intelligent (because they are getting closer to God). Swedenborg describes this transcendental feat of radically removing constraints from principles of action in his *Doctrine of Forms*. He builds a hierarchical ladder to describe more elevated forms and trajectories and to discover new powers of infinity (beyond the spiral) that ultimately lead back to a pre-geometrical realm and the Creator (again, I will say more about this in Chapter Five).

Swedenborg's Doctrine of Forms can help modern science turn quantum physics into a more geometrical theory. He uses curvature (non-linear expressions) rather than vector spaces to describe the physical idea of quantum superposition. This unique challenge will be addressed in the next chapter.

▶ Advancing Science Through New Insights About Love

There is nothing essential in motion or process except endeavor, effort and disposition. Endeavor is a spiritual dynamic and a derivative of Love. God consciously creates structure and law by adapting motion and form to His Divine disposition, which is Love. Self-organization in the universe can only emerge as an incarnate form of Love, that is, structured utility and cosmic usefulness.

A powerful theological theme of Love underlies all the emergent properties of matter and biological complexity in evolution. Biological systems are simply an extension of the theme of differentiation perfecting unity through *reciprocal utility*. Both physics and biological systems are under the same reciprocal laws of flux and reflux, which have their source in the original Divine disposition of Love operating within the endeavor of singularities.

Creation is not the result of a mindless Big Bang explosion (or even a mindful one). The universe exists because of its essential disposition to generate and promote *coexistence*. Nothing exists unless it coexists. Existence is relationship. There is no union without reciprocation and no reciprocation without some act of goodness (sharing). The universe is fine-tuned for reciprocating action and promoting profound relationship. Order and organized structure based on the reciprocation of action provide powerful evidence that dynamical process follows a non-spatial, non-temporal, principle of unity—the spiritual principle of Love. As mentioned previously, this is why the universe is so bio-friendly.

The Creator can only be in relationship with creation and spacetime from the things in it that reflect God. This means that God can only create by conjunction (which is the big, important missing ingredient in the current view of Intelligent Design). The conceptual universe is based on Conjunctive Design (CD).

Conjunctive Design and the exaltation of Love explain more than creation. CD also explains why the universe is expanding and why this expansion is accelerating. God's Infinite prolific principle continues to produce more space, matter and intelligent structure. As this process of creation and evolution continues, the universe becomes a more perfect recipient of God's Infinite Life and Divine influence (increased dynamics through increased correspondence).

Swedenborg's spiritually-based approach to science can explain some of the most difficult modern issues of the New Physics. In Swedenborg's vision, as we have just seen, spacetime is not a backdrop. It is active information flowing in discrete parcels. This creates the first units of space and matter directly from dynamical process. Spacetime structure consists entirely of relative relationships and holistic process.

From his recognition of pre-space principles striving to become actual measurement outcomes in time and space, Swedenborg anticipated Einstein's discovery that energy and matter are related. But he took this idea much further. In his cosmological model, dynamical process and kinetic wholeness results when active information springs out from a theological disposition and Divine consciousness. On the deepest, non-spatial and non-temporal level, the fabric of reality is Infinite Love, which causes the kinetic universe to form a spiritual tapestry consisting of irreducible usefulness and goodness, not the irreducible chance that some interpretations of quantum theory posit.

Swedenborg offers an interesting spiritual reason for the emergence of complexity and why the flow of time (arrow of time) is asymmetrical and irreversible. To make time go backwards is to undo Divine purpose and goodness and *de-sanctify* dynamical process.

The essence of Love is to unify. This provides us with a theological reason for why everything emerges from one (non-physical) formative substance and why everything is non-locally connected simultaneously. Order, derived from Love, is only universal if it takes place in all the minutest parts, creating a holographic universe. According to Swedenborg, such a universal order cannot have one of its parts touched or excited without producing an effect on everything else.[17] This instantaneous transfer of signals is given scientific support in Bell's theorem and the Aspect experiment, which shows that there cannot be a real world out there, independent of an observer, without faster-than-light signal propagation.

Swedenborg, like Einstein and David Bohm, believed that there was a real objective world out there (unlike the Copenhagen interpretation of quantum mechanics that depends on an observer), and that signals could travel faster than light (unlike Einstein's special theory of relativity). Even as I am writing this, reports have reached me that two German scientists are proving that faster-than-light speeds exist in the universe. I suspect more such discoveries and verifications will be made in the future. Faster-than-light velocities will challenge modern science's current view of causal spacetime structure and its cosmological notion of invariants (constants of law). Chapters Four and Five will address this.

Another modern scientific notion derives when God's nature is recognized in creation and dynamical process. In terms of the theologically consistent consequences of *differentiation perfecting unity*, wholeness (holonomy) is intrinsic to action. One of the main features of quantum physics is that action comes in whole units (quanta).

Now comes something more intriguing. As you can see from the illustrations provided in this chapter, Swedenborg used forms of *curvature* to define units of action and build spacetime geometry from dynamical functions. Remember that in this background-independent model the kinetic action is the "space," and action and space are consubstantial (the same stuff). One of the main features of general relativity theory has been to redefine the curvature geometry of spacetime as *gravity*. Amazingly, Swedenborg's concepts concerning space, time, energy, and matter, embrace both features contained in our most important modern theories in physics—curvature of space and quanta—with *theological* considerations.

Swedenborg's science was not merely a product of his own era and therefore quaint. (Scientists reading the information of this chapter will even see some similarity between Swedenborg's concepts of spacetime and the background-independent theory of *loop quantum gravity*, a main alternative to string theory.) His discoveries would be even more relevant to 21st Century paradigms if he had actually solved the great mystery of quantum gravity.

What remains to be teased out of Swedenborg's thinking is a way to express the *superposition* principle of quantum theory in terms of spacetime curvature. It is worth a try, which is the challenge of the next chapter.

SUMMARY

▶ Spacetime is derived (an outcome of a deeper, pre-space principle).

▶ The universe is conceptual.

▶ There is no spacetime until it has coherent structure.

▶ Swedenborg's singularity is a force-carrier of God's universal patterning principles.

▶ The first essential of motion is endeavor.

▶ Spacetime and matter are physical analogs of the various qualities of Divine Love (unity) and Wisdom (distinction).

▶ The emergence of self-organization and complexity involves differentiation perfecting unity through reciprocal action.

▶ The Infinite is real.

▶ The nexus between the Infinite and the finite is itself Infinite and operates by correspondence (similitude).

▶ The Infinite is in finite things according to reception (and a form's ability to change states).

▶ There was no Big Bang beginning. The universe is expanding because it is growing and displaying God's infinite nature. God is creating more matter and space.

PREDICTION

In this century it will be conceded that
physical laws and forces are actually spiritual laws and forces
constrained by what exists in time and space
and, therefore, correspond.

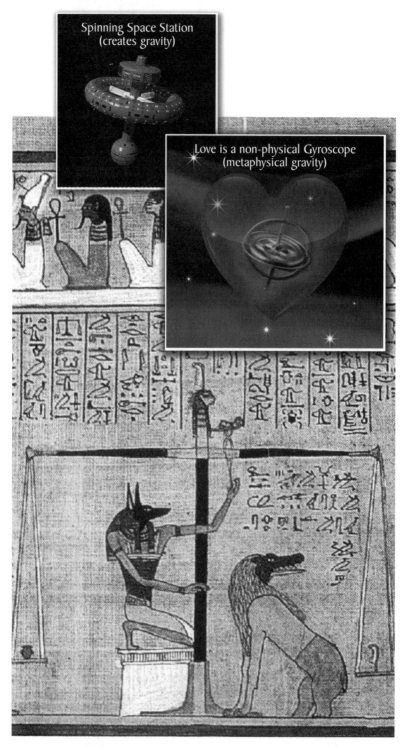

Spinning Space Station
(creates gravity)

Love is a non-physical Gyroscope
(metaphysical gravity)

Weighing the heart against the feather from *The Egyptian Book of the Dead*

LOVE AND QUANTUM GRAVITY

> "Science without religion is lame, and religion
> without science is blind."
> — Albert Einstein (1947)

God did not create the world with a snap of the fingers. Nor was the world created all at once. The process of creation is not continuous from God but is a top-down sequence of distinct events and successive stages moving from a non-material realm to a realm of space, time and matter. This is not to say that God's creative action is not constant (*creatio continua*) throughout the created universe; rather, it says there are discrete jumps between the steps by which the world takes shape over time in a lawful, evolutionary process (the idea of discreteness is in keeping with the discoveries of quantum theory). Swedenborg offers us a theological reason for why this must be so: if creation were continuous with God, the universe would be God rather than a finite, mirror image of God's Divine nature. That would make the act of creation an act of self-love, that is, God making Himself the focus of Divine Love.

Swedenborg claimed that self-love is diametrically opposed to true, spiritual Love, which seeks to love "some one outside of self and by whom one can be loved in return."[1] Therefore, Divine Love necessitates the creation of a distinct finite world with finite beings in whom there is nothing of the Divine itself (but who can love the Divine).

This is why there must be discrete or discontinuous "jumps" of activity in the unfolding sequence by which God created the physical world. Quantum physics, which embraces discontinuity in its fundamental picture of nature, may well have its origins in theological dynamics. This may seem far-fetched, but the fact is that no one knows where the laws of quantum mechanics come from (bear with me—the difficult issue of quantum gravity requires some setup).

Ironically, even though quantum theory promotes *discreteness* or *quanta*, the Schrodinger equation, which is the mathematical description of how a quantum event evolves over time, describes a *continuous* process. I will spare you the math. What is important to know is that the Schrodinger equation doesn't portray any "jump" to support the idea that the fundamental world of possibilities collapses into real measurable outcomes, which physicists assume take place in a quantum event. So there is a big problem with our fundamental understanding of the microworld and how it correlates with the behavior of the larger world we can experience. This opens the door to new interpretations.

I propose that the current formulation of quantum physics is wrong (some physicists already sense this). While the mathematics of quantum physics is very successful in portraying the idea of superposition (that a subatomic particle can exist in a mixed state of multiple locations simultaneously) it does not describe what is really going on in the microworld. The reason for this is that a particle existing in the ambiguous state of superposition (also called wave/particle duality) is a magical condition never observed in the real world (Figure 4.1).

This is why quantum theory finds itself faced with what it calls the measurement problem. The measurement problem is this: how does the fundamental realm of quantum potentials "choose" one actual outcome over the others? Stated differently, how does something that acts like a murky wave of probabilities collapse into a clear, single actuality?

On the one hand, physicists assume there is a wave of probabilities. On the other hand, they assume that this wave function "collapses" at the very location where they happen to find the particle at the moment of making a measurement (Figure 4.2). But the second assumption in this scenario is never described in the Schroedinger equation.

In the Schrodinger equation, wave functions do not collapse. The notion of collapsing was added on later in a forceful attempt to make the mathematics fit the observations. Even when scientists accept the rude mathematics of all this, they are still left with the puzzle of how a wave function actually "chooses" to be in one particular location out of all its many possible locations. Most physicists believe that the act of measurement or some other extraneous thing in the environment causes quantum potentials to "choose" and collapse together into a real, measurable location.

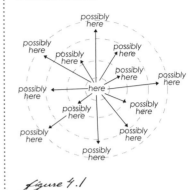

Wave Function Distribution of Possible Outcomes

figure 4.1

Some physicists have tried to get around this by proposing theories where the wave function never has to collapse. Hugh Everett, in 1957, suggested a *Many Worlds interpretation*, in which the wave function consists of particles in "parallel universes." This allows every quantum probability of the wave function to lead to a real outcome in one of those universes—without collapsing. This interpretation is gaining in popularity.

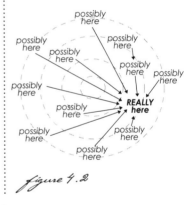

Collapse of Wave Function

figure 4.2

Also in the 1950s, David Bohm proposed that waves and particles maintain their independent reality as in classical physics, but that quantum possibilities, rather than collapsing, operated as hidden-variables within what he called "pilot waves." These waves contain information and spread out in space as possibilities until they reach an actual particle and guide it in a particular way. Pilot waves get their information from various influences in the universe, even distant influences. For this idea to work, quantum particles must possess subtle and complex inner structure (like a radio) and pilot waves must be *instantaneously supersensitive* to all the conditions of the universe. This instantaneous transfer of information is called *non-locality*.

While Bohm's interpretation does not require a wavefunction to collapse, he preserves the Schrodinger equation and non-locality by adding this complex ability of instantaneous change to the force of the wavefunction, called the quantum potential. Individuals who are spiritually minded prefer Bohm's model because it provides a way to see the world holistically, where everything is causally connected into an entire system. However, while there's no measurement problem in Bohm's theory it is unclear exactly how wave influences lead to the ubiquitous principle of least action found in the larger, classical world of Newtonian physics.

Before we turn to the specifics of Swedenborg's solution to how the microworld meshes with the macroworld, we need to look at Nobel Laureate Richard Feynman's efforts to show how the classical laws of least action (nature's economy of action) arise from the erratic dynamics of quantum superposition. It is called a "path-integral" or "sum over histories" approach.

Schrodinger's wave equation assumes that a quantum particle has no specific physical trajectory. Instead, the function implies that the particle just spreads out like a wave (Figure 4.1). Feynman, however, bypassed the Schrodinger equation and described the superposition of a quantum particle as one that takes *every possible trajectory* simultaneously, to get from one place to another.

The different potential paths or possible histories depicted in Figure 4.3 take place in imaginary time and interfere with each other in a way that cancels out some and reinforces other alternatives. Eventually, they arrive at a final combination and trajectory that is more classical (Figure 4.4), and best upholds the principle of least action. Scientists call this a *correspondence principle* that *recovers the action principle of classical mechanics.*

One problem of Feynman's path-integral approach is that a particle like an electron has to consider every possible path at the same time as it also interacts in complex ways with other things, like photons. This complicates the situation and results in an endless increment of probabilities, leaving scientists faced with infinities (which they hate).

Another problem is that this path-integral

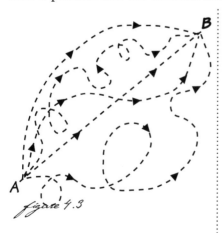

figure 4.3

approach is purely a math-ematical technique and convenience. It goes to great lengths to create a system that assigns amplitudes of great magnitude to a range of *sensible and absurd* trajectories. These trajectories then gain significance through mutual interference. Their possibilities strengthen or negate each other until there is one observed integrated outcome. Yet none of these "paths" requires a physical basis in reality. So we

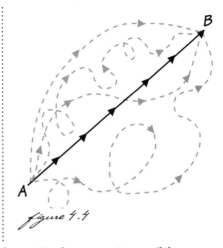

figure 4.4

are once again left with a mathematical convenience (like the Schrodinger equation) that tells us nothing about what is really going on in this microscopic world.

Recall from Chapter Three that Swedenborg embraced both infinity and the principle of correspondence or self-representation (pp. 34-35). In suggesting that *infinite direction* produces the principle of correspondence by the deterministic endeavor of possibilities, Swedenborg demonstrates his unique and timeless genius. He provides a more rational and even spiritual approach to depicting infinite possibilities and their distribution instead of depending on absurd trajectories that depict mathematical contrivances. In other words, he demonstrates, geometrically, how a process moving in all possible directions generates a preferred spatial direction by means of the correspondence principle. We are about to see why this physical process is an analog of Love.

▶ A Non-Linear Approach to Superposition

One major reason why quantum physics is an incomplete theory and needs to be reformulated is that it does not satisfy the correspondence principle. Quantum indeterminacy does not correspond to the determinate processes and order we see on larger scales. Fundamental randomness at one end of the spectrum does not produce orderliness at the other end. A true correspondence principle involves more than a mathematical gimmick to recover the classical laws that operate in large systems of the universe. A true science of correspondences would show that inherent within the mixed states of quantum superposition there is a dynamic or determining mechanism by which potentials seek relationship and coalesce into a preferred measurement outcome through analog and self-representation.

Swedenborg's spiritual approach to mixed states of potentials in the microworld is based on the unifying principle of Love. Rather than relying on mathematical convenience, he offers up a geometrical approach to multiple path histories that require virtual particles (singularities) to actually form mutual relationships and follow *real paths* that describe not only infinite direction but a *coherent* and *stabilized* unit of infinite direction.

What is a stabilized unit consisting of simultaneous infinite direction? It is a complex of reciprocal actions taking form within a coherent unit of curvature. Simply put, Swedenborg believed potentialities could tend toward a common equilibrium and form a microsystem. Nature has an unceasing compulsion for complexity and self-organization at its most fundamental level.

Swedenborg gives this *simultaneous superposition of coexisting paths* a real physical interpretation through complex spiraling curvatures or vortices. This curvature allows the

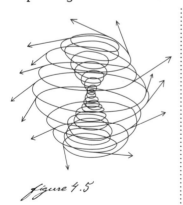

figure 4.5

action of singularities to move in every direction whereby all possibilities find their equilibrium in one unanimous coherent form of ordered flux and reflux (Figure 4.5). The superposition principle of possibility outcomes is built into the very structure and dynamics of Swedenborg's model of a quantum particle or field entity (without smearing the particle throughout space like a spreading wave).

Swedenborg views infinite trajectories not as random dispersal, scattered outwardly in all directions, but as ordered flux and reflux. This geometrical approach has another benefit: it gives us something resembling a "collapse," whereby a state of potentialities (represented by singularities) makes a sudden discontinuous jump into a specific, permanent outcome (quantum particle). The correspondence principle emerges because the quantum particle or field entity forms a likeness of the effort in each singularity to move from a center to a periphery and a periphery to a center (flux and reflux).

Something even more astonishing can be teased out of these ideas. Swedenborg has wedded Einstein's ideas about gravity as spacetime curvature to the concept of quantum discreteness. Remember, the emergence of spacetime structure in Swedenborg's cosmological theory of causal processes was background independent. Spacetime structure emerged from the operation of dispositions described by *functions* expressing complex geometries (flows of force moving in coherent, infinite directions by means of a complex ordered curvature). In other words, in this non-linear treatment of superposition, quantum mechanical principles are built into his theory of creation and the constituents of space.

By linking quantum possibilities (path histories) to coherent units of spacetime curvature, he makes *quantum gravity* a natural outcome of his theory. Science, by contrast, has yet to formulate a satisfactory theory of quantum gravity.

▶ Quantum Gravity

World-renowned mathematical physicist Roger Penrose, who is never afraid of swimming against the current, believes that the key to finding a correct theory of quantum gravity is not to modify general relativity theory, but to use Einstein's ideas to modify quantum mechanics. Most physicists take the opposite view. This is a bold and direct challenge to the Schrodinger equation and its use of a linear picture of superposition (a complex mixture of arrow-like vectors) to portray quantum possibilities mathematically.

Swedenborg would side with Penrose, who prefers to bring the principles of curved spacetime geometry to bear on the rules of quantum theory.[2] Swedenborg did not need a wavefunction or a complex linear notion of superposition to describe the concept of possibilities in the microworld. Instead, he envisioned the distribution of possibility within a dynamic *geometric* theory, consisting of virtual objects whose trajectories represent infinite direction through complex curvature.

Wavefunctions make it difficult to interpret quantum mechanics from geometrical principles. I have even heard quantum mechanics described as non-geometrical. Since gravity is tied to spacetime curvature, the anti-geometric nature of a wavefunction challenges the ability to visualize what quantum gravity might be.

To correct quantum theory, Swedenborg would back up Penrose and suggest that we get rid of quantum theory's complex linear superposition principle and exchange it for a non-linear superposition principle. Curvature geometrizes quantum possibilities, as shown in Figure 4.5.

The best way to illustrate Swedenborg's unique approach to the microworld of possibilities and how this might lead to a correct theory of quantum gravity is to return to the theme of the emergence of spacetime structure discussed in the previous chapter. Recall that in Swedenborg's pre-space and pregeometric realm of singularities, momentum and position had no coherent existence until the individual singularities came into relationship and became dependent upon each other. Before that, they represented mere potentialities and virtual events.

In Swedenborg's model, the so-called "collapse" of potentials into an actual distinct outcome is deterministic and consists of something akin to cooperation and collective behavior. A new principle of constraint or physical law emerges from this collective behavior as each singularity adjusts to and tempers the other.

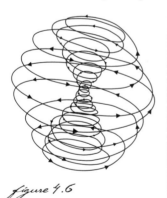

figure 4.6

If we again reproduce Swedenborg's first finite particle (field-entity) discussed in the last chapter, we can see that its topology or surface consists of an organized swarm of virtual particles.

Figure 4.6 gives us a quantum entity (energy-bundle) generated from a *simultaneous superposition of coexisting paths*. The vortex above is formed by a number of curved paths produced by the flux and reflux of individual virtual particles. In the kinetic medium of pre-space potentials, the singularities have found their equilibrium "end-state" in the form of a continuously spiralling field-entity. This *unified action* is the first example of gravitational order and gravitating systems spawning self-organization.

Here, Swedenborg unifies the idea of infinite direction with the idea of potentials non-linearly, because these curved paths have a trajectory *that is moving in no single direction more than any other*. Where modern quantum theory envisions a fuzzy "cloud" of probability outcomes, Swedenborg's virtual particles or singularities find coherence and unity by moving in a geometrical form that represents *all possible directions!* This kinetic system of complex coherent curvature is a tidy way of solving superposition and the mathematical problems of infinite possibilities that bedevil the calculations of scientists. Again, all the potential outcomes are built into the dynamical structure of this smallest unit of spacetime curvature and matter. The emergence of spacetime structure and matter is the outcome of gravitational order. Where there is spacetime curvature there is also gravity. This geometrical insight allows us to conceive of a unit of gravity (called a graviton).

I admit that these ideas are difficult to express, but if Swedenborg is correct, we have a solution to the measurement problem in quantum theory, that is, how potentials "choose" an actual, measurable outcome. Also keep in mind that he incorporates possibilities into the internal structure of quantum entities from a geometrical principle and dynamic analogous to *Love*. That is, *potentials are deterministic*, and form networks of relationships that conspire towards an actual, measurable outcome.

Swedenborg's field-entity will offer us a way to visualize the deterministic mechanism behind potential directions conspiring towards one actual direction. In this way all states in superposition endure and contribute to the outcome. Swedenborg devised a way by which potentials could interact and be in continuous relationship as they make discontinuous "jumps" into more easily defined trajectories and new units of action. As you are about to see, the potentials *determine* the outcome through correspondence (similitude).

Even those who are inclined to reject the premise that a theological principle of Love operates in the microworld have to admit that quantum theory, as it is currently portrayed, is a theory lacking principles of any kind. Instead it concerns itself with only what is phenomenological (observable). The lack of an identifiable principle stands in the way of understanding how quantum potentials lead to the orderly laws of classical Newtonian physics. It further leaves open the more confounding question of how quantum potentials lead to the finely tuned laws that make our universe so conducive to profound self-organization and are so bio-friendly. Let's now take a look at Swedenborg's theistic solution to how the process of creation makes further discontinuous jumps into new orders of structure in the physical world of space and time through the *correspondence principle*.

If the creation and governance of the physical universe proceeded from the non-stop activity of an Infinite Designer, then action is more fundamental than rest. Therefore, the first created things, by necessity, had to take form in continuous, self-sustained motion, which necessitates a vortex. The mutual flux and reflux of each singularity not only creates the organized structure and "dressed" mass of our first finite particle or field-entity through constant action, it allows them to generate, through their unified path histories, a common geometrical center. This geometrical center also becomes a *center of gravity* for the entire micro-entity, shown here.

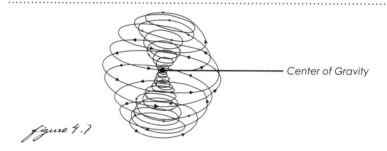

Center of Gravity

figure 4.7

This sets up a mechanism by which the kinetic energy of the singularities can become the *potential energy field* for generating a new *discrete* function and level of action. The disposition and trajectories of the singularities will now be able to impart to the whole field-entity a *similar* and corresponding disposition to move externally.

For the process of creation to continue "downward" through a sequence of distinct steps into more tangible material entities, the spiraling trajectories of the singularities must now make a discontinuous and unified "jump" into *an action on another scale*. What makes this possible is that even though the curved trajectories of the singularities permit them to encompass all directions, they share the same chirality, or right and left-handedness. (For what it is worth, experiments in modern physics, have shown that chirality characterizes nuclear particle reactions.[3]) This cooperative relationship allows a new disposition and force to build up with a *preferred potential direction*. In order for this force, which is generated by the singularities, to produce a new kinetic outcome, it must be able to displace its center of gravity and accelerate the entire "mass" of the field-entity they comprise in an outward direction.

At this point, Swedenborg says another dynamic enters the process. If circumstances permit and there are no hindrances, the spiraling motion of the singularities eventually creates an axillary or circular compound motion as shown in Figure 4.8. Contemporary science calls this motion *particle-spin.**

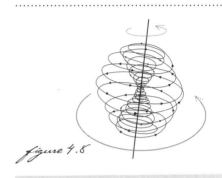

figure 4.8

This axillary spin turns the field-entity into a flywheel and "packages" momentum. It also puts the dynamics of the field-entity into two distinct geometrical situations—a spiral motion of its parts (the singularities) and a circular motion of its compound form. To appreciate what is going on here it is important to grasp that the center of a spiral is *not* in the same place as the center of a circle. Figure 4.9 represents a birds-eye view of Figure 4.8 and depicts the two different geometrical centers of this primitive entity. The

✱ **To appreciate the importance of this, consider that the modern notion of particle-spin and subatomic particles was "discovered" by twentieth-century physicist Paul Dirac in the development of his mathematical basis for Relativistic Quantum Mechanics.** Swedenborg anticipated the standard model of modern particle physics, which identifies two major types of fundamental particle-matter constituents (called fermions) and force carriers (called bosons). Swedenborg identified these two types of fundamental particle as "actives" and "passives." Spin plays a big role in Swedenborg's model in whether a particle is a passive finite entity or a force carrier.

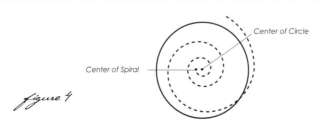

Center of Circle

Center of Spiral

figure 4

consequence of this double-dynamic within the structure of the field-entity is that it allows the axillary spin to impart a force to the center of gravity, which gives it the power to perpetually impel and *accelerate* the whole field-entity to a motion that is extraneous—in other words, to local motion. This new dynamic of action between two distinct geometrical centers is not unlike two dance partners swinging themselves around across a dance floor.

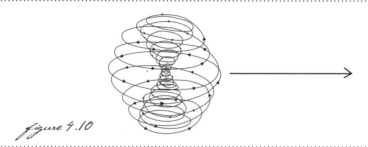

figure 4.10

This, according to Swedenborg, is the deterministic and geometrical mechanism of how potentialities in the microworld cooperate to make a discontinuous "jump" into a preferred outcome. The outcome is deterministic because the preferred direction and new progressive motion of the finite field-entity shown in Figure 4.11 will actually start to take a curved trajectory that mirrors and *corresponds* to the spiraling flow of singularities or virtual processes within its internal structure. In other words, the virtual particles find their preferred direction by creating a new image of themselves on another scale. The only difference is that one more constraint (law) has been added to the original action of the individual singularities. The "jump" results in action from one boundary condition recommencing in a distinct new boundary. This creates a new principle of action that describes a new kind of space. (Remember, because Swedenborg's theory is background-independent, a new principle of action will yield its own kind of space.)

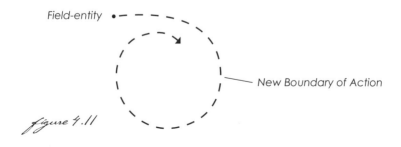

Field-entity

New Boundary of Action

figure 4.11

Once put into motion, this primal field-entity, acting as a distinct new action principle, will find its equilibrium not only in a different structure of "space," but it can form new relationships with other finite particles. This leads to increased mass and more tangible material entities.

Figures 4.10 and 4.11 illustrate in a most simple way how an actual event results from, and is an analog of, a deeper prior order and causal principle. Similitude is the *causal link* and the only means by which successive derivative events can be consistent with a prior causal principle and promote some original "end." In fact, a top-down sequence of distinct (contiguous) events always consists of a *triune order* moving from potentials, through intermediary causes, and into some concrete result (end, cause, and effect).*

In this cosmological model, new relationships and the growth of structure mirror the dynamics of Love to unite all action and process. Swedenborg offers us a real and full-blown correspondence principle, in fact, a *science of correspondences*, with explanatory and predictive powers.

Several important conclusions emerge from this top-down sequence of generative events. First, all action has its origins from perpetual endeavor (and perpetual endeavor has its origins in the Divine Will).

Second, since the act of creation is discontinuous, action principles become layered (hierarchical) and generate different levels of spacetime structure. The geometry of space not only has discrete structure (and is therefore, not a continuum as Einstein described it) but a multi-leveled structure. In fact, because different principles of action must manifest as stabilized scales, their fluxes and trajectories must form coherent "loops." So spacetime, like matter, not only comes in parcels, but in parcels on different levels and scales.

The third conclusion involves a whole new way of looking at gravity. The "jump" from one discrete level to another, illustrated above by the displacement in the center of gravity of Swedenborg's first finite entity, represents a *gravity transition*

* **In the top-down order of creation, virtual particles (singularities) act as potentials and first causes.** Cooperating singularities act as the intermediary cause, relative to any single virtual particle. The reconciliation of both produce the concrete effect of new structure. Next, the effect (or result), such as a new finite particle or field-entity, becomes a new first principle for another corresponding series of end, cause and effect. In each series, the effect is always an analog or similitude of its first principle.

to a distinct new function and action principle. This transition from one action principle to another is motion (and mass) finding its function or *new direction* in a distinct, new equilibrium "end state." Furthermore, this new direction and function bends the geometry of space into an additional layer and species of curvature. All top-down causation consists of gravity transitions whereby action makes discontinuous jumps through multiple levels in its endeavor to find new equilibrium every step of the way.

In this model, the laws of gravity and quantum discontinuity do not work independently. One might even say that Swedenborg gives support to Roger Penrose's notion that quantum gravity is involved in the collapse of the superposition principle into a specific event—providing that "collapse" means a gravity transition to a new principle of action and structure of space. In the next chapter, the different forms of curvature defining these distinct orders of gravity and the multi-dimensional structure of space will be discussed. These ideas may well be more advanced than those proposed by modern string theory or loop quantum gravity, because they involve gravity taking on different *qualities*.

▶ The Origin of Physical Mass and Law

The creation process of the universe continues as new levels of matter and spacetime structure become units for additional new levels. In this way, new levels of order emerge from the matrix of previous levels. At each generative level, forms become more compounded, so matter continues to have more of its energy tied up in aggregations of mass and physical constraints until, at the bottom of the totem pole, is inert matter (radically fixed energy), which finds its equilibrium in a state of complete rest. Again, as pointed out in Chapter Two, Swedenborg knew even before Einstein that energy and matter were interchangeable.

The reason why "mass" exists in the universe is that creation requires non-physical spiritual force and energy to be fixed in time and space. Mass is derived from both accretion (Divine Love's power to unite) and increased resistance to change through new constraints faced at each step, which leads to inertia and fixedness. Natural laws are the outcome of these constraints placed on action, for all law is the introduction of a passive principle in nature (Divine Truth's power to quantify Love).

Within this downward descent, the gravitational arrange-ment of action produces coherent structure, order and orientation on different scales. This endeavor and instanta-neous ability of the universe to self-organize in profound ways is still a mystery to the natural sciences and the philosophy of materialism. But Swedenborg believed he had found the key—a spiritual teleological key.

The laws of physics today (which deny purpose in the uni-verse) lack a "supervising" organizing principle. This is why there is as yet no satisfactory explanation for the universe's ability to self-organize and grow structure—whether in the formation of galaxies or in higher-level systems like the human brain. The key to understanding the concept of agency and of gravitational order is that all things in the universe emerge out from the spiritual dynamic of Love (which, as a first causal principle unites things and breathes "fire" into the equation). The capacity of a quantum world of potentialities to become coherent and stabilized matter mirrors Love's endeavor to become a "concrete" reality in spacetime. Potentialities are formative, non-material substances, because Love's action, rather than matter, is basic.

When you consider Swedenborg's theory that space is not simply a backdrop for events to happen, but that its structure and curvature results from the kinetic event, this allows grav-ity to become equivalent to any dynamic function—including brain function. This brings us back to another unique idea about gravity transitions hinted at earlier—gravity not only expresses *quanta* but also changes its own quality! The notion that grav-ity changes its *quality* goes beyond the current thinking of 21st century science (even for those physicists willing to embrace the concept of gravity variability). According to Swedenborg, the capacity of gravity to change its quality allows gravitational order to describe discrete and different geometries of space, each with distinctly different *dispositions* for organizing action into systems of varying complexity—from galaxies to the human brain and even the belief-systems of non-material mind.

▶ The "Gravitational" Organization of the Universe

The seemingly miraculous tendency and disposition in the universe to grow structure is directly related to gravitational systems. But Swedenborg augmented the notion of gravity by linking it to a more theological model of reality in which Love is the form-giving creative agent and unifying principle of the cosmos.

The gravitational power of self-organization in the universe comes from Love's ability to grow structure by adapting form (information) to its own disposition (the correspondence principle). This process of self-representation, by necessity, always finds equilibrium in some unified action and design mirroring the dynamics of Love (otherwise structure would be disconnected from its essential substance). All coherent structure and organization in the universe consists of unified action, which informs us about the nature of first causes. You cannot have cooperation in nature without an underlying (nonlocal) principle that *communicates* and *promotes* cooperation. In this Divine design, the emergence of complexity creates the pattern for the exaltation of Love through higher forms of utility (useful unified action).

Gravitational order is the outcome (epiphenomenon) and expression of Love's unifying disposition operating within the constraints of time and space. And as I pointed out in Chapter One, because Love is psycho-energetic (mind-substance), it is the only unifying principle that could emerge from a non-spatial and non-temporal domain and describe the initial conditions of an un-manifested universe. Divine Love is a pregeometric substance and grand unified field that encompasses all possible spacetime curvatures, gravitational orders and potential structures of the universe.

Swedenborg is not alone in linking gravity to Love. Peter Russell (who studied under Stephen Hawking) makes a similar observation in *Waking Up in Time*, where he argues that gravity and Love are similar in that they are forces pulling things towards unity. Russell quotes Buckminster Fuller: "Love is metaphysical gravity."[5]

For gravity to be so mutable that it can represent a physical force on one level and a psychical force on another requires that spacetime geometry be equally mutable. It may seem that Swedenborg has generalized the concept of gravity and space almost beyond recognition, but his *Doctrine of Forms*, discussed in the next chapter, will show, from geometrical principles, a multi-level model of spacetime that allows for gravitational organization and complexity to arise on all levels of reality—from the material to the spiritual.

Each step by which the dynamics of Love descend into the spacetime arena (top-down causation) and take on new constraints, represents a gravity transition, and each gravity transition not only creates a new function and trajectory in *another kind of space* but brings a *distinct organizing principle* into the picture. The same rules that organize the universe into physical systems apply to an individual's ability to organize his or her ideas into a belief system.

Swedenborg's notion of the true nature of gravity is much richer than Newton or Einstein's description. Gravity is not just one simple physical force. Instead it can morph into new qualities. Swedenborg states:

"All gravity is from the active forces which determine it."[6]

This means that gravity and principles of action are equivalent, whether we are talking about the trajectories of physical objects or the trajectories of our thoughts (remember, thoughts do have a directionality and find their equilibrium end-state in some goal-structure). Therefore, gravitational order has analogs (correspondences) even in *mental* and *spiritual spaces*.

All gravitational order consists of distinct principles of action seeking their equilibrium in some layer of this multi-leveled spacetime hierarchy. Swedenborg's contribution is that at each substratum, gravity enjoys *qualitative* new properties, allowing creative novelty to emerge in the universe.

Gravity is not simply a warping of spacetime structure. Gravity organizes spacetime structure into rational order and form because it is the realization of God's original disposition and cosmic scheme—Love. The deepest strata of reality consist of higher orders of activity that generate civil, moral and spiritual "spaces." Religion is God's wise strategy for helping humans apply the correct gravitational order to their inner, spiritual *reality*. This Divine restructuring of the human heart and mind allows the biosphere to evolve complexity adapted for a non-material world (called heaven).

▶ Reformulating both Relativity and Quantum Theories

This "layered" model of reality is extraordinary for several reasons. Firstly, it ties quantum discontinuity to the Divine need to create a universe through discrete steps and contiguity.

Secondly, it is a real quantum gravity theory. Each of these levels represents a distinct *gravitational order*.

Third, it also provides the causal link between God and creation through the correspondence principle. There could be no Divine governance in a contiguous physical world unless gravity and self-organization emerged out of some equivalent metaphysical, unifying first principle residing in a non-spatial and non-temporal realm—a principle like Divine Love.

Fourth, it offers a new interperetation of quantum theory. It generates spacetime structure from a new non-linear geometrical interpretation of the superposition principle, with complex curvatures. And, since these kinetic fluxes of form are consubstantial with spacetime structure (the same stuff) this model is background-independent and therefore extremely relevant to modern scientific thinking.

Swedenborg's work cannot be easily dismissed. He offers new insights towards unifying relativity theory with quantum mechanics, which is exactly what a theory of quantum gravity would lead to. But his hierarchical model will also challenge science to unify quantum and relativistic theories for every level of the scaffolding of spacetime structure. This will ultimately include unifying science with theology, because on the deepest levels of reality, spaces and times find measurement in moral and spiritual values!

Swedenborg's concepts compel a modification of *both* relativity and quantum theories. He dismisses the speed of light as a maximum velocity and the notion of spacetime as a continuum. As we have seen above, he portrays the superposition principle geometrically and deterministically, as complex curvature defining a unit of action by the cooperation of potentials in common equilibrium.

In relativity theory, the *constancy of the speed of light* is the cosmological foundation and thesis for understanding the causal structure of space. But Swedenborg's search for invariants led him to identify a top-down *family of constants* (absolutes), so that universal laws and action are the same for all observers even when these laws extend into other, distinct classifications of space that are not continuous with Einstein's space. Whereas Einstein made spacetime elastic, Swedenborg made it radically morphic. All these distinct "spaces" contain their own distinct geometrical principles of action, the *qualities* of which remain invariant relative to each other.

These distinct qualities of space represent a "family" of hierarchical invariants, because they are all generated by the original disposition that Swedenborg identifies as the Divine Will. These distinct levels are discontinuous analogs of God's essential nature. So *endeavor in nature* corresponds to *human will*, which corresponds to *Divine Providence*. They are analogs that operate in different spheres and are therefore invariant qualities relative to each other. The laws of physics are derived from spiritual laws.

While all this may all seem totally alien as a concept, science may actually be moving in this direction. Professor of Philosophy and Vatican cosmologist, Michael Heller, like Swedenborg, is playing around with ideas that do not impose a limiting velocity for signal propagation.[7] He is currently seeking the possible unification of quantum mechanics and relativity through the concept of pregeometry. He believes that the laws of physics have their origin in a pregeometric realm where even "points" have no meaning. This implies that nature has its *equivalence* in non-spatial and non-temporal measurement.

A prespace measurement (metric) would be one that describes "distance" between *states* instead of difference between *points* in space. What would remain of causality if it were freed from its involvement with time and space and consisted solely of states?

▶ Multi-Dimensional Spaces and Forms derived from Measures of Love

What, indeed, would remain of causality? Swedenborg's answer is: the dynamics of the spiritual world. As evidence, he would point out that we each experience this miraculous prespace activity in mundane ways—for example, in the daily operations of our mind, which *consists solely of states*! We can feel close to or far from other people according to the difference between our hearts and beliefs. States of mind also affect how we experience time. The mind is our internal reality and spirit. The causal structure of the spiritual world is based on *states of Love*, and its metrics consist of thoughts (the measure of what we Love and intend).

Love is spiritual substance consisting of real qualities and quantities. Spiritual *values for living* can therefore be put into real scientific language, building bridges between science and theology. The physical ratios and proportions of the manifest universe have their origins and analogs in the non-physical metrics of Divine Love (substance) and Divine Wisdom (form). The evidence for this is that all order and orientation has reference to universal utility, which is the extension of spiritual Love and its dynamical magnitudes into physical forms and processes for the purpose of exalting unity through increased complexity and cooperation. This *correspondence principle* helps us understand how God can act in the world and do so without circumventing the constants of natural law.

Swedenborg was able to impose strict conditions on the laws of nature not with just one cosmological invariant but by describing *multiple* invariant values within the causal structure of spacetime and gravitational order. This allows the universe and the constancy of its laws to be so finely tuned that human-kind and human consciousness can make their appearance in creation. In this multi-level model, the causal structure of spacetime and its constants even have equivalents in the higher leveled structures of bio-complexity, including the human brain and the self-organization of our worldviews out of *values*. (This will be important for our later discussion on evolution).

The main difference between Swedenborg's multi-dimensional view of the world and modern string theory (or membrane theory) is that his extra dimensions (in the order of abstraction) take us lawfully, by steps, all the way to a pregeometric spiritual realm. Modern string theory, despite all its dimensions, never takes us out of the physical world (many scientists will admit that their job description is to explain nature purely by natural laws).

Swedenborg accomplishes this transcendental feat in his *Doctrine of Forms*. Interestingly, Penrose believes that the physics governing our brains may well involve some new kind of *physical* action.[8] Swedenborg introduces new kinds of actions and their distinct geometries with his *Doctrine of Forms*.

He describes the action principles (concepts of flow) that occupy each level of spacetime structure as expressions of a new principle of infinity within curvature. Each species of curvature represents the operation of a disposition (substance) whose function defines a direction in "another kind" of space within the hierarchical scaffolding of the cosmos. This order of abstraction eventually moves us to forms and structures taken completely out of relation to spaces and times (pregeometries).

Swedenborg's formalism is important to modern physics, because it offers insights into the true geometry of deeper dimensions. Of equal importance to modern theology, Swedenborg's Doctrine of Forms illustrates how, by removing physical constraints from dynamical process, we can lawfully arrive at the non-temporal and non-spatial gates of heaven.

SUMMARY

▶ Swedenborg was the first to propose particle spin in the microworld.

▶ He offered models that could reformulate both relativity and quantum theories (and provide a background-independent alternative to string theory).

▶ He believed the speed of light was not the maximum velocity by which information could travel.

▶ He believed that spacetime curvature could be applied to the rules of the microworld and its potentials.

▶ Potentials in prior realms form relationships (analogs of Love), creating preferred directions and outcomes in a posterior realm.

▶ Nothing happens by chance. Quantum potentials are deterministic and tend towards a common equilibrium (relationship-building).

▶ Spacetime structure and gravity are not only discrete, they can embrace different qualities. This gives rise to different levels and orders of complexity.

▶ The universe is hierarchical.

PREDICTION

Quantum Mechanics and General Relativity
will both have to be modified
before the laws of physics can be unified.

THE DOCTRINE OF FORMS

"As a man who has devoted his whole life to the most clear-headed science, to the study of matter, I can tell you as a result of my research about atoms, this much: There is no matter as such! All matter originates and exists only by virtue of a force which brings the particles of an atom to vibration and holds this most minute solar system of the atom together ... We must assume behind this force the existence of a conscious and intelligent Mind. This Mind is the matrix of all matter."

— Max Planck

The human mind cannot comprehend anything unless it can be given some form to focus its attention on. Indeed, nothing can exist unless it has form, including God. Form is the consort and predicate of *essence*, or being. An Infinite God consists of all form, and these forms represent the various ratios, proportions, and extensiveness of Divine Love and Truth. The physical world of time and space and its organizational complexity emerged from these non-physical and theological quantities and qualities.

Depicting ontological, mechanic agency for such a top-down operation and progression from the *supra*natural to the natural requires a new geometrical interpretation of forms changing through distinct *qualities* by their transition from spiritual to physical forms. Such an interpretation is beyond the current ability of modern science, which still focuses attention on quantities as opposed to qualities.

Physicists will readily admit that they don't have a satisfying geometrical grasp of the microworld, either through quantum mechanics or superstring theory. As mentioned in the last chapter, Roger Penrose believes we may need to discover a new kind of physical action. Lee Smolin echoes this sentiment when he suggests in *Three Roads to Quantum Gravity*, that, "It may be time for us to add another layer of insight to our understanding of what motion is."[1]

Swedenborg's Doctrine of Forms offers such insights. He believed that dynamics and action are not limited to the dimensions of ordinary space. His radical insight is that trajectories (fluxes of energy) can have *qualitative* differences and describe directions in completely different *kinds of space*. This idea of space as consisting of discrete structures challenges Einstein's idea of spacetime being one continuous medium.

Swedenborg's Doctrine of Forms is a cosmological model that depicts how different principles of action build the sandwiched layers of spacetime by spanning the physical and spiritual worlds. His model of a multi-leveled universe is superior to that of superstring theory, which holds that action is one thing and space another.

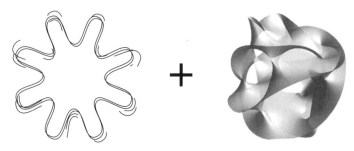

Vibrating String plus Calabi-Yau Space

figure 5.1

The current model of superstring theory shown above makes use of a classical filament of vibrating energy (which is not quantum mechanical) and a "Calabi-Yau" shape. The Calabi-Yau form is a mathematical depiction of the warped multi-dimensional "pretzel" of spacetime geometry, upon which the strings of energy vibrate, supposedly creating all the laws, forces and particle properties of the universe. God and consciousness play no role in the foundations of this model, because no matter how many dimensions scientists add, by bending space, it still remains a physical theory.

String theory has two problems. Physicists would prefer that their model start from a quantum mechanical framework instead of using classical strings (Newtonian physics) to vibrate on a warped background (background dependent). They also admit having no way of really knowing the exact geometrical form in which space can bend into extra dimensions. The Calabi-Yau shape represented in Figure 5.1 is just one example of the many ways that extra dimensions can be mathematically made to curl up.

In *The Fabric of the Cosmos*, Brian Greene states: "Many researchers consider the development of a background independent formulation to be the greatest unsolved problem facing string theory today."[2] Swedenborg had an answer. Rather than fluxes of energy wrapping around the geometry, Swedenborg saw the fluxes creating their own spacetime structure. Different spaces curl up from the operation of different dispositions and their functions.

Swedenborg's Doctrine of Forms may well provide the key to finding the proper geometrical forms of extra dimensions, because it is background independent and describes action principles far more abstract and energetic than those of classical strings. By allowing spacetime geometry and action to be consubstantial—the same stuff—Swedenborg not only has parsimony (Occam's razor) on his side, but he also gains the theoretical flexibility to take the concept of form and action back to their initial pre-space theological conditions by distinct "jumps." Swedenborg accomplishes this by lawfully removing physical constraints from action and applying new principles of infinity to forms of curvature, until dynamics no longer have relation to physical spaces and times. This means that curvature is not only a physical property of spacetime structure, its type of bending is intimately tied to *distinct* kinds of spacetime structure, all of which have their origins in the operation of Divine Intelligence and purposeful action.

Again, Swedenborg believed that the foundation of physics lies in a pregeometrical world that is entirely spiritual and consists ultimately of Divine, conscious action. The purpose of the created universe is to give *fixedness* to the infinitely complex flow of spiritual forces, which have their domain in pre-space. Spiritual forces gain fixedness by constraints and increased resistance to change, making quantum uncertainty simply action with fewer constraints—action that is therefore more expanded, nonlocal and indeterminate. These constraints on action are the laws of the universe that structure the universe into layers of existence. All action obeys laws, which are the parameters of the operations and functions of various substances, forces and their dispositional properties. The more constraints, the more classical (Newtonian) action becomes.

▶ God Geometrizes

Plato is said to have coined the phrase "God geometrizes." But Swedenborg boldly shows how. He brings a geometrical and topological approach that straddles both classical and nonclassical physics by showing that they are lawful extensions of each other. His Doctrine of Forms not only gives us "another layer of insight to our understanding of what motion is," but adds layers of insight.

He identifies seven distinct "genera" of form, which define the overall hierarchical scaffolding of the universe and the geometrical framework by which the laws of the universe are to be stated. Swedenborg's cosmological model involves laws and dynamics that go far beyond those on which contemporary theory is based. These seven forms are *invariant* (constant) relative to each other and represent the lawful restrictions placed upon the top-down order of creation from pre-space first causes into the physical world of inert matter.

Swedenborg also uses these forms as quantum steps that describe the ascent or return back to the Creator through the emergence of increased biological complexity and the nonspatial operations of the human mind (which employ more subtle forms of structure and qualitatively distinct principles of action). These seven constants of form and dynamics apply to both physics and biological structure. In effect, they span God's creative involvement into the world of matter and evolution from matter back to spiritual realities. Swedenborg called this completing process the *Circle of Life*. (This has new implications for how we can envision the arrow of time as holistic process and as goal-oriented.)

Throughout this book I have maintained that natural law is derived from spiritual law. Our discussion of forms will bear this out. For the purpose of illustrating Swedenborg's Doctrine of Forms in the clearest way, it will be best to start our progressive analysis from the bottom up, with forms that are most familiar to us, that is, at the point where disposition, force, and energy have come under the greatest geometrical constraint and where nature finds her greatest fixedness and inertia. Then we will follow Swedenborg's epistemic ascent and proceed in the order of abstraction to the hidden geometries deep within nature and beyond to a realm where number, measurement, and ratio, transition into qualities of goodness, truth and holiness. Swedenborg accomplishes this ascent by logical geometrical rules, that is, by showing how the dynamics of substances and their action can be lawfully removed from physical constraints and become more unbounded and non-local.

▶ Building a Geometrical Ladder to God

The first or lowest geometrical structure on nature's cosmic totem pole in Swedenborg's Doctrine of Forms is the *angular* form. This includes squares, rectangles, rhomboids, parallelograms and triangles. Swedenborg calls these forms "terrestrial" geometries and they represent the forms most associated with solid substances. In nature, these forms can be found in large geological formations, in minerals, crystals, even in the highly directional bonding of molecular structure.

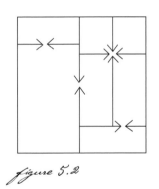

figure 5.2

Angular forms are poorly suited for motion, because their angles hinder any natural rotation around an axis, as you can easily observe when you throw dice or try to roll a six-sided box. Furthermore, in such a geometrical situation, the forces of substances can only be in opposition to each other at every point of the structure (Figure 5.2).

The angular form is subjected to the greatest constraints of physical law in nature and therefore is not geometrically suitable for motion. Opposition to motion rules this geometrical form. This gives it the quality of fixedness and inertia that is conducive to "solids." As the predominant form in the mineral kingdom and in crystals, the angular form plays a fundamental part in the evolution of biological order. In his anatomical work, Swedenborg makes the point that the angular form is most suitable for skeletal structure, enabling it to serve as the rigid foundation and support for the entire human physiology. (In simpler forms of life angular forms play a crucial role in molecular chemistry.)

Swedenborg believed mathematics and geometry played a fundamental role in living systems. This allowed him to tie the discipline of physics (both classical and non-classical) to biology. His geometrical approach to biological science uniquely provides mathematical insight to universal order and biological evolution that microbiologists and evolutionists have overlooked: organic forms have geometrical properties, and principles of geometry have everything to do with organic complexity and bio-dynamics. This may sound "alien" to many readers, particularly scientists, but biological complexity and higher-level organization (reverse entropy) are directly related to the addition of some new qualitatively different principle of geometry, *which reveals the presence of a distinctly new dispositional property and action principle within a system.*

By tying geometric principles to the organic functioning, architecture and scaffolding of the human anatomy, Swedenborg sets the stage for Divine purposefulness in both creation and evolution. His Doctrine of Forms ultimately attempts to unify physics, geometry, and biology with *theology*, especially by postulating the theological role geometry plays in the evolution of organic complexity and the capacity for intelligence and consciousness. His working thesis is based on the assumption that the hierarchical structure of the universe and the hierarchical structure of the human organism, including the human brain and mind, are built on the same geometrical principles and constants of law.

In terms of physics, Swedenborg's Doctrine of Forms offers a geometrical approach back to first causal principles and the Divine Creator. Concerning biological science, it explains how evolution can create forms of life with increased intelligence and consciousness as principles of action and their forms are freed from the shackles and constraints of natural law. In terms of religion, increased intelligence and consciousness offers humans a means by which they can continue to evolve inwardly, whereby bio-complexity is organized around spiritual substances and adapted to live in a non-spatial and non-temporal environment (called heaven).

But how would we know that Swedenborg's Doctrine of Forms actually represented God's way of doing things? How could one determine that a jump from one geometrical principle to another, whether it were through the layers of spacetime structure or layers of the human brain and mind, had God's fingerprints on it?

Swedenborg offers us a credible answer through an approach that is self-consistent with God's character—God's *infinite* character. His multi-dimensional Doctrine of Forms gives each cosmic step between discrete geometrical forms an infinite "difference." Certainly, a progression of distinct geometrical forms representing some new portrayal and characteristic of infinity would provide the proper measure for illustrating God's hand in creation.

The geometrical contingency by which nature sheds off the shackles of physical constraint comes from adding some new "infinite" principle to the rectilinear or angular form described above. What infinite principle can be applied to an angular form? This is where curvature comes into play. Curvature takes us from an inert form to a more dynamical and superior geometrical form.

▶ The Circular Form

Next in order of geometrical perfection is the *circular* form. Swedenborg states that this form is *infinitely* superior to the angular form:

> "The other or spherical, is more perfect than the former in this, that its surface resembles an infinite angle, and relates to only one fixed point, opposite to all the points of the surface, which is called the centre; it is therefore accommodated to motion and variations of form."[3]

Swedenborg also shrewdly realized that the way to remove constraints lawfully from forms was to eliminate, one by one, the static qualities of geometry. While the angular form is completely static at every point, the circular or spherical form is static only in its center and its radius.

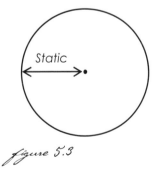

figure 5.3

The circular, or *perpetuo-angular* form has the new geometrical property of infinitely continuous variation. The circular form is not under the same geometrical constraints that angular forms are. Nature employs the circular form for continuous motion and the promotion of angular momentum, such as the revolving of planets around the sun. The circular form also provides Nature with a new principle for *organization*.

But Swedenborg is zeroing-in on something else. Remember, the difference between physical and spiritual action is that the former represents *change of place* while the latter represents *change of state.* He is looking for forms that can progressively increase their capacity for modification, their ability to change states. Not only is the circular form more suitable for continuous motion, it is capable of adapting itself to greater variations of form, that is, *changes of state.* This increased capacity for changing states explains why human physiology utilizes angular forms for rigid skeletal structure and spherical forms for its internal organs. A square stomach, for example, would be unable to change its modes sufficiently to provide the proper churning functions for digestion.

The circular form is geometrically "exalted" because a static property of the angular form has been eliminated and something *infinite* added. Swedenborg calls the spherical form *perpetuo-angular,* because it represents an uplifting of the angular form by an infinite principle. By systematically raising-up geometry through new infinite principles, *we can make predictions about the next form* (which is important for evaluating any theory).

The next degree of geometrical perfection will require an elevation of the circular form by a new principle of infinity. In other words, the circular form is subjected to some new expression of infinity and a new concept of flow, that is, some new type of infinitely continuous variation. This geometrical upgrade of the circular form and its flux of forces into a new order of complexity and change of direction, gives us the *perpetuo-circular form,* or a spiral.

▶ The Spiral Form

We have seen that the rectilinear or angular form is static, a geometrical circumstance in which forces are always in opposition to each other, and that the circular form is geometrically static only across the radius, which is the distance between its center and circumference. In a spiral form, yet *another geometrical constraint is removed.* Now, even the radius is no longer static.

The spiral's perpetually circular nature allows it not only to rotate but to undulate as well and undergo even more variations of form in the process. The spiral form and spiral gyration give nature a new potency, since an action once started, can be

promoted almost instanta-
neously. Swedenborg says
this is clearly apparent
in "the helix and screw of
mechanics."[4]

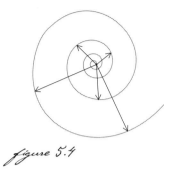

figure 5.4

This geometrical cir-
cumstance makes the spiral
form especially suited for
signal propagation. It allows
a mode or modification of
its form to be instantaneously transferred, making it a perfect
wave-guide to receive, accommodate, and promote the transfer
and flow of information. Swedenborg calls the spiral form a
sensitive form. His anatomical research led him to believe
that nature utilizes this twisting form in biological structure
wherever there is need for signal transfer. He observed from
dissections that the medullary nerve fibers and the whole
cerebrum was folded around itself and drawn into spirals,
describing surfaces, planes, axes and centers. Without these
geometrical properties and organization, the brain's activity
would become unordered. Today's advanced investigative tools
have found this same biological coiling in DNA molecules,
microtubules and in the complex folding of proteins that carry
information and transmit signals. In fact, protein folding constitutes an astronomical number of "proper" conformational
states in order to transfer signals correctly.

Sensation is the reception of forms—constantly changing
forms. The human nervous system must be geometrically
adapted to receive and transfer a variety of changes in the
ambient world to the cerebrum. This reception and transfer
requires organic forms to be able to *change their state* according to the variations in these influences. Since the spiral form
is better suited for variation and changes of state than the
circular form, it enjoys an enhanced geometrical capacity for
receiving the modifications of forms (active information) from
the surrounding world.

In Swedenborg's hierarchical model of form and order, biology makes use of nature's disposition to "curl up" into different
spacetime structures. This is similar to the Kaluza-Klein idea of
multi-dimensional space used in string theory (except it does
not rely on space being continuous). While physicists today are
puzzled why space would curl up, Swedenborg believed, that
the curling of space resulted from action principles removing

resistance and physical constraints as they move closer towards first principles and God. Removing physical constraints makes action not only more energetic, but also more intelligent! Keep in mind that these jumps into superior forms represents a jump into new qualities of action. Swedenborg's multi-leveled theory of existence allowed him to describe how action can transcend physical laws, something string theory, in spite of all its dimensions, is unable to do. Swedenborg's approach is unique to contemporary science because biology, including molecular biology, employs no geometrical principles into its discipline.

This is where consciousness enters the geometrical picture. It is not that the spiral form is sentient in itself. Instead, it provides a geometrical mechanism and contingency by which organic life can become aware of its surroundings. It is the form in which sensitivity becomes embodied in nature. Intelligence increases as forms gain greater capacity for variation. There is a theological reason for this as well. The more a form is able to vary its states and modes, the more perfectly God's Infinite influence can be conjoined to the form.

Again, Swedenborg's Doctrine of Forms addresses Lee Smolin's suggestion that we may need to gain a new understanding of what motion is. The deeper we probe motion, the more remarkable it becomes. The three geometrical forms discussed so far covered inertia, physical motion, and sensitive motion. These three forms consist of geometries that are *qualitatively* different. This allows the forces and kinematics that determine these forms to be qualitatively different as well.

The Doctrine of Forms takes us, by discontinuous steps, out of the physical and closer to God and first causal principles. *Qualitatively* speaking, the geometries of a square, a circle, and a spiral have no finite ratio relative to each other. This idea may seem odd if we merely draw a circle next to a square and show that there is a *quantitative* ratio between the two. But Swedenborg sees these forms as wholly different "animals." Simply put, there is no finite ratio between a form with angles and a form that is perpetually angular, as there is no finite ratio between a circle and a form that is perpetually circular. Each form gives us a new expression of infinity. Viewed as functions representing dispositional operations, these forms describe different *orders* of action. Action and form could not become qualitatively different if there were only finite ratios between them (an idea missing in Einstein's spacetime continuum). There would be no geometrical contingency allowing dynamics to jump from physical process, to sensitive process, to intelligent process. Each new order allows for greater change of state and enables forms to undergo increased modification within their dynamical functions. The more a form can undergo variations, the more it can accommodate psychical forces and take on states that articulate the more subtle relationships and ratios of intelligent information. That is, the more mutations a form can undergo, the more effectively a mental disposition can supervene its influence and functional property on that form.

Science does not yet know how the brain actually processes information or thinks. Swedenborg suggested that human brain cells (neurons) consisted of layered structures consisting of forms that could undergo greater modification. When we sensate something brain cells react by changing the state of their form. This activity turns information into mental pictures (we see it in our heads). Deeper *meaning* can be distilled from the cognitive function of sensation, when changes of state caused by the sensory data can be elevated to a new form of a superior order and degree. Meaning emerges from the recognition of new relationships that are detected concerning the information. The mind (which is spiritual) could not do this unless the brain consisted of subtle forms and substrates that had greater capacity for variation and literally reproduce these deeper relationships through *shape-shifting*. All relationship consists of ratio, proportion, analogy and order. All apprehension comes from the mind's ability to reproduce this order and difference on the brain's multi-layered structures. Thought forms or mental pictures are the results of the variability of modifications in neural structure that produce analogs of these new relationships. Swedenborg's Doctrine of Forms provides the geometrical mechanism necessary for a modification to pass into a contiguous realm of a higher order. This is what allows sensory data to be elevated into abstract thought.

▶ Why Curvature?

There is a form more subtle than the spiral that nature adopts to create the deeper neural substrate in the brain that performs more exalted cognitive functions. But before moving on to this next elevated form, we need to give more consideration to why increased curvature provides nature with a mechanism and potency by which process and change can jump into new orders of activity. In *A Tour of the Calculus*, author, mathematician and philosopher, David Berlinski calls curvature a "creature of change."[5] In *Newton's Gift*, he states, "Change beyond change requires a curve."[6] In other words, curvature can change as it is *changing*. This is how action and motion can increase its dynamics, which are expressions of change. Increased curvature is also how nature creates new orders of organization.

Observation clearly shows that the world of nature prefers curvature to a straight line (a straight line is a mathematical tool used to simplify making measurements—like vectors in the Hilbert space of quantum mechanics—but is not to be mistaken for what the world is actually like). The preference of nature for curvature is significant, since nature consists of processes that are organized. This includes the phenomena of the microworld as discussed in Chapter Four.

In *When Science Meets Religion*, nuclear physicist and theologian Ian G. Barbour refers to Ilya Prigogine's Nobel Prize-winning work on dynamic systems which shows examples of the sudden appearance of order from disorder by way of curvature—like the vortex that emerges from the turbulence of

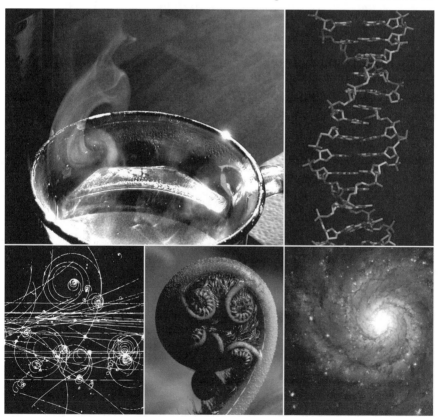

a flowing river.[7] Steam from a cup of hot coffee or tea will rise up in spiral patterns. Anyone watching a weather forecaster over the years knows that storms organize through rotation and even increase in organization as winds increase their curvature, transitioning from circular rotation to a spiral rotation as in tornados and hurricanes. Star systems seem to adopt this pattern of organization as well.

David Bohm speculated that higher orders of organization might be described by increased curvature, whereby "differences themselves become different," such as in spirals.[8] This is exactly the idea on which Swedenborg based his Doctrine of Forms more than 250 years ago. He successively increased the curvature of each of his forms through a new *expression of infinity*, a process by which the static qualities of classical geometry were removed, allowing action to expand and become less "material." Each new species of curvature was a constant of law, a cosmological invariant, and each described directions in "other kinds of spaces." This sense is lost when space is looked at from Cartesian coordinates and as a continuum.

We are moving into deeper levels of nature's structure and closer to first causal principles and God. To ultimately ascend to this non-physical world beyond time and space, forms must continue to become free from the static properties of geometrical constraints. Again Swedenborg adds a new principle of infinity to both *form* and its *function*, increasing its capacity to undergo more subtle changes of state and become more receptive to intelligent action.

> "In each degree, when forms are carried up by this ladder, something earthly, material, and finite is cut away and put off, and a certain celestial, perpetual, and infinite is superadded and put on" ... "*Until at last nothing except what is perpetual, infinite, eternal, pure, holy, that is, Divine remain.*"[9]

Swedenborg's Doctrine of Forms (in the order of abstraction) now approaches a realm where nature and her phenomena become removed from direct observation and hidden from empirical science. The next superior geometrical form in this cosmological ladder is infinitely superior to the spiral or helix. It is called the *vortical* or *perpetual-spiral* form.

▶ The Vortical Form

Swedenborg admits that his next form reaches such a degree of abstraction and subtlety that it is hard to visualize. Yet, by applying the same rules of removing geometrical constraints, it can still be rationally understood. The angular form is completely static, the circular form static in its radius, and the spiral form static only at its center. In this new form, the center of the *vortical* or *perpetuo-spiral* lawfully possesses a radically new degree of freedom.

Unlike the spiral form, which gyrates around a static center, the vortical form gyrates around a center that is perpetually moving and dynamic. By successively removing geometrical constraints we get a form so far removed from geometrical extension that it brings us to non-classical physics and closer to a place beyond time and space. As limitations are removed and curvature becomes more extreme, we find less distinction between the center, radius, and periphery of the form, which are classical qualities of extension.

Swedenborg refers to the vortical form as the *magnetic* form since its fluxion of force creates poles. Again he ties physics and cosmology to anatomical design. The vortical form not only describes an interior layer of spacetime structure but also the brain structure! Nature's intelligence and human intelligence are hierarchically layered by a similar design and blueprint, making the human brain a micro-version of the macroworld. The human brain adopts this superior form to weave the structural substrata that lie within the neuron itself. Swedenborg describes this deeper organic substrate as a *superior cortex* and *nervous system* that function as an interior and higher sensory organ.

Swedenborg understood that a single-leveled neural network was not sufficient for explaining higher cognitive functions or abstract thought, let alone mystical experience. The vortical form and its increased ability to take on subtle variations and mutations of shape provide the mind with a deeper substrate in the brain to absorb information from the world and delve into the more intricate ratios, computational analogies, and equations that raise sensory data to higher powers—that is, to abstract thought and meaning. The vortical form and its enhanced mutability enables the non-physical mind and its cognitive function of *imagination* to flow into and be embodied within the physical structure of the brain.

The concept of the neuron having its own interior nervous system may once have seemed absurd. But science is actually moving in Swedenborg's direction. One area of current brain research is seriously looking into the possibility that structures called *microtubules* represent such an interior system within the neuron. This research is also inquiring into the possibility that connections exist between the ability of microtubules to *rapidly change their shape* and the cognitive functions of learning and memory. Swedenborg believed that cognitive functions, such as those that produce thought, were embodied in changes of states within the subtle organic forms of the brain. In *The Quantum Brain*, psychiatrist Jeffery Satinover reports that the power of self-organization in microtubules may be analogous to a kind of magnetic system called "spin glasses" that exhibits unified behavior.[10] This magnetic organization in the neuron is exactly the claim made by Swedenborg.

The two main ways by which scientific models are judged is by their abilities to explain and to predict—in other words, their fruitfulness. Swedenborg's Doctrine of Forms makes definite predictions concerning forms and their orders of abstraction— that each successive form represents a discrete, discontinuous degree and that each evolves by the successive removal of static geometrical properties and the addition of a new principle of infinity. (Keep in mind that this is the reverse of creation, whereby action and form pick up constraints on their descent from God into inert matter.)

In terms of its explanatory powers, we will see in the next chapter that the Doctrine of Forms provides a means to construct the comprehensive, multi-level cognitive theory that modern neuroscience lacks, and the neural basis for religion that modern theology lacks. Swedenborg also presents a spiritually-based framework that supports the concept of discontinuous morphological evolution that Stephan Jay Gould and Niles Eldredge call "punctuated equilibrium." They challenge the Darwinian paradigm because it is clear to them that the fossil evidence did not support the idea that evolution proceeded on a smooth and gradual path. Swedenborg would add that natural selection could not bring about "jumps" in complexity and intelligence, because environmental pressures do not cause organisms to adopt new and infinitely superior geometries.

Swedenborg's Doctrine of Forms supplies the blueprint needed to account for the radical jumps in bio-complexity in the observed world. These jumps of structure and intelligence are jumps of activity into *different principles of action* that describe and find their equilibrium in qualitatively different gravitational orders of structured spaces. In other words, Swedenborg's idea of different or *implicate* orders consists of distinct orders of action and equalization. Each evolutionary jump in living structure and function requires that the appropriate geometrical principles be built into the bio-complexity of the organism.

In the next chapter we will see how various cognitive functions are hierarchically layered. Human cognitive architecture could not emerge unless nature could fine-tune herself through a layered scheme of cosmological invariants (with no finite ratio between them). This is why the Doctrine of Forms is a multi-leveled *invariance theory*.

In Swedenborg's spiritually-based approach to physics, action is not only a more fundamental state than is the state of rest, but action is fundamentally conscious, stemming from God's principle of Love that is the agency found in nature. When you increase the curvature of flux and reflux within a whole unit of action (as does the Doctrine of Forms), coherent, self-referable action increases in complexity and abstraction by infinite degrees. This allows *conscious force* to become more manifest and embodied in dynamical process.* Science fails to lead us to first principles and causal agency because it attempts to describe the fundamental nature of reality only in terms of a *physical* theory, rejecting purposefulness in its cosmology. In spite of the fact that quantum mechanics hints at a *non-material* realm where consciousness seems to play a part in quantum events and the fact that a Big Bang singularity hints that nature could not have had a physical cause, few scientists have recognized that the inner core of reality is *psychical*.

> ✳ *Since mind is not located in physical space, it must consist of forms devoid of physical extension and, therefore, non-material.* The mental function of abstract thought requires biostructure to consist of units devoid of geometrical description.

Swedenborg's cosmological model challenges the material paradigm that consciousness is an epiphenomenon, or a fluky outcome of inert matter. Instead, it posits that consciousness is primary and ultimately theological. In fact, a material view of consciousness has no strict experimental basis and represents scientism rather than true science.

There can be no real *Theory of Everything* that fails to demonstrate a nexus between God and nature, religion and science, or mind and body. And this depends on recognizing that as action and process become free from geometrical hindrances, they become less physical and more *psychical*. Since mind is not located in physical space, it must consist of forms devoid of physical extension and, therefore, be non-material. The non-physical human mind can *supervene* on the physical brain because its more interior structures, which are freed-up from geometrical constraints, consist of forms capable of changing their states from new infinite principles to produce the subtle ratios and analogs of abstract thought.

Even more fascinating, Swedenborg's cosmological model of a layered universe is the same as his model for the cognitive architecture of the human brain. The multi-leveled design of the human brain takes advantage of nature's hierarchical order by which geometrical hindrances are removed from forms and forces as you move toward first principles.

Swedeborg's multi-rung ladder of forms, substances, and kinetic fluxes, takes us, by successive steps, to a dynamical and non-spatial realm. As geometrical hindrances are removed, action and its curvature fold in on themselves until they become a form completely devoid of extension. So let us continue our ascent into deeper realms of reality and its nonphysical forms.

▶ The First Natural Form or Singularity

The eminent physicist John Wheeler was the first to suggest using the notion of "pregeometry" to formulate theories of causal processes that were not based upon grids, that is, a system of Cartesian coordinates, or manifolds, for spacetime. As we have seen, Swedenborg had already described the structure of spaces, not in terms of grids, but in terms of forms generated out of *functions* that represent the manifestation of dispositions. These include pregeometric "spaces" with dispositional operations consisting of pure endeavors (potentials) that define the special, initial conditions of the prenatal universe.

Swedenborg's Doctrine of Forms next describes a supernatural entity with curvature and fluxions infinitely more complex than the vortical form. Spacetime, process, and form all curl up into an invisible dimension of reality. Again, this resembles the Kaluza-Klein notion of extra dimensions that are curled into a tiny ball, except it is action folding in on itself. By systematically removing constraints from action and its form, Swedenborg brings us to a pregeometric realm where substance, action, and form exist as a "state" of *pure effort* and *possibility.*

This supernatural or pre-geometrical form is Swedenborg's essential singularity (mentioned in Chapter Three), a zero-dimensional point of infinite curvature that contains *all natural forms* and all of nature's forces in potential. The prenatal universe was solely comprised of these virtual entities, making its initial condition a pregnant void.

Contemporary physics calls this invisible realm the "quantum vacuum," which consists purely of "tendencies to exist." Swedenborg maintained that there was nothing essential in motion and its measurement outcomes except *endeavor* (tendency). So Swedenborg and the New Physics are both in agreement that primal nature represents a state of pure endeavor and potentiality. Where they disagree is that science sees this invisible realm as a probabilistic froth, whereas Swedenborg sees it as a realm saturated with non-dimensional entities or singularities sustained by God. Modern science describes this realm as one of irreducible chance, while Swedenborg sees it as one of irreducible order and purposefulness.

Swedenborg's view of this realm of pure possibilities is that qualities like endeavor, disposition, and tendency are primal states of self-referable action. Disposition in primal nature is the physical equivalent and analog of God's *Will* because it is the first *natural* form to receive the flow of information from Divine consciousness. The supernatural singularity is a medium by which God's will and intellignece can transition to a state of pure tendencies in the prenatal world. This form is the interface between metaphysics and physics.

When Divine omniscience flows into and takes on this prenatal form of pure tendencies it results in a perfectly self-referable and self-interacting endeavor that Swedenborg calls a *mathematical intuition of ends* because it is a *force carrier* for God's cosmic goals and administration. This supernatural form derives its dynamics from infinite Divine Love and Wisdom. It instantly comprehends in itself all form, inorganic and organic, and intrinsicly meets the most severe restrictions in creating a universe that is fine-tuned for the Cosmological Anthropic Principle, a bio-friendly universe that anticipates the arrival of humankind. This self-referable action, which contains at its core heavenly intelligence, holds the patterning principles which generate all series of connected productions whereby first principles and destinies are simultaneously present. The organic version of this form resides in the deepest natural substrates of neural structure in the human brain. Its operation is involuntary and instinctive (more on this in the next chapter).

Swedenborg calls this perpetuo-vortical form the primal and supreme form of nature and geometry itself. It is pregeometric and represents the unique concept of a cosmic singularity that *exists without a physical cause*. Compared to the vortical form previously described, its fluxes have been raised to a new, infinite principle of self-referable action, resulting in a curvature so extreme that its center occupies every place of its periphery, and its periphery consists wholly of its center. With so many geometrical limitations removed from this supernatural form it enjoys perfect spontaneity and conductivity to Divine consciousness.

All physical restraints and geometrical oppositions are thus removed so that the perpetuo-vortical form enjoys *frictionless flow*. Infinite in its flux and reflux, it has no extension, figure, magnitude, gravity or levity. There is no up or down, left or right, in its internal action. Here are Swedenborg's own words concerning this paradoxical and supernatural entity:

> "Thus it lacks resistance, and is something perpetual, self-moving, spontaneous, most perfectly geometrical, the principle figure of potencies and forces, and as it were a force figured, or motion in figure. Consequently it is the figure of figures, and the sole figure to which nature finally betakes herself, and from which she brings forth all the geometry and mechanism of her world. In this figure she knows nothing of being resisted, for she feels nothing resisting. Therefore, by its aid she returns at every change to her own equilibrium."[11]

This pregeometric form, or first natural point, is Swedenborg's version of an essential and non-material singularity that has infinite density but no physical volume or extension. It is a true quantum entity because its "dressed mass" and density consists of pure active information seeking a concrete result. This density of information contains a provident design, because the dynamical state of pure striving in the singularity comes from Divine agency. Again, the infinite curvature and frictionless flow of Swedenborg's singularity allows it to possess infinite conductivity to God's influence. As a medium between God and nature, it is a force carrier for the theological cosmic code.

Swedenborg describes this primal entity as an "exemplar representative of the universe" containing all the potentialities and contingencies necessary for the topology-changing transitions of creation and evolution. The laws of the universe are enfolded in the dynamics of the singularity, so all the things of the future universe and its contingencies already exist non-temporally and non-spatially. This interior level of reality is infinitely complex! Swedenborg presents us with the interesting and novel notion about the emergence of complexity in the universe: physical complexity emerges from non-physical complexity, that is, the innumerable aspects of God's purpose.

Swedenborg anticipated physicist John Wheeler's desire to formulate theories of causal processes through the concept of pregeometry. Such theories would show how physical laws and forces emerge out of the non-material realm. The physical universe can arise from a non-physical cause only if there is a causal link between both realms through *correspondences*. Everything in the physical world must be an analog of something in the invisible and nonmaterial realm of spirit.

In the previous chapter, I mentioned that current scientific research is moving toward non-physical dynamics. Using the recently discovered *noncommutative geometry* of French mathematician Alain Connes, philosopher and Vatican cosmologist Michael Heller and his coworkers are trying to show that the physics ruling the universe at the most fundamental level springs from a realm where there is no spacetime. This requires an extreme conception of "spaces" where even "points" become meaningless and dynamical magnitudes are *removed from their involvement with time and space*.[12] Heller realizes that a noncommutative space can be described not by distances between points, but by the differences between *states*.

This is exactly the direction Swedenborg's Doctrine of Forms leads us—to a realm where non-spatial and non-temporal process are recognized as *changes of state*. And Swedenborg is not afraid to say that he had felt a Divine hand in its formulation:

"Wherefore, that I might not be held in things so extremely ultimated and finited by the Lord, there was given me a notion of forms entirely transcending all geometrical forms, for geometry is terminated in the circle, or in curves referring themselves to the circle, which are merely terrestrial, and do not embrace even the lowest atmospheric and aqueous forms. From these lowest or

terrestrial forms, it was given, by the removal of imperfections, such as the causes of gravity, rest, cold, and so
on, to perceive forms which were free from the operation
of such causes; and that then there remained forms still
more free from them, and others freer still, till at length
forms were given in which nothing could be conceived
but centers in every point, so that they consisted of mere
centers from whence all circles and peripheries, each of
the points of which represented centers, and from these
centers still had respect to similars, till the lower form
being removed, in which were those termini signifying
the boundaries of space and time, I saw myself carried
forward to forms almost entirely void of limits and thus
taken out of relation to spaces and times."[13]

Swedenborg's Doctrine of Forms does not stop with singularities. Instead, it moves into a realm where even "points"
have no meaning. So then, what *corresponding* qualities or
forms is one left with when physical forms, speed and process
are removed from their involvement with space and time?
Swedenborg claims that in such a pregeometric world, everything is measured by a theological metric—the measurement
of Love and spiritual values!

▶ Spiritual Initial Conditions of the Universe

Swedenborg's Doctrine of Forms includes two more exalted
forms, both of which are pregeometric. These are the *Spiritual
form* and the *Divine form* itself. The Spiritual form is the actual
emanation (not radiation) of God's infinite Love and Truth. It is
the first created form from God. Its form encompasses the non-
material complexity of the Divine Holy Word, or *Logo*sphere. Put
in a more scientific language, this emanation acts as a Grand
Unified Field, a spiritual aura, consisting of conscious *living
force* that generates the goal-directed and *anthropic* patterning
principles for all creation, evolution and even the salvation of
the human race. (The Divine form will be discussed below.)

The non-physical world of pregeometric spaces and spiritual
forms embodies the countless representations of God's Love and
Truth. In Swedenborg's theologically-based cosmogony and
cosmology, Love is spiritual-creative *substance* and truth is the
form and measure (metrics) of that Love.

Spiritual measurement requires a third aspect. That is why
Divine Love and Truth take part in a third function that unifies

the two in some *activity* (theologically called the "Holy Spirit"). The true measure of Love can only be determined by its utility or cosmic usefulness and the part it plays in promoting holistic relationalism. Creation is formation through relationship. Relationship is formed by shared utility. Shared utility is Love. So all ratios, proportions, and organization in the physical universe are measurements of usefulness. All created things have utility. The fruitfulness and variety of the natural world come from the living force of spiritual forms which are in perpetual endeavor to promote *goodness*.

A correspondence principle between the physical and spiritual world results from the various measurements of Love's value and utility to transition to their analog and equivalent in time and space. Since all relationship is ratio and proportion, even various qualities of Love consist of quantities and represent a standard of non-physical measurement. In a universe with manifest structures organized and arranged to promote unity through utility, the concepts of extensiveness, ratios and proportions originate in and can only be understood in terms of the non-physical properties and dynamical magnitudes of God's Love and Wisdom. Love is the nonmaterial origin of complexity in nature. The preeminent spiritual force is the reason why physical nature is in continuous endeavor towards self-organization.

The "tendencies to exist" operating on the most fundamental level of reality are not quantum probabilities seeking some random measurement outcome in physical nature; they are *living* spiritual forces (determined Loves) seeking measurement and concreteness in purposeful forms of utility or "ends." For this reason the future with God is always present. Again, this correspondence with Love is why the universe is unified and everything in it is interrelated, interconnected and interdependent.

Swedenborg called the *causal link* between the distinct metrics of spiritual and physical worlds the *science of correspondences*. Divine Love cannot generate or maintain any form or process in the physical universe unless God's nature can be represented within that form or process. Love has no meaning or measure unless it can take the form of usefulness. All coherent form and matter finds its stability in utility. The universe is proof of a powerful disposition to organize and unify, which itself is a derivative of Love. The physical world and its processes, when looked at from the point of view of interconnected and mutual usefulness, is a reflection of God's Love. Divine Love is the non-physical principle behind holonomic continuity and whole-part causality.

Correspondence is why God can be present in all created things. In fact, the nonmaterial human mind has its bio-complexity from a most exalted organization of spiritual substances and therefore has been created to serve as a means for perfecting God's Conjunctive Design (mentioned in Chapter Three). Without spiritual forms and substances, men and women could not think or become wise.

As noted earlier, some current researchers are attempting to reinterpret dynamical magnitudes and the laws of physics in a way that is beyond time and space through the application of noncommutative geometry. This seems to be fertile ground, but it will succeed only if physical laws and forces are proven to correspond to non-physical laws and forces. Swedenborg believed that non-physical dynamics were spiritual laws and forces. In other words, if you free ordered process from its association with time and space, you get its non-physical or theological equivalent, which is Love.

Scientists would certainly pay special attention if it could be shown that stripping coherent process from its involvement with time and space yields a "noncommutative" result. And theologians would pay special attention if it yields a measurement of *Love* and *purpose*. In *The Reflexive Universe*,[14] philosopher and inventor Arthur M. Young believed that what remains of causality from the breakdown of commutation in quantum theory is a nonmaterial unit of action, one that implies purpose within universal process.

Young shows that in a *commutative* operation we get the following equation:

$$(1) \quad ab - ba = 0$$

But in the *noncommutative* operation of quantum physics we get something quite different, something greater than zero:

$$(2) \quad ab - ba = ih$$

In physical terms, equation 1 above is the mathematical equivalent of a person going to a store (*ab*) then returning home

(-*ba*) and ending up where he or she started (0). But something more than a physical event took place. The person went to the store for a *purpose*—to buy something. Stripping this event from its involvement with time and space leaves a measure of purpose and intention (which is spiritual measurement). Young states that the "*i*" in *ih* is *imaginary* and represents a nonmaterial element hidden in the world of quantized physical process ("*h*" represents Planck's constant or unit of action).

God created a conceptual universe based on the nonmaterial dynamics of Love, one that can be viewed (according to Young's explanation) as a noncommutative equation, revealing a unit of purpose within physical process. In other words, if you strip *all* ordered and coherent process in the universe from its involvement with time and space, you get its non-physical or theological equivalent—holistic units of usefulness and utility (Love and its derivative good).

The physical and conceptual universe becomes more God-like as Love exalts utility and relationship through its embodiment in the evolution of complexity and higher ordered bio-structure. Love is the causal agency where the power in nature for self-organization comes from. This allows for novelty to appear in the universe. Humankind's important role in God's cosmic scheme of creation is to help the universe reach a "Divine end."

God's holistic activity and conceptual universe—the expression of a "grand" unit of action—is both a downward process and an involvement with matter (*ab*) and a return (-*ab*) or evolvement from matter, back to the non-physical realm. This cycle is completed through the spiritual evolution of the human race by means of a sacred covenant implicit in the two big religious tenets of loving God and *loving the neighbor.* This return of God's Love is greater than "zero," because it is more than the return of Love to its starting point. It includes the creation of a heaven from the human race. I have referred to this transcendental cycle earlier (Figure 3.5) as Conjunctive Design (CD).

▶ The Divine Form and Ultimate Purpose

The ultimate, ineffable form or *Divine form* is God, the Eternal Being and Divine Personality who is uncreated and ground of all reality. God is Infinite Love, which finds its form in Infinite Wisdom. Infinite Love and Infinite Wisdom are the ultimate expressions of free will and reasoning; hence, they are infinitely human. God's Divine form is the origin of humanness.

God creates because *Love needs a subject*. Not even a physical world is enough. Since perfect Love is reciprocal and not one-sided, God must create a creature who can be the object of this Love, a creature who can recognize and respond to that Love.

The evolution of the human brain and consciousness provides the cosmic possibility for this intimate relationship between the created and the Creator. This intimacy would have little meaning and not be worth the expenditure of all the Divine energy if eternal life were not a part of the equation of this relationship. Conjunctive Design is the means by which the Lord God can bestow on others all the blessings and delights that belong to Love and wisdom. In this way Love exercises its fullness.

Not only has the universe been fine-tuned for human life to appear on earth, but it has been designed to allow human evolution to continue in such a way that it extends the biosphere into the pregeometric realm. After the death of their physical bodies, humans can be safe from entropy and the tyranny of time and continue to exist in a domain of unimaginable Love and eternal life. This interior realm is called *heaven*. The human mind and spirit are non-physical organic forms created for just that glorious purpose.

In the following chapters we will continue to see why scientists must change their assumptions about physics. We will also learn why theologians will have to change their assumptions about Holy Scripture. If Holy Scripture is the Lord God's Divine Word, and all things were created from the Word (as stated by the apostle John, 1:1-3) then its various lessons must also contain higher-level meanings and interpretations removed from their involvement with time and space. At the same time, these sacred stories must display top-down causality that accords with the science of correspondences.

SUMMARY

▶ Biology and complexity are profoundly connected to distinct geometrical forms and principles of action.

▶ As constraints are removed from action, process becomes more abstract and intelligent.

▶ Human intelligence and the intelligence of the universe consist of similarly layered structures and geometries.

▶ There are no finite ratios between the distinct layers of reality's architecture. Jumps between each layer of reality are separated by a new principle of infinity governing both form and action. At the lowest end is inert matter. At the highest end is a God of Love.

▶ All coherent and ordered physical process is driven by a nonmaterial (spiritual) goal.

▶ Scripture must also contain this top-down layered scheme hidden deep within in the literal sense of its narratives.

PREDICTION

In this century it will be determined that the neuron has its own interior nervous system. And deeper still, is the human soul.

Chapter Six

LOVE, NEUROSCIENCE AND HIERARCHICAL DESIGN

> *"And he dreamed, and behold a ladder set up on earth, and the top of it reached to heaven: and behold the angels of God ascending and descending on it."*
>
> Jacob's Ladder, Genesis 28:12

In most of the world's religions, humanity is the crown of creation and the key player in the cosmic drama of the universe. A few post-modern, relativistic, and progressive approaches to theology reject all notions of hierarchy and make worms as important as people. Nevertheless, the human race forms an important feature of a universe in which Creator and the created can be in reciprocal relationship through the agency of free will and first-person phenomenal experience. I have referred to this Divine scheme as Conjunctive Design.

By contrast, science and its materialistic philosophy of the last several hundred years has challenged the spiritual nature of the universe. Not only is God not needed to explain the world and its operations, but poor Homo sapiens, rather than being the special creation of the Divine hand, is merely the by-product of a cold and indifferent universe. Consciousness is an *epiphenomenon* and therefore a pointless coincidence in the grand scheme of the cosmos. Humanity is just a quirk of quarks in a universe that is going nowhere special.

Recently however, some physicists—including Ian Thompson, Brian Josephson, Fred Allen Wolf, William Tiller, Amit Goswami, and John Hagelin—have proposed that science needs to be founded on different assumptions. There are real signs that science has started moving toward belief that the emergence of life forms that are conscious and intelligent enough to ask questions about the universe is part of a grand, purposeful design in which the prime principle of existence is consciousness. This is called the Cosmological Anthropic Principle. Even staunch materialists and atheists marvel that the universe is so finely tuned and precise that it allows for a human brain to appear that it can recognize and seek to explain this very precision.

Ironically, it is the very success of science and mathematics to describe many of the laws of nature that drives this new shift toward the conclusion that consciousness is fundamental to the cosmic design. Nobel Prize-winning physicist Eugene Wigner was fascinated over why the universe is ordered in such a way that science and mathematics can work so "unreasonably well" in describing the laws of nature. How is it that the laws of nature agree so well with human reasoning? How is it that the universe is conceptual at all?

Another Nobel Prize-winning physicist, Louis de Broglie, strongly believed that "the structure of the material universe has something in common with the laws that govern the workings of the human mind."[1] The success of mathematics in describing the laws of the universe has prompted other serious thinkers, like astrophysicist Paul Davies, to wonder if there isn't "a deep and meaningful resonance between the human mind and the underlying organization of the natural world."[2] Davies brings up another intriguing implication of this notion when he asks, "would it be true to say that the world is a manifestation of reason?"[3]

Does human reasoning and cognition reflect the structure and rational order of a hierarchical, multi-dimensional universe? Does the human mind work the way the universe works? If so, this would have profound consequences for our understanding of the origins of the cosmos, the constancy of law, universal symmetries and the foundation of physics.

Certainly, a Big Bang explosion by itself could not account for such a super-rational model of a world that meets the necessary and rigorous conditions for the human brain and consciousness to emerge. A conceptual universe that anticipated the arrival of the human species would require special initial conditions beyond those that would generate a mindless and violently spreading fireball. A conceptual universe must be conceptual from the very beginning and require consciousness to be a fundamental rather than a random consequence. Since consciousness is a living quality that perceives and intends, it follows that creation must be the action of a *living* designer. This would also explain why the laws of nature are so bio-friendly.

Swedenborg's spiritually-based cosmological model, as discussed in the previous chapters, was premised on the notion that nature and geometry had a common origin in Divine Love and Intelligence. Even Einstein believed that the universe was conceptual and that a full understanding of the laws of nature would enable us to "know how God thinks."

While Einstein was trying to fathom how God thinks, he apparently forgot about his famous comment that "science without religion is lame." Religion tells us that God is Love. Yet Love played no role in Einstein's noble quest for the laws of nature. This attitude reflects a scientific and *masculine* bias towards the material constituents of reality. This bias is the main reason why the natural sciences fail to grasp the true nature of agency. There is great need in the academic world for the God-guided heart to inform the head.

Swedenborg maintained that substance and reality are dispositional. Not just computational. Love is the hidden dynamic behind the mathematical order and rational structure of the cosmos.

A spiritually-based theory of the universe, such as Swedenborg's, is built on the theological premise that *God is Love.* God thinks and acts out of principles of Love. This is wholly self-consistent with God's character. Love is not some wispy emotion or mere passion that humans get mixed up in. It is the fundamental organizing principle of a dynamic and volitional universe that *strives.* Love is the grand hidden variable that ultimately generates spacetime geometry, matter, bio-complexity, human brain structure, and ultimately the worldviews and paradigms of the human *mind* that emerge from subjective experience and dispositional leanings. Swedenborg treated the study of Love as an *exact science*, one that was fundamental to any complete theory of the universe.

In Swedenborg's model of the universe, the human mind and the physical brain form a kind of effigy of the universe and its forces. In other words, human intelligence and the intelligence of nature share the same hierarchical structure and order. The implications of this shared, top-down pattern is that the universe and the human brain are both *biocentric* in a way that mirrors God. The infrastructures of the universe and human brain can be looked at as sensitive membranes operating in the different and discrete structures of space. All structure, no matter what kind of space it occupies, is formed to receive Divine influence, which is why the universe is a dynamical and unified process. Love, as a living conscious force, generates a unified field or sphere consisting of *Divine active information*, the flow of which produces everything from the bio-friendly geometry of spacetime to the complexity of sentient life forms.

Quantum physicist John Hagelin believes there is a strong correspondence between pure consciousness and the unified field of modern physics. For instance, he points out that by definition, both the unified field and pure consciousness are perfectly self-interacting, that is, self-referable functions. Since it is not consistent with common sense to have more than one fundamental unified field, he concludes that they must be one and the same thing.[4]

Hagelin bemoans the fact that there is no "comprehensive theory of consciousness comparable to the unified field theory of physics."[5] He speculates that human intelligence and nature's intelligence are hierarchically structured into similar layers, "from gross to subtle, from excited to de-excited, from localized to unlocalized or field-like, and from diversified to unified."[6]

Psychiatrist Stanislav Grof echoes Hagelin's conclusion that we need a comprehensive theory of consciousness. He states, "We need to find a new map of the human psyche which corresponds to the hierarchical forces and dimensions of hyperspace."[7]

Hyperspace means extra dimensions beyond Einstein's 4-D spacetime continuity. This multi-dimensional map will not be found in Einstein's concept of geometric space as a continuum. In other words, it will not materialize when ordinary space is rotated into additional dimensions (as string theory does). It will be based on non-continuous *functions* and new principles of action that describe the structure of other "kinds" of spaces, with *qualitative* differences. The geometric structure of these other, non-continuous spaces will also be found in the layering of brain structure and its neural substrates. This enables the human mind to direct mental process and cognitive function in new and expanded directions according to universal laws and mechanisms.

Swedenborg provided such a multiple level map of the human psyche in his Doctrine of Forms and his later work, *Rational Psychology*. His map shows that the deepest levels of reality are indeed levels of the mind.

▶ A New Comprehensive Theory of Mind

Many of Swedenborg's unprecedented discoveries came during his studies of the human brain and human psychology. He developed a sophisticated, multi-level cognitive theory that included a hierarchy of emotions or loves, addressed subjective human experience, and articulated a neural basis for religion—actually a spiritual basis of neural substrates.

This deserves attention for several reasons. First, as Fraser Watts puts it, "There is no generally accepted, comprehensive theory of cognitive architecture and its neural substrate."[8] Second, Swedenborg's theory was broad enough to include a personal-level framework, which presents a big problem for modern *physical* theories. Third, it goes deeper than the current synaptic mapping of neural connections and speculates on additional orders of structures within the neuron itself. Without this, there would be no suitable substrate for explaining the higher functioning of mind and abstract thought. Fourth, it incorporates new physics and geometrical principles into the biology and organic structures of the brain, neuron and mind. And most important to the theme of this book, his cognitive theory encompasses God and salvation.

Certainly, the above paragraph will create some skepticism among readers concerning Swedenborg's accomplishments in brain research during the early 1740s. But Swedenborgian scholar Dr. Rev. George F. Dole points out that in spite of the fact that scientific methodology was in its infancy 250 years ago, "We are still discovering his discoveries." Dole goes on to say, "It does seem, perhaps strangely, that in the two areas of his primary concern, science and religion, Swedenborg did not find his way into the mainstream of Western thought. There is a change now going on in both science and religion, however, and the climate is favorable for a fresh assessment of Swedenborg's contributions in these areas."[9]

▶ Way, Way, Ahead of his Time

Such a fresh assessment would recognize Swedenborg as the father of Neuron Theory, as this 1968 article from the Journal of the American Medical Association (JAMA) will attest:

> "Swedenborg's contributions in the medical sciences, prepared in Latin, lay unnoticed in the library of the Swedish Royal Academy of Sciences until the 1880s when Tafel translated into English the four-volume treatise, *The Brain*. Included in this monograph were observations on cortical localization, the somatotropic arrangement of the motor cortex, reference to integrative action of the nervous system, the significance of the pituitary gland, the formation of the cerebrospinal fluid, and a pronouncement on what is now known as the neuron theory."[10]

More recently, Harald Fodstad wrote in the introduction to his article, *"The Neuron Theory"*:

> "In 1740 the Swedish scientist and philosopher Emanuel Swedenborg described what is the first known anticipation of the neuron (a nerve cell with its processes)."[11]

That Swedenborg also addressed cortical localization, both specific sensory representation areas and the somatotropic arrangement of the motor cortex is nothing short of extraordinary. *In Emanuel Swedenborg: A Neuroscientist Before His Time*, Charles G. Gross states:

> "Remarkably, Swedenborg had the idea of the somatotropic organization of motor function in the cerebral cortex. He correctly localized control of the foot in the dorsal cortex (he calls the 'highest lobe'), the trunk in an intermediate sight and the face and head in the ventral cortex (his 'third lobe')."[12]

Gross adds:

> "There is no other suggestion of the somatotropic organization of the motor cortex until the experiments of Fitsch and Hitzig in 1870."[13]

Swedenborg was the first to identify the frontal lobes as the center for our higher reasoning powers. He recognized that the left hemisphere was more intellectual while the right brain was more emotional and creative. In spite of this new assessment of his ideas, there is still much more insight he can bring to neuroscience.

Today's scientists know that neurons process information in the brain but they do not know how. I believe neuroscience has come to a dead end in trying to explain mental processes in terms of physical processes, such as the synaptic operations of neurotransmitters or the flows of ions through the neuron membrane.

Furthermore, explaining brain function neurophysiologically as the firing patterns of neuron groups does not really address the deeper problem of how states of mind correlate to physical brain states or help solve the "hard problem" of human inner experience. This perplexing situation has caused some modern researchers to speculate that the neuron might have its own nervous system.

Swedenborg had already anticipated this fairly recent idea of neuroscience within his own comprehensive theory of human cognitive architecture. His theory portrays the neuron as a "little brain" with its own hierarchical structures and discrete dispositional functions. Various cognitive functions have their sphere of activity in their own unique spacetime geometries, which form the interior and more subtle networks deep within the neuron.

In a way that anticipated Planck's discovery that action comes in whole units or quanta, Swedenborg's multi-level cognitive theory describes various cognitive functions, such as *sensation, memory/imagination, rational thought, intuition*, and *spiritual revelation* as qualitatively distinct units of activity. He organized these particular operations of the human intellect into a unique *hierarchy* of discrete cognitive functions and bio-structure—like discrete rungs on a ladder, in which each rung widens to reveal a more unbounded realm of the universe and its increasing consciousness.

▶ Discrete Cognitive Functions

We tend to look at all of our mental activity as a continuous operation. But Swedenborg showed that cognitive functions like *sensation, imagination,* and *rational thought* are as different from each other as a square is to its root, a general to its particular, or a compound to its simple. These discrete functions of mind operate in different kinds of *structured* spaces; that is, each of these mental processes represents a new class of determined action or trajectory, each finding its greatest economy of action in different species of spacetime geometry. Human cognitive architecture is based on this cosmic scaffolding.

Swedenborg pushed the boundaries of neuroscience and neuron theory to integrate body, mind and soul. He confessed:

"I have pursued this (brain) anatomy solely for the purpose of discovering the soul. If I shall have furnished anything of use to the anatomic or medical world it will be gratifying, but still more so if I shall have thrown light upon the discovery of the soul."[14]

In fact, Swedenborg provided humanity with a unique theory capable of both advancing modern neuroscience and of providing rational evidence for the workings of an infinite and Divine guiding hand, a hand that leads evolution towards a positive cosmic goal. Each of the human brain's layered cognitive networks allows for the embodiment of a new and expanded principle of Love to operate within consciousness. This further illustrates the purposefulness of the universe, an idea that is considered taboo in the natural sciences, and particularly, in the biological sciences.

▶ From Outermost to the Innermost Human Experience of Existence

We live in an ocean of vibratory influences. These modifications and changes of form represent *flows of information.* The universe is a chatter bug!

The information that is carried by all these vibratory patterns throughout the universe would be wasted if evolution did not create creatures capable of receiving and interpreting them. This implies that biological intelligence and consciousness evolved for the purpose of increasing the capacity of living creatures to receive and process influences from the environment. This increased capacity is easily confirmed by the distinct differences among the senses of touch, taste, smelling, hearing, and seeing through which the function of sensing expands its reach and acquires new subtleness.

All sensation is the reception of form, which is information. Sensation then gives birth to *affection* (the affect or reaction to information). This reaction, whether it is to touch, taste, smell, sound or light creates a physical change of state in the organism to indicate either *like* or *dislike*. In other words, the influence is detected as harmonious or hurtful. Swedenborg's great insight is that all these different degrees of awareness and recognition are based on the operation of some principle of Love.

Even inert substances react and change their states in a similar fashion to the way that Love does. When metals are affected by external influences like heat or cold, they expand or contract accordingly. This same law can be extended to the human heart, where Love can grow warm or cold depending on how various influences affect us. Affection and repulsion (like and dislike) are changes of state in the mind and the *psychical* equivalence of expansion and contraction. When we like something we open up to it. When we dislike something we close up.

This uniformity of behavior between the physical and the mental arises from the fact that Love is the ultimate substance, the *substantia prima*, that preserves this correspondence through a transcendental family and spectrum of substances from the spiritual to the material. The only difference is the degree of complexity and subtlety by which substances and forms change their state and are affected by influences. This is evidence that nature and her processes is founded on conscious, living, first principles.

In animals, the instinctive principles of Love revolve around mere self-preservation and reproduction of their species. They *cannot transcend their natural level of Love*. Their cognitive architecture lacks the deeper layers of structure where the "like" and "dislike" take on the moral and spiritual values of "good" and "evil." (This also implies that the evolution of complexity and intelligence has a spiritual goal.)

In humans, the principle of Love is not instinctive but *chosen* through one's life experience. It is intimately tied to free will. Of interest to bio-physics, a person's chosen Love, his likes and dislikes, actually has its *center of gravity* (equilibrium) in one of the layered cognitive networks in the hierarchy of the human intellect. In other words, humans can transcend their biological selves by gravitating to a more expanded principle of Love that operates in its own unique spacetime geometry of bio-structure. We instinctively recognize that different mental dispositions operate in distinct kinds of "space" when we say that a person who is talking to us is coming from "another place."

▶ Metaphysical Digestion

In terms of evolutionary theory, what the neo-Darwinian synthesis misses is that the human brain and its intelligence not only adapts to the physical environment, but actually *internalizes* it. As Loren Eiseley puts it, concerning the main difference between the evolution of life and human evolution, "They were starting to occupy, not a niche in nature, but an invisible niche carved into thought..."[15] Human complexity has developed higher-level structures capable of foraging information. Unlike other life forms, the entire universe and all its endless transmission of information nurtures human beings, like a cosmic breast. The universe "mothers" humanity in a most intimate way.

In the same way that we take physical food from the environment and turn it into "you" and "me," we also digest information and turn it into the fabric of who we are. We internalize information from the ambient world according to our likes and dislikes. The mental processes of this cosmic scheme correspond to digestive processes in the sense that we absorb certain information and eliminate others. This psychic gastrulation enables us to distill meaning out of information based on what harmonizes with our personal values.

Science does not deal with values. Yet, in Swedenborg's model of cognitive architecture, the engine of evolution moves into the realm of values. First-person phenomenal experience allows humankind to take part in its own further evolution. This is where religion comes into the equation of natural science.

Religion and altruism are not simply evolutionary ploys by selfish genes to improve survival chances. They are part of God's strategy within the conceptual universe to prepare us for a reality beyond the physical world. The human brain and its neurons consist of *interior sensories* by which our cognitive powers can make quantum jumps between sensing, imagining, thinking, intuitiveness and spiritual revelation.

Proof that these cognitive operations point to the existence of superior senses is that each represents a distinct and qualitatively different species of *seeing*. Only the bottom rung, or ocular sight, requires physical light for determining likes and dislikes. The higher cognitive functions of seeing can take place in our head—the mind's eye. They see deeper things within the things our physical eyes see. They see ratios of ratios and qualities of qualities. Whereas one cognitive function sees physical objects, a deeper function sees ideas. And they all operate organically in distinct kinds of structured spaces or dimensions.

Swedenborg demonstrated that human cognitive architecture was designed first to recognize the topological features of the physical world, and then, by a process of abstraction, to remove them from their involvement with space and time. The interior cognitive functions of humans allow for such things, and because of this, they also provide the necessary mechanism for personal transcendence.

We cannot tap into these higher functions of human cognition unless we appropriate newer, nobler principles of love into our lives. Religion fills this role. Human evolution has nothing to do with natural selection and everything to do with the selection of values. God's conceptual universe is directed towards the evolution of the heart.

▶ The Evolution of Consciousness through Values

Because humans enjoy subjective experience, they can take part in their own further evolution. Humans can choose what they Love, and more importantly, they can choose to Love what is wise and leads to greater goodness.

Swedenborg's model of human cognitive architecture starts from the premise that at its deepest level resides the soul, which consists of purely *spiritual substances*. While the soul has profound, intrinsic intelligence, it still has to learn what the changes of states of the brain and its neural substrates mean when their substances are affected by external stimuli from an alien world of physical data and spacetime. The non-material soul forms and perfects the human intellect over time and through discrete steps. Let us explore these steps—the ascendency of love.

▶ Sensation

Awareness begins with the function of *sensation*. Humans have five senses, which at birth are under the rule of corporeal Love. We are born completely corporeal creatures, whose likes and dislikes are limited to creature comforts. These are evaluations of the physical things in the world that bring us pleasure, and those that bring us discomfort.

All attention and recognition is based on the harmony or disharmony between an influence and the reaction of organic substances and forms in our brains. As information enters the eye and is directed inwardly towards the brain and neurons, the organic structure reacts, changing its state in a way that produces certain ratios and proportions that it associates with the visible object. Images from the world are transformed into ideas of the mind from these changes of state in its sensitive substances. All ideas are changes of state in the brain. If we like something, it will be represented in the ratios and proportions that the substances of the brain have learned to most readily assume, called recollection.

In order to preserve free will at all times, the soul supervenes and changes brain states according to a person's acquired disposition, or Love. This operation is the origin of human, first-person, phenomenal experience, or subjective consciousness.

▶ Memory

Memory is internal sight, the mind's eye. In memory, the soul and mind supervene on brain and neuron structure from within to recreate those same changes of states and reproduce the object as a mental image. Recall from the previous chapter (p. 93) that some researchers speculate that the shape-shifting abilities of microtubules may be the dynamic behind learning and memory. Swedenborg's model solves several modern neurological mysteries, including the nomadic and holographic nature of memory and Alzheimer's disease.

Stanford University neurophysiologist Karl Pribram is credited with the theory that memory has holographic properties. His research has shown that even after surgery or head injuries patients experienced no gaps in their memories. Memory can be *distributed*, rather than tied to specific regions of neuron clusters in the brain.

How is this possible? Swedenborg would answer that a memory can exist anywhere a neuron can learn how to change its state to recreate a mental image and its unique ratios. Alzheimer's disease blocks the neuron from being able to run through its normal states (probably through the build-up of plaque and abnormalities of microtubules).

The cognitive function of memory, in which the images of our experienced world are transformed into ideas of the mind, prepares for us a whole *new field of vision*. This field of vision is the foundation of our inner world. It encompasses a different kind of "space," one that is not continuous with ordinary space. In Swedenborg's model, humans differ from other creatures because evolution can continue in this inner world. By means of the cognitive function of memory, the spontaneous power of self-organization moves into this non-physical, inner space of the mind. New complexities of structure arise as various mental images and information are organized into intelligent systems. Keep in mind that the conception of a system intrinsically implies *simultaneous* constituents as wholes—whether it be a system of stars (galaxy) or a belief system (worldview).

▶ Imagination

Active memory, called *imagination*, is a new qualitative level of seeing. Whereas ocular sight has its field of vision in the physical world, the imagination scans the world of our memory. The imagination can behold simultaneously the things that had entered into the memory *sequentially* from the senses. This gives imagination the power both to see new patterns and find novel relationships and harmonies between the various ideas of our experience and to reconstitute them into new orders or homogenous unities. This superior cognitive function is creativity and ingenuity.

Relative to the eye, which takes in images sequentially, the cognitive function of imagination beholds all the things in the memory "at once." *There can be no finite ratio* between these two distinct cognitive functions of seeing: one beholds the image of trees, rocks, mountains, and stars, while the other harmonizes the idea of a circle with solid material and envisions the wheel. Imagination is infinitely superior to ocular sight.

Because these cognitive functions are qualitatively different, they require structures in the brain and neuron that are layered into distinct networks, each under the operation of different dispositions that describe different geometrical principles. This may be Swedenborg's biggest contribution to neuroscience. If visual images in the brain are caused by the changes of states in its structure and substances, then the imagination or abstraction of these images requires a substrate whose forms have an infinite new power of varying themselves. This increased ability to change state allows the imagination to distill particular ideas out of the more general information from the senses.

We can now begin to see that consciousness is multiplexed. It is in the awareness of our awareness and deeper sight into the things we see that we find meaning. Human consciousness and the subjective self rely on more than one kind of "observer" within the human brain and psyche. We see the physical world in one frame of reference, and we see our ideas in a second frame of reference. One "observer" in our psyche lets us acquire knowledge from the external world. Another, more interior observer lets us form abstractions from what we know by discovering new harmonies (homogenous unities) among the things gathered by the first observer. This ordering of experience is the non-physical equivalence of nature's spontaneous power to self-organize into orders of complexity. Furthermore, organization and order always come from a superior power of *simultaneity.**

✳ It is incorrect to think that complexity theory is a notion of modern science. Swedenborg not only anticipated this theory, he extended it to the non-material mind and organized structures of the spiritual world!

What goes unnoticed in computational models of brain activity is that Love—our likes and dislikes—focuses our mind and shapes our attention.

Our corporeal Love focuses on the harmonies we perceive from individual physical things, while the imagination is perfected from a *different and more elevated species of Love* that focuses on the harmonies among ideas.

If there were yet another discrete observer or more interior sensory operation in the brain that provided a frame of reference beyond seeing and imagining, this would add another dimension to our consciousness. This new level of seeing would expand our awareness of reality by recognizing and perceiving a new order of harmonies that would otherwise remain hidden.

Consciousness increases whenever more observers and frames of reference are superadded to the subjective self. Each new observer operates in a different dimension and represents a new value of Love. This has obvious theological implications for the growth of mind.

To awaken the next cognitive power in human potentiality requires adopting a more noble Love or value into our lives. This Love seeks even deeper and more profound harmonies in the universe. *It is the Love of seeking truth.*

▶ Human Reasoning: Another Non-Continuous Jump of Mind

The next discrete mental operations in the ascent of the human intellect, or observer, are *thought, reasoning* and *judgment*, which Swedenborg described as dispositional operations infinitely superior to imagination. Here again, Swedenborg takes us to another level of *simultaneous order*, a new discrete frame of reference and geometrical substrate that requires another principle of infinity to increase its power to change states.

Thought is another species of seeing. Unlike imagination, which creates new abstract relationships from images of sight that enter the memory from the physical plane, *thought* works from the field of ideas and abstractions already provided by the imagination. Human reasoning uses the imagination as its field of vision. As a new cognitive function, reasoning surveys and looks down upon the field of vision produced by the memory and imagination and creates a wholly new world of order. Rational thought simultaneously comprehends the things that are produced by the imagination and creates a new plane or *rational field of vision* by finding deeper harmonies between abstract ideas.

Swedenborg makes it quite clear that rational thought is discrete from imagination. Human thought starts its operation with an abstraction. That is, it takes creative ideas formed in the imagination and combines them into new unities and harmonies of another order. For this reason, Swedenborg states in his *Rational Psychology*, real thought "procures for itself and brings forth a new idea never before presented to the sight ..."[16] Such refined and distilled ideas include *justice, ethics* and *morality*. These novel values cannot be detected by the five senses because they are not direct products of mere sensory data; if they were, they would have been developed in other life forms, many of which enjoy keener organs of sense than humans do.

Some people incorrectly assume that imagination is superior to thought because it seems more creative. But that wrongly compares imagination to associative, reflexive thinking instead of comparing it to the deeper operation of *contemplation* and *pondering*. True thought is reasoning; it is not associative or reflexive. Rational thought and reasoning, according to Swedenborg, creates more rarified analogies and ratios from ideas of the imagination by means of:

> "...an analysis not unlike infinitesimal calculus, that is to say, by the rules of natural philosophy and by a mode of reduction, transposition, and equation. The equation itself, which is formed by the sole help of the mind, is called an idea of thought. Thus, an idea of the imagination is an idea insinuated through the gates of the senses; while an idea of thought is one that is formed from the ideas of the imagination, which are like the numbers in calculus, by a force proper to the mind itself. These ideas of thought, which are called rational, intellectual, and immaterial, when once formed, even though somewhat compounded, are yet in their turn regarded as simple ideas, not much unlike entire equations in algebra, and entire analogies in geometry and arithmetic, assumed as unity. The mind in turn disposes and distributes these its ideas into a certain rational order or form, and from them in turn it then brings forth a new analysis and equation. From this is born and formed an idea that is still more perfectly rational and intellectual."[17]

Swedenborg further adds:

> "Imagination, therefore, grasps only the form of an object or objects, and the nature thereof according to the order, situation, and connection of the parts or ideas, while thought does not grasp the material form of the parts, but from such form, or from like forms compared together, it elicits some meaning not visible in the parts and the connection of parts, but lying deeply concealed. Wherefore, thought is said to understand, and imagination to perceive; and an idea of thought is called immaterial, and an idea of imagination material ..."[18]

Philosophy, ethics, morality, and the codes of justice, are all examples of *immaterial* ideas that deal not with mere ingenuity but with *judgment*. (Animals, even with the capacity for ingenuity, cannot create immaterial ideas.) Reasoning and rationality are therefore infinitely superior to imagination. There is no finite ratio between the invention of the wheel and the visionary formulation of justice. Imagination is cleverness (a worldly trait) and not dependent upon personal conduct or morals. True reasoning, however, ponders and becomes wise, which is more than an intellectual exercise, for *it is directed towards the heart*. The mental inquiries of truth lead us to inquiries and choices of good.

> "To understand and be wise are two altogether distinctive things, for we may understand and still not be wise; but one leads to the other, namely science to the cognition of truth (*veri*), and truth (*veritas*) to the cognition of good, and it is the good which is to be sought for."[19]

True rationality is where spiritual consciousness begins to enter Swedenborg's multi-dimensional cognitive theory, because contemplation moves us beyond the domain of mere ego-reasoning and the self-love of the habitual mind. Swedenborg's multi-leveled model of cognitive function offers us powerful clues to God's grand cosmic scheme. As we elevate human cognitive function, the soul (which is under God's governance) more perfectly supervenes on brain structure and its changes of state to reflect spiritual values. So Swedenborg believed that "doing good" had everything to do with brain science!

Our corporeal, imaginative, and rational minds are made, not born. That is, they are acquired and arise out of the Love that reflects our subjective dispositions. In each upward step, our capacity for Love expands and becomes more inclusive. Each quality of Love creates a *new gravitational order* to our inner world that reconstitutes our mental material into real organized structure, thereby embodying our evolving values.

We become what we Love, literally. The human soul, through the medium of the mind, actually fashions a spiritual body to adapt its bio-complexity to the conditions necessary for eternal life in a non-physical realm. Darwin's theory of evolution did not take this eternal realm into account. It is spiritual selection of values, not natural selection, that drives human evolution. Religion is a crucial part of God's evolutionary strategy.

▶ Quantum Relativity in Cognitive Functions

Swedenborg's ideas for unifying neuroscience with religion would be interesting enough on their own, but he may also have unexpectedly applied the principles of relativity theory and quantum discontinuity to his multi-level model of human cognitive architecture.

Einstein's special theory of relativity gave us two interesting concepts: the *constancy of the speed of light* and the *relativity of simultaneity*. Both concern the way observers experience space and time, their frames of reference. In the first concept, no matter how fast an observer is traveling, the geometry of spacetime merely compensates itself so that light always speeds past the observer at the same rate, as if one is standing still.

In the second concept, the speed of light is invariant to all observers but different observers moving in relative motion will have a different view of simultaneity. For instance, if an observer looks down upon a city from a mountain, he or she will see everything going on in the city *all at once*, or simultaneously. However, a person walking down below along one of the city's streets will take in all the different sights sequentially. This dynamic is reversed when the city dweller looks upwards and sees everything that is happening on top of the mountain at a mere glance. Their different frames of reference let observers experience simultaneity differently.

Swedenborg took the idea of observers and tied their frames of reference to different cognitive functions. A higher-level mental function could look down on a lower mental function just as an observer from a mountaintop could look down at a person below. From the point of view of our imagination and memory, we can simultaneously behold things that had entered the physical eye sequentially. Higher up, the rational mind, with its loftier frame of reference, can view simultaneously everything that has ever been imagined by us.

This is relativity theory on steroids! Each distinct cognitive function views simultaneity differently, but differently in a qualitative way! One observer sees physical objects, another sees concepts and still another sees rational truths.

Since the operation of imagination is simultaneous relative to the sequential operation of physical eyesight, and since rational contemplation is simultaneous relative to the imagination, each distinct operation is always a *constant* and an invariable, relative to each other. We can change the things we look at in our field of vision in the physical world, or change what we imagine in the field of vision of our memory, or change what we contemplate from the field of things imagined. But these operations always remain discrete and invariant relative to each other and share no finite ratio. They operate from *distinct principles of Love* in qualitatively different spaces. This brings quantum gravity back into the picture!

Each discrete cognitive function has its center of gravity or equilibrium end-state in a species of curvature and structured space that is non-continuous with Einstein's spacetime continuum. Again, the layered structure of the human brain follows the same pattern of the layered structure of the universe. To make the quantum jump from one discrete cognitive function to another requires a change in the center of gravity of being, a jump from a Love of sensory things, to a Love of ideas, to a Love for truth.

Since each of these distinct minds arises out of some new simultaneous function and level of Love, some higher, sacred intelligence must secretly be operating within us and working on our behalf. Where does this conscious power of simultaneity come from?

According to Swedenborg, on the deepest levels of our being we have a connate intelligence that acts as an *omnipresent intention*. This inmost intelligence is the human soul. It fashions our consciousness and our discrete cognitive functions from our life-choices and Loves. (Modern science already knows that new synaptic connections are made between neurons from a person's changing intentions and interests.) The soul accomplishes this through a supernatural mind, consisting of the first forms and substances in nature, and a spiritual mind, consisting of non-physical substances (described in Swedenborg's Doctrine of Forms).

All humans have these superior minds fully functioning at birth. The more developed our reasoning, judgment, and choices, the more perfectly these higher minds and their powers are flowing into and influencing changes of state in the brain. The role of religion is to help us tap into these influences so that human reasoning becomes more *spiritually* rational, which is the basis of true wisdom.

Love occupies the cockpit of human cognition. We all have the freedom to choose noble principles of Love to live by. When we do, the soul flows naturally into and redesigns our *inner world* in a way that is analogous to the way gravity spontaneously grows structure and spacetime orderliness in the physical universe. In humans, the operation of the soul allows bio-structure to continue to self-organize and evolve into a non-spatial and non-temporal realm, where it becomes immune to the second law of thermodynamics and entropy. Without being seen, the soul engages in a perpetual effort to create a spiritual body that will enable individuality to survive the death of the physical body. God has designed a universe where the biosphere, through humanity's spiritual evolution, extends itself into a realm totally removed from the constraints of time and space.

To construct new organic forms of complexity that embody distinct levels of Love requires that the soul be furnished with an intelligence that instinctively knows the most secret laws of science—all at once. Evidence that profound intelligence lies hidden and supervenes upon nature is that even the simplest forms of life show a complete mastery of science.

▶ We all have a Supermind!

Beyond the discrete mental operations of sensation, imagination, and reasoning, there is an intelligence that sports an infinitely higher energy level and is infinitely more spontaneous in its operation. This mental operation *comprehends simultaneously what judgement and reasoning comprehended successively.* According to Swedenborg, it grasps at once the premises, principles and consequences of all things instantaneously!

Swedenborg called it the *pure intellect*; I refer to it as "supermind." It is the innate and intuitive power to discern all the constants, contingencies and symmetries of natural law. These are the principles that govern the entire universe! This *paralogical* intelligence can be described as a perpetual intuition of causes, and causes into effects. This intellect anticipates the strict mathematical requirements for maintaining perfect self-consistency (correspondence) between first causes and ultimate physical effects within the framework of top-down causality. Swedenborg states:

> "Moreover, it [the pure intellect] contemplates all things past as present and, at the same time, as future, being things which flow from the connection and according to the natural order of things. Therefore, moments and degrees cannot be predicated of the operations of this intellect, as neither can time, space, place, motion, speed, and other predicates of which presupposes succession and distance." [20]

This supermind instantly comprehends the totality of natural law as it relates to universal holistic processes. Recall from the previous chapter that this universal intelligence was also described by Swedenborg as a *mathematical anticipation of ends*. Its operation of cohesion meets the severe mathematical tolerances required for supporting an anthropic principle in the universe. It sees things as "already done." Its functioning is so holistic that its every creative act instantaneously mirrors the whole-part patterning scheme of the creation of the universe, creating self-similar wholes and complexes on all scales. This intelligence operates both within the deepest substructures of the human mind and in the deepest substructures of nature at what today's physicists would call the quantum level. I will provide a schematic for this universal whole-part process or *mathematical philosophy of universals* in the final chapter.

This super-intelligence allows for a holographic universe. It obeys a higher, holistic spiritual intelligence, and it therefore provides the mechanism by which everything in the physical world can be non-locally connected and orchestrated simultaneously from pre-space, theological principles.

Swedenborg realized that many people, especially those in academia, would deny the existence of this deep intelligence. Nevertheless, we can catch glimpses of its action in our daily mental functioning. Evidence for this connate, *a priori* super-intelligence operating within human experience is the power to *instinctively* prepare our learned knowledge in an instant, resulting in further analysis. In other words, when we contemplate something, supportive ideas are brought *immediately* to mind. This is how it is possible for assorted ideas, facts, and information *harmonious* to the direction of one's *intention* to instantly rush in, forming the coherent structure of our thoughts and the words of our sentences. This power lets us, from time to time, be "quick on our feet." We have all experienced "strokes of genius" and "flashes of brilliance" that seem to take place in spite of ourselves. They happen quickly and spontaneously, without our taking the time to apply the successive operations of our analytical powers. Solutions to difficult problems can, and do, sometimes simply "pop" into our heads.

Without such an infinitely superior intelligence, we could never recognize the universal patterns, beauty, and deeper symmetries in the world around us. It gives us the aptitude for leaps of instantaneous creative insight. This preeminent mind works on the level of the quantum vacuum as a fundamental and universal formative force in nature. It is the origin from which the laws of physics emerge. This intelligence operates within our deepest neural substrates and cognitive architecture. Swedenborg makes the bold claim that, "Inmostly within ourselves, we possess a most perfect knowledge of all natural things..." [21]

Imagine that! We have a super intelligence whose operation encompasses the Theory of Everything, or to be more precise, a universal theory of cohesion and ordered process! This super intelligence has its seat in our unconscious and involuntary mind.

According to Swedenborg, the operation of this supermind is most consistently and directly experienced when we are asleep and our usual conscious mind is at rest, because it creates the fanciful images and symbols of our dreams. Many dream researchers (including the noted analytical psychologist Carl Jung) have observed that dreams are formed from a higher level of intelligence than that of our ordinary, everyday waking consciousness. Dreams are more objective; they come from a level that is beyond and above our subjective and personal apprehension of things.

This intelligence expresses itself allegorically. As an intelligence infinitely superior to our ordinary reasoning mind, it contains ideas so totally abstracted from "worldly meanings" that it can only communicate holotropically, through analogy and highly abstract symbols. Psychiatrist Stanislav Grof describes this holotropic language as *double bookkeeping*, in which the non-ordinary states of a higher dimensional mind manifest themselves on lower planes and create metaphor and allegory. According to Swedenborg, it was in these non-ordinary states of consciousness that the prophets of Holy Scripture received their visions. It is also the reason why Jesus *only* spoke in parables, as attested in *Mark* 4:34. He was communicating on several levels in accordance to the principles of His own Divine hierarchical order.

For thousands of years, indigenous and shamanistic cultures all over the world have tapped into this super intelligence through dreams, visions, and trance-states, either by following ancient rituals or using powerful narcotics. Under the influence of these various holotropic dream-states, a shaman receives information about certain plants that provide specific medicinal cures. It is a mystery to western science how so-called "primitive" people can acquire this knowledge. Often, shamans work with plants that are so toxic there's no room for trial and error!

There is also a very famous case of a western scientist, Friedrich August Kekule von Stradonitz, who had been struggling to fathom the structure of the benzene molecule. One night he dreamed of snakes chasing each other and forming a ring. The circular motion of the snakes gave him the insight to discover the benzene ring! There is also a similar story that the Periodic Table of Elements came to Mendeleev, in a dream.

Both Einstein and the early 20th Century inventor Nikola Tesla reportedly took small naps and upon waking up found *fully formed* solutions to the problems each was contemplating. We recognize this implicitly in our common language when we instinctively decide to "sleep on" problems that have no immediate solutions.

Swedenborg maintained that if we were more perfectly connected to our supermind, we would not need teachers. There is some evidence for this in those individuals known as *savants*, who seem to have certain brain functions directly connected to the powers of a supermind. These rare individuals possess superhuman abilities that allow entire musical symphonies and mathematical solutions to pour spontaneously out of their heads. Some possess seemingly superhuman memories. Swedenborg points out that because this supermind is connate and primal, it is perfectly developed and running in newborns, as well as in those who suffer retardation or brain injury that has resulted in poorly formed, damaged or obstructed neural connections.

While each of us enjoys a more or less perfect connection to this super-intelligence, our unconscious or involuntary lives are quite another matter. We could not live for one minute without its providence. It gives our bodies the intelligence to grow and organize from a single cell, heal itself, digest food, create chemical compounds better than any chemist, regulate body temperature, and subordinate and coordinate the myriad of organic processes that need to work as a unified whole. All this requires a holistic intelligence that has mastery over all the physical sciences and laws of the universe. It grasps the Theory of Everything instantaneously.

In God's all-wise providential plan, no scientist will be granted access to a true Theory of Everything that ignores the spiritual dimension. The physical world cannot be fully explained by physical processes. The physical world and its laws emerged out of the spiritual laws of Love.

▶ The Spiritual Mind

Continuing his climb up the hierarchical ladder of mind and consciousness, Swedenborg took his study of the human intellect even further, all the way into the spiritual realm. Above the supermind is yet another, infinitely superior form of intelligence, attended by an infinitely superior principle of action. Even the supermind is subservient to the spirit, for it produces brain bio-chemicals in response to the intentions, moods and loves that are our spirit. Researcher Candace B. Pert has described neurotransmitters, steroids, and peptides as *molecules of emotion* in her book of the same title. There is scientific evidence that Love strengthens the immune system. The supermind also seems subservient to a higher mind, because the dreams it creates during sleep not only communicate in symbolic ways, but in ways that often contain urgent messages concerning our spiritual well-being.

Swedenborg maintained that since we are all spiritual beings clothed in material bodies and that we are destined to become angels, we all must possess this ineffable mind as well. At the same time, he acknowledged that this higher mind is remote from our normal conscious experience.

This higher cognitive function is the *spiritual* mind. From its infinitely superior vantage point, it can simultaneously comprehend all the physical laws and symmetries of the universe, and from that, distill new harmonies infinitely more potent, such as God's *eternal* or *spiritual* truths.

Eternal truths are heavenly harmonies. They represent the direct perception and consciousness of God's infinite Love and wisdom in all its various forms. The spiritual mind's *field of view* is angelic and heavenly. It presents us with the deepest and ultimate beauty within all of creation. The spiritual mind would not merely comprehend the physical world *physically* (like us earth-dwellers) but *spiritually.* For instance, its "likes and dislikes" have transcended entirely to the issues of good and evil. It looks at all natural things and relationships abstracted into their spiritual *equivalents*. It is the consciousness of form totally abstracted from physical measurement.

So, if we were to make a sudden cognitive leap into our spiritual mind we would see everything in the universe as an effigy of God's Love and truth. Angelic perception represents the ultimate abstraction of human thought. In this rarified field of vision, topological invariants are removed from their involvement with time and space. Dynamical magnitudes and action now find their orderliness and maximum efficiency in forms of *goodness*.

In Swedenborg's multi-level cognitive model, religion has a neural basis, because its deepest substrate is non-material and operates in pre-space. The neuron has its own superior cortex consisting of a hierarchical layering of bio-networks and non-continuous cognitive functions that ascend all the way to the soul. Each distinct cognitive function represents a "species" of Love—from the corporeal to the spiritual.

Love focuses our attention and shapes our inner world. This many-layered design implies that God wants to make contact with us and enter into a covenant with us, but in a way that does not interfere with human free will.

Swedenborg believed that the primary means by which God attempts to communicate with us is through the revealed truths of the Holy Word. But if human cognitive function is multiplexed or layered, does Scripture accommodate all the potentials of human consciousness as it makes quantum leaps into new frames of reference? Would an angel derive the same meaning from the stories of Scripture as those of us on earth? Could the Holy Word be a multi-dimensional document, designed for a multi-leveled intellect and contain the patterning principles of a hierarchical universe?

We now turn to these questions.

SUMMARY

▶ Consciousness is multiplexed. The levels of awareness represent the power of "seeing" raised to higher cognitive functions.

▶ There is no finite ratio between the discontinuous cognitive functions of sensing, imagination, reasoning, intuition and spiritual revelation.

▶ Swedenborg anticipated the science of complexity theory and took it all the way into the spiritual realm.

▶ Swedenborg unified quantum discreteness with gravity on all levels of mind and the brain's hierarchical structure.

▶ All cognitive function represents some agency and principle of Love.

▶ The neuron's layered substrates have their origins in the spiritual substances of the soul.

▶ Angels may interpret God's Holy Word in ways far abstracted from the way terrestrial humans do.

PREDICTION

The next upheaval in science and religion
will spring from the realization that
everything in the manifest universe emerges from
the Holy patterns and spiritual dynamics of Scripture.

DID NOAH'S ARK TRAVEL ACROSS POSSIBILITY WAVES?

"In the beginning was the Word, and the Word was with God, and the Word was God. The same was in the beginning with God. All things were made by him; and without him not any thing was made."
— John 1:1-3

"There are more sure marks of the authenticity of the Bible than in any profane history."
— Isaac Newton

Swedenborg's scientific research had a strong mystical component. As we saw in the previous chapter, his multi-level cognitive theory included mental operations that consisted of *holotropic*, that is, non-ordinary, states of consciousness. In these rarified states, the mind speaks in symbols and can even perceive God's spiritual influences directly.

What makes Swedenborg's discussion of these non-ordinary states of consciousness unique is that throughout his life, he was able to actually tap into them, and with increasing intensity. This process started when he was very young. Recall from Chapter Two that he would read the Bible and ponder its meaning while noticing a change in his breathing. He then became aware of a more interior and tacit type of breathing that he believed came from the *spirit*. He claimed that this tacit form of breathing was indispensable for entering into a mental state of more profound ideas and thought. This practice also prepared him for the supersensory consciousness that he would later enjoy over the last three decades of his life, when he was a prodigious writer of systematic theology.

So there is good reason to believe that his scientific works were also a result of efforts to tap into higher states of consciousness. As he moved from one scientific topic to another over a period of 34 years, he was determined to solve the difficult problem of how God acts in the world. This involved finding nothing less than the causal nexus between God and nature, the infinite and the finite, the soul and the body. He was a theist (as opposed to a deist) and therefore believed God not only created the universe but also perpetually maintained it (*creatio continua*).

Theism requires top-down causality. Top-down causality between the Creator and creation requires a nexus between qualities *that have no finite ratio between each other* (such as the force of Love and the force of gravity). A truly theistic science must lead to the discovery of a causal link that permits relationship between discrete (non-continuous) qualities if it is to succeed in explaining how an infinite God acts in the finite world. This all-embracing science was the ultimate goal of Swedenborg's search. Unfortunately, this science cannot be obtained through the activities of the normal reasoning mind or through academia. This science must be the product of higher mind, which cannot be obtained without a sincere and vigorous approach to God. This science would ultimately have to support the sacredness and authority of God's Holy Word.

Swedenborg's first efforts to find symmetries between science and Scripture hit a brick wall. Early in the 1720s, he realized that the biblical flood of Genesis did not square with geological evidence, which supported the idea that nature's processes had required long periods of time to unfold. Steadfast in his belief that Scripture was relevant to science, he found himself forced to take another approach.

If God was Infinitely wise, how could the sacredness and authority of Scripture be based solely on the accuracy of its historical data? Could scripture contain God's infinite intelligence? Perhaps there was more to God's Holy Word than meets the eye.

In 1734, Swedenborg became curious about hieroglyphics as symbols. By 1736, he had made his first attempts at writing down his dreams and interpreting them. In that same year, he began to be influenced by remarkable inner experiences. Periodically, his mind made discrete jumps into higher states of consciousness in which the mundane world began to convey the miraculous. By the early 1740s, he was fully convinced that inferior orders, such as those representing the physical world of phenomena "regard the superior as their analogues and eminences..."[1] In other words, higher levels of reality correspond to lower levels. He believed he had found the science of sciences. This universal science would address the symmetries and causal links between God and nature, thus unifying religion with the natural sciences. It was a method for interpreting all physical process abstracted from spacetime measurement.

In 1744 Swedenborg began a work entitled *A Hieroglyphic Key To Natural and Spiritual Arcana By Way of Representations and Correspondences*. It demonstrated that non-continuous qualities of actions that shared no finite ratio with each other could be arranged into families consisting of three distinct classes. Concerning these classifications he states, "In the first class are contained all those things that are purely natural; in the second those that are rational and intellectual, and consequently also moral—or those that pertain to the human mind; and in the third, things theological and Divine."[2] Examples of these three distinct classes include:

> Motion...Action...Operation
>
> Conatus...Will...Providence
>
> Nature...Human Mind...Divine Mind
>
> Faculty...Benevolence...Grace
>
> Direction...Intention...Divine Disposition
>
> Principle of Effecting...Intuition of Goals...Salvation
>
> Physical Pleasure...Happiness and Joy...Heaven

Each progression of the terms represents a quantum "jump" between qualities of action. Activity can take place in physical spaces, mental spaces and Divine spaces. You will notice that the terms used to illustrate these three distinct classes of dynamics form groups that are *similar in their differences*, that is, they correspond. Physical action has analogs in mental action and Divine action. This correspondence permits there to be causal links between realms that seemingly have no finite relation to each other. Each term is invariant relative to the others in its triad, yet the terms form kindred and *non-continuous families of constants*. Without the principle of correspondence, God would not be able to govern and maintain the world as a unified whole.

Swedenborg found these top-down symmetries, not from customary states of consciousness, but from higher states of consciousness in which terrestrial and localized events observed in nature take on a corresponding *nonlocalized* and *expanded* quality. If the laws of nature have their origins in spiritual laws, we can look at the inverse of this situation and see that nature mirrors God's non-local and non-physical qualities. Every quality of the natural world and its scientific truth will therefore have an *inverse* (*a priori*) meaning that corresponds to a spiritual truth. With this new insight, the universe instantly changed before Swedenborg's very eyes. He set out to systematize these findings:

> "In our Doctrine of Representations and Correspondences, we shall treat of both these symbolical and typical representations, and of the astonishing things which occur, I will not say in the living body only, but throughout nature, and which correspond so entirely to supreme and spiritual things, that one would swear that the physical world was purely symbolical of the spiritual world: insomuch that if we choose to express any natural truth in physical and definite vocal terms, and to convert these terms only into the corresponding spiritual terms, we shall by this means elicit a spiritual truth or theological dogma, in place of the physical truth or precept; although no mortal would have predicted that anything of the kind could possibly arise by bare literal transposition; inasmuch as the one precept, considered separately from the other, appears to have absolutely no relation to it."[3]

Swedenborg began to suspect that God's Holy Word could be treated as more than a simple linear or historical document, describing localized events in space and time. Instead, he thought there might be a similar three-tiered cosmological framework hidden within its stories—containing Divine, mental and physical levels of expression. In the apostolic teaching of Scripture (John 1:1-3) quoted at the beginning of this chapter, we read that everything in creation emerged from the Holy Word. This could only mean that God's word contained the blueprint for the created universe. For God to act in a perfectly self-consistent manner with His Divine nature, the world could not have been created in one way and the Sacred Text another.

Dissimilarity means imperfection. The laws of top-down causation must be intrinsic to the Bible's Divine architecture. The universe is multi-dimensional in the same way the Holy Word is multi-dimensional. Both are hierarchically stacked.

Does Scripture contain stories with inverse or *a priori* levels of meaning? Can the stories of Scripture contain interpretations that include the three distinct classes of activity mentioned above (physical, mental, and Divine) and give us wholly different but coherent narratives that depict events in qualitatively different kinds of *spaces*?

The effectiveness of a scientific model is in its repeatability of results and in its power of prediction. Since the three classes of dynamics identified by Swedenborg depict universal constants, there are strict rules for expanding the meaning of physical terms and qualities. These rules of abstraction are the same as those portrayed in the Doctrine of Forms whereby *constraints are removed*. Symbolic meaning, if it is to satisfy the correspondence principle, cannot be assigned to words arbitrarily. Symbolism is a process by which constraints are lawfully removed from meaning. Physical terms and qualities become "de-materialized." Scientifically speaking, abstraction must form non-continuous families or analogs, giving us their proper mental and spiritual equivalents. In other words, each layer of meaning in Scripture, while distinct, is a *similitude* of the other. Therefore Swedenborg's theory of correspondences can be tested.

Will it show us how Scripture exemplifies the physics governing the fundamental world of nature? Will it show that the laws of nature are patterned after the spiritual laws of an invisible, non-material realm?

▶ A "De-Materialized" and "Non-Local" Flood

Physicists with strong religious convictions, even those who acknowledge from their faith that all things were created by God, would doubt that they could learn anything about quantum theory from Scripture. They certainly will not find the Schrodinger equation (which is a flawed concept and does not fulfill the correspondence principle) hidden in the pages of the Holy Word. However, if they embrace the ideas of *non-locality, discreteness,* or *mixed states* or the idea that a quantum entity acts like a *non-material* wave then they may be surprised.

I posed a question in the title of this chapter: *Did Noah's Ark Travel Over Possibility Waves?* Is it possible to interpret this famous biblical story in a way that does not require the Ark to travel across *physical* waves? Could there be a deeper quantum narrative that is timeless and relevant to the difficulties we face in the modern world?

The wavefunction of quantum physics does not describe physical waves, but *waves of possibilities*. These "tendencies to exist," that is, the non-physical endeavors to become actual measurement outcomes in time and space, are similar to our inner world of volitions and thoughts. Our volitions and thoughts "jump" into outcomes of measurement when we act on them and they then become localized (and non-reversible). This top-down activity in human experience is similar to a quantum discontinuous "jump" from the realm of potentials into a realized outcome in which one possibility is chosen over the others.

Many physicists believe that consciousness plays an important role in the process whereby quantum events become actual events. Some suspect that quantum process is also involved in the interaction between mind and brain. It is! In humans, however, this process has *eternal* consequences. Our choices among possibilities can produce either hurtful or beneficial results. Our choices also shape who we are on the most fundamental level of existence, even enabling evolution to continue beyond a person's biological self.

It is the quantum microworld of *fluctuation* and the uncertainty principle of possibilities and tendencies that God's Holy Word inwardly addresses. This is where religion must take hold within human consciousness to prevent harmful tendencies from making the discontinuous leap into physical reality.

Temptations and proclivities are tendencies to exist. They represent non-physical forces seeking to become measurement outcomes, that is, non-reversible events through human acts and deeds. The strategy of religion (and God's wisdom) is to address the quantum potentials operating within the deepest recesses of the human psyche. Religious tenets supply a conscious vehicle, like a lifeboat, on which we can ride out our *personal* storms and emotional *fluctuations* when our minds are being inundated by harmful tendencies and compulsions.

Where is this non-physical message actually addressed in the historical narratives of Scripture? Noah's Ark seems to portray a physical event. But could it represent something deeper and more profound? Perhaps, like the quantum world, Scripture contains an *a priori* level of meaning, with an *expanded* quality and non-local message.

If the story of Noah's Ark contains an inner narrative or quantum language, then all its physical terms must be perfectly transferable to a nonlocalized or spiritual value. So as an experiment, consider the word "water," which is a significant feature in this biblical story. Let us see if its meaning can be expanded in a way that *reveals* another, deeper level to God's Word. In this case the "water" of the biblical flood would represent a spiritual threat rather than a natural disaster.

If we apply Swedenborg's universal science of correspondence to the word "water," we find that all its physical qualities easily de-materialize into the psychological qualities of *knowledge* and *memory-data*. For instance, information streams through our senses as waves from an ocean of frequencies and collects in our memory, just as water collects into pools, lakes and oceans. We *pool* our ideas.

The human mind has an instinctive ability to form such analogies, yet science cannot tell us how nature is conducive to such things. The ocean of information we collect throughout our lives, for instance, is equally prone to *all the conditions* we find with water. Our memory can be *muddied* by false, trivial, or unorganized data and even contain *polluted* and poisonous information. We can become *swamped* under a *deluge* of data to the point we *drown* in it. We can get into *hot water*. Our anger can reach a *boiling point*. Or we can be *ice cold*. If we *harbor* negative things in our mind, they can *stagnate* and become *putrid*. Getting into a conflict is often described by the phrase, getting into *troubled water*.

On the more positive side, the memory can also contain elements of untainted truth, particularly for those who seek it. That is why water, when purified or distilled, can also correspond to the clarity of truth. Truth makes things transparent to us.

Water quenches the thirst, knowledge quenches the mind and truth quenches the soul. Put another way, water is refreshing and revitalizing to the body, while truth is refreshing and revitalizing to the soul. Without water the body becomes parched and dies. Without truth our inner being withers and dies. In the same way that water bathes and keeps the body clean, truth cleans and purifies the soul.

The academic world may see such metaphors as sloppy science, but this comes from its materialistic and reductive views of reality. Swedenborg insists that metaphor has onto-logical reality within higher levels of the hierarchical universe.

▶ Applying the Science of Multi-dimensional Order to Exegesis

Swedenborg's study of the "layered" structure of nature and the human psyche led him to conclude that Scriptural architecture also contained the same scaffolding. He theorized a multi-dimensional universe and applied the assumption to his systematic approach to theology and the structure of Scripture. Through the correspondence princicple, he began to apply more rarified meanings to the words of the Bible. And he discovered that a wholly new dimension of God's revealed wisdom opened up to him.

Let us apply this technique of abstraction to the larger story of Noah's Ark. In this famous biblical story Noah tried to ride out a great flood. If we take the term "water" and *de-material-ize* it to its psychical and spiritual equivalent as we did above, we gain more profound insights into this story. In fact, we get a story that has quantum components! Water corresponds to memory-data. A "spiritual" flood represents both an inunda-tion of information and a chaotic state within an individual who is experiencing inner conflicts. This conflict is represented by the stormy *fluctuations* that rise up from our memory and threaten to drown the soul. The non-physical waves of such troubled waters are a *superposition*—a mixed state of tempta-tion pushing us in all kinds of bad directions at once.

God helped Noah by giving him the specifications for build-ing an Ark. This too has a spiritual meaning. Such an Ark could be built only from God's exact specifications because they are not for the purpose of building an actual boat from timbers. Instead, they are spiritual instructions to help us construct something that will allow us to remain buoyant and ride out our agitated and mixed state of threatening possibilities before they become non-reversible and measurable consequences. These specifications are the tenets of religion.

The story of Noah's Ark conveys two distinct types of navigation that share no finite ratio. One level conveys a physical event, the other, a purely spiritual event. Yet as non-continuous constants, they are perfect mirror images of each other. In the modern world, however, only the one physical interpretation is grasped and promoted to the faithful. Swedenborg believed that to deny deeper levels of meaning within Scripture is to deny God's infinite intelligence. Even worse, it leads others towards disbelief, viewing these "fantastic" stories as being hokey or contrived and driving them towards atheism and secular reasoning.

To test the theory that Scripture contains a quantum language even further, water will have to give us a similar spiritual meaning wherever it is used. In each case these higher meanings will provide us with deeper revelations about human interiority (the human heart and mind) and the subject of salvation.

The *separating of the waters* into higher and lower realms in the Creation Story of Genesis can be understood spiritually as the stage in our inner development where we gain new cognitive powers of discernment and the ability to discriminate what is really important in life from the clutter of our everyday knowledge. The mutable meaning of the words frees the creation story from scientific criticism, because it now pertains to the deeper story of our spiritual rebirth and new creation, our *epigenesis*.

Interestingly, many of the miracles of the Bible alluded to Scripture's quantum language. Jesus *walked on water*, which signified that God's knowledge is of a higher order that can easily rise above the natural and disturbed, stormy thoughts of worldly men and women. The Lord turned *water into wine*, which symbolized God's ability to transform natural and mundane knowledge into something with more "kick" and potency.

The Lord directed His disciples to look for a *water-bearer* who would lead them to an "upper room" for their special meeting place. This symbolized the challenge and necessity for His followers to take their knowledge to a higher level. The lower cognitive functions of the habitual human mind are not able to *digest* the quantum language embedded within the strange rituals that took place at the Last Supper. One purpose of the Lord's ministry on earth was to invite all followers to drink and absorb lessons of a whole new cognitive order. Hence, the Lord spoke *only* in parables.

I will open my mouth in parables; I will utter things that have been kept secret from the foundation of the world. (Matt. 13:35)

If a terrestrial understanding of the words of Holy Scripture were sufficient, the Lord would not have had to make any extra efforts to aid the comprehension of His disciples:

He opened their minds to understand the scriptures. (Luke 24:45)

Open their minds to what? The answer is that the Lord sought to reveal this quantum language of correspondences to His disciples so that they could probe deeper into the mysteries of faith through their powers of reason. The human race needs to activate its higher, God-given mental facilities to get out of its current predicament. Here, religion meets science dead-on. The upper room where the Last Supper took place and the *three levels* of Noah's Ark address both the top-down order of the hierarchical universe and the multi-leveled order of the human cognitive scaffolding. Scripture embraces cosmology and neuroscience! The Holy Word contains three distinct narratives. They address physical events on the lowest level, psycho-spiritual events on a higher level, and Divine events on its highest level. These distinct levels of meaning in Scripture refer to various cognitive levels operating within the multi-level structure of the human brain and mind. Scripture serves as God's guide for helping humans tap into higher levels of mind and allowing them to interpret physical events in a non-physical way. Each level allows God to enter into a more profound relationship and covenant with humankind.

Swedenborg had discovered much more than clever plays on words or poetic license. These metaphors in Scripture were ontological, representing objective realities in higher realms. This deeper "language" came from *spiritual* cognition, and since this capacity lies within all of us, it seeps even into our ordinary lexicons. Again, using the term "water" as an example, this seepage of one cognitive level into another is evidenced when say that we are "thirsting for knowledge," or, when we know something is not true, that "it doesn't hold any water." These common phrases of our language betray the existence of an unconscious and connate intelligence within us that instinctively knows the ontological correspondences between distinct realms. We intuitively know that physical water lawfully symbolizes knowledge and truth, which are non-physical.

The great mythologist Joseph Campbell would describe this lock-and-key relationship between water and knowledge as an *innate releasing mechanism* (IRM) of the unconscious. But he did not know how such relationships were established in the human psyche. Campbell did not know that relationships of these kinds represented the causal structure of reality that the human mind and brain are patterned after.

If abstraction of physical terms leads us to entities with real ontological status, where all things are removed from their involvement with spacetime, then there must actually be a world whose landscape and topological features consist of forms derived from the qualities of heart and mind. Religion calls this pregeometric world heaven. If heaven is non-material, then it cannot be a place you go to after the death of the physical body. Rather, it is something you become. God's kingdom is a kingdom of Love. It can only be obtained through a sincere love for God and neighbor.

Before we look at the dynamical landscape of this non-physical kingdom, it is important to know why the human race plays a critical role in evolution. Humankind is the means by which God extends the biosphere into the pre-space of heaven through the perfection of Love.

▶ Evolution as Conjunctive Design

Swedenborg embraced the idea of evolution. But he brought to the debate about evolutionary science the concept of *Conjunctive Design*. It takes the Intelligent Design concept that the complexity of living organisms is purposeful one important step further. It identifies the Divine purpose and mechanism that is directing evolution. It takes us all the way to God's covenant with humanity.

The neo-Darwinian synthesis argues in favor of randomness and fortuitous factors in evolution. However, William Dembski, the author of *No Free Lunch*, argues that "specified complexity" and utility in biological structure provides reliable empirical evidence for intelligent design. Conjunctive Design adds an important theological ingredient to all this. It brings specified complexity under one holistic umbrella and grand scheme. Swedenborg tied "utility" to God's original and eternal purpose. Conjunctive Design also extends Michael Behe's idea that biological systems are "irreducibly complex" into a pre-geometrical realm of non-physical bio-complexity!

Earlier in this book (pp. 59, 103), we discussed that a God of Love needs something to Love. The universe was created to serve this purpose and requirement of Love. However, God cannot create or sustain anything unless it mirrors some quality of Love (usefulness). God creates through self-representation and similitude. So the eternal unites the time-bound to itself by means of these correspondences. This allows temporal matters to coincide with eternal ones and all created things to be recipient forms of Divine influence. Discussing the created universe Swedenborg says:

> "For having been created in God from God, it is adapted to conjunction; and because it has been so created, it is an analog, and through such conjunction it is like an image of God in a mirror." [4]

The Infinite is in finite things according to reception, that is, *conjunction*. If God's infinity consists of infinite distinctiveness infinitely unified, then God is more perfectly conjoined to and more perfectly acts in the world through the emergence of complexity and higher-order bio-structures. The main component of organic complexity is division of labor, which contributes to the integrity of the *whole*. Evolution is a miraculous process by which differentiation becomes more perfectly integrated. "Individuation" is perfected by promoting community *through utility*. The unifying principle behind this holistic relationalism is found in the dynamics of spiritual Love. Love is a non-physical, formative substance that is present in everything. This non-spatial and non-temporal substance is the reason why everything in the world is non-locally connected simultaneously (which is a feature of quantum theory). Swedenborg states:

> "... that there can be no harmony between natural entities, without a principle of harmony in superior nature, which conjoins single things universally, and the universe singularly." [5]

Science does not deny that the universe is unified. But it has no satisfying "physical" answer for this unity. Even Intelligent Design suggests only that the universe is the result of an Intelligent Designer without identifying the theological principle guiding *all* design and creation's ultimate goal.

In everything created by God, Who is Life, there is reaction or response. In Life alone there is action and determination. All physical process in the universe is caused by a response to the action of life. It is based on a Divinely living principle (which is why the laws of the universe are so bio-friendly). Matter does possess endeavor, but not from itself. Nature by itself is dead and it is not self-organizing. Studying matter or trying to find material explanations for everything will not uncover the key to agency. Scientists who insist that anything going beyond explaining nature by purely physical laws is not science will have to remain content to explain reality without first causes. This view is not science but *scientism*.

Only Life can create. Love is living creative force and primal substance. All novelty in the universe emerges from a profound, unifying essence. As Swedenborg states, "For new things to exist, previous things must coexist." Whereas Intelligent Design simply points to organized utility, Conjunctive Design points out that utility is *goodness* and reflects God's involvement with nature in different degrees of complexity. In other words, Divine Love combines matter with utility and purposeful design.

But Love is imperfect if that Love is not returned. Evolution is not driven by natural selection, but by the exaltation of Love. This requires the emergence of a being in nature that has the complexity and the higher-order cognitive structures to *consciously* recognize God and love Him in return (conjunction). Swedenborg maintained that the purpose of creation was to *create a heaven from the human race*. Divine Love would be impotent if it did not seek its fullness in an eternal reciprocal relationship.

▶ **Evolution is Spiritually Driven**

Since evolution is spiritual and driven by God's wish to have an eternal relationship with humanity, there must then be a lawful contingency by which bio-complexity can be extended into an indeterminate, non-localized, and non-temporal realm, far beyond the ravishes of entropy. This contingency does not come into play by humans who are merely adapting to a physical environment or finding a niche to exploit in the physical world. Nor does it come from new reproductive strategies to ensure genetic survival, as the neo-Darwinian synthesis would suggest. It comes from religion and following God's tenets.

We were all created by God to live in a blessed state forever. Love wants to share what it has, its blessedness, with others. But Love is also primal, formative substance in a perpetual effort to create organized structure through the subordination and coordination of successive and simultaneous order. Therefore, the human mind, which consists of the things we love and think about, consists of spiritual substances. So the mind, which is actually our spirit, has non-material bio-complexity that is tied to the civil, moral and spiritual principles we adopt in life. In the same way animal bio-complexity on earth is adapted to various physical environments, the bio-complexity of the human spirit will be adapted for some specific environment in a non-physical world.

Unfortunately, what humans inwardly love, intend, and think about can be far removed from anything resembling a blessed state. Because of free will, one's spiritual biology is self-organized according to one's values and life choices.

Religion deals with values and life-choices because it is part of the real science behind the bigger picture of evolution. When God's tenets are adopted, concrete changes take place in our lives. While this change may be manifested on earth as increased good will and helpfulness towards others, a new complexity and self-organization is occurring inwardly in the spiritual substances of our hearts and minds. This is a startling idea. And it is crucial for humanity to acknowledge.

Since our ideas and values are real substances and not "wispy" or ephemeral things, then we must truly be careful about what we wish for. What we call our belief-system or our worldly paradigm and its supporting ideas, are actually arranged into the non-physical organs, tissues and bio-structure of our *spiritual bodies*. Unlike our physical bodies, which are the results of inherited genes, our spiritual body gets its features from what we love and is therefore even more perfectly constituted to do our bidding. *This inner body is who we really are.* The soul fashions this spiritual body from our life choices. It is what survives death and allows us to gain access to non-terrestrial real estate.

The stories of Scripture, when translated into a deeper, quantum language, address nothing else but the evolution of the spirit and the way its development coincides with and receives God's blessedness. So when we contemplate the ritual of Baptism and de-materialize its meaning, we see that it serves as a lesson reminding us to think about the *inner* cleansing and purification of our mind and spirit. To the degree that we remove unsavory things from our lives, we organize our inner being to receive more of God's blessedness and Love.

Heaven is acquired. It is a *co-established* harmony with the Divine Creator. This is why the Creator seeks a covenant with humanity. The human race, through spiritual principles, allows Conjunctive Design to reach its fullness and perfection.

▶ A Paradigm Shift to a More Adequate Theology

The conflict between modernity and Scripture is caused by the fact that many biblical stories no longer seem relevant to today's world. Even worse, some of the passages in Scripture seem to approve slavery, patriarchy, the colonization, oppression, and slaughter of other peoples, and even discrimination against those who do not conform to the mainstream lifestyle. Because of this, even some of the faithful have accepted Scripture not as an absolute and infallible deposit of truth, but as a document that is better interpreted through the *relativism* of its truth.

All this is cleared up when the Holy Word is seen as a multi-leveled document with distinct narratives. The quantum language hidden within the literal sense of Scripture gives entirely new and unexpected meanings. These are profoundly relevant to the post-modern world, where scientific discovery, archeology, and religious relativity have challenged the "inerrancy" of the biblical text. The higher meanings conveyed by Swedenborg's science of correspondences also abolish the different worldly interpretations of Scripture that are divisive to religion, lead to intolerance, and even to atheism.

Philosopher Ashok K. Gangadean, author of *Between Worlds: The Emergence of Global Reason*, believes that a holistic dialectic is needed to move humanity beyond its fragmented worldviews. He realizes that this exalted grammar will emerge only from "the recognition of the bankruptcy and failure of egocentric reasoning."[6] He believes that such a universal linguistics can only be approached through *ontological relativity*, where words, thoughts, and meaning gain new degrees of freedom and objective potency.

This is exactly the approach Swedenborg took. His science of correspondences is a universal language in that it adds new degrees of freedom to the terrestrial meaning of words. These higher meanings will be the same for everyone, no matter what language they start from.

Scripture contains three orders of narrative corresponding to the three distinct orders of process in the universe—physical, mental and Divine. The lowest level of narrative is the literal sense of the words. Above that, the narrative represents *deep psychology* and is purely *soteriological*, that is, it addresses our inner journey and our personal salvation. Higher still, the *Christocentric* narrative speaks about how the Lord made the Word "flesh" and glorified His human nature with His Divine nature to become the Alpha and Omega.* (Chapter Nine addresses this topic.)

✳ While this assertion may alienate non-Christian readers, Swedenborg points out that this three-tiered language was originally incorporated in all symbols used by more ancient religions and that it can even be teased out of Greek mythology.

In the higher orders of interpretation, we can begin to experience the depth of God's wisdom. The stories of the Bible are not the fanciful tales of goat herders sitting around a campfire. In fact, Swedenborg would later learn from his contact with the spiritual world that the failure of these "quantum" interpretations to be adopted into the hermeneutics or methodology of human inquiry is one of the major reasons why the Lord promised He would return.

To borrow a phrase from chef Emeril Lagasse, God wants to "kick it up a notch." This means that humanity must also rise above its biological self and tap into a higher cognitive level, one that is closed off to humans who immerse themselves in merely corporeal-sensual worldviews.

Knowledge of these higher levels within the architecture and scaffolding of Scripture helps us gain powerful insights for unifying science and religion. For instance, these inner levels are analogous to Bohm's notion of implicate orders. Each distinct level of Scripture represents a narrative whose continuity is discretely different and invariant relative to the others. They are God's fundamental constants, and they represent three major boundary conditions of reality and their particular dynamics. Each level of Scripture was generated directly out from the level immediately above it, in a way that set the finely tuned pattern for all top-down causation in the universe. They are non-continuous, but similar in their differences from the principle of correspondence. At the upper levels, words and the actions they describe have expanded meanings that portray non-local events and represent our possibilities for spiritual growth.

Not only did Noah's Ark travel across non-physical waves, but when the totality of Scripture is viewed from a higher cognitive level, all its stories can be seen to be non-locally connected simultaneously. Each word, each sequence of events, is orientated to and contains the eternal goal, much like a hologram where everything contains an image of all. In this way everything in scripture portrays God's infinite wisdom, containing both previous things and future destinies as simultaneously present.

According to the theory of Conjunctive Design the *arrow of time*—the directionality of process and change—moves in a transcendental circle. Creation and evolution move out from God and the spiritual realm into constrained spacetime matter. Then they return back to God and the spiritual realm through the proper development of human minds and hearts (See Figure 3.5). Time is asymmetrical: it has a "one-wayness" that is the result of Love's eternal disposition towards goals. Time does not move backwards. If it did, Love, which is the principle of agency in the universe, would have to act contrary to itself. Such a reversal would undo Love's action to self-organize into complexity. *Unity through utility*, or "goodness," would perish and thwart God's plan. The show must go on! As discussed in Chapter Three, time and process are actually the same things and are inseparable. You cannot turn time backwards while process in the world—and God's influx—is moving everything forward.

Conjunctive Design, unlike other theories of creation and evolution, requires our having real knowledge of the spiritual world and its topological features. What strange ontology emerges from a non-local and non-temporal realm where spiritual values define the structure of its "spaces"? How does one understand a world where measurement, dynamical magnitudes, and topologies exist according to the states of the human heart and mind? Can there be trees, animals, roads, mountains, seas and cities in this non-physical environment?

Swedenborg found out—through firsthand experience.

SUMMARY

▶ There are three distinct categories of action in the universe—physical action, mental action and Divine action.

▶ The sacredness and authority of Scripture is contained within its deeper levels of meaning.

▶ The evidence that the laws and boundary conditions of the universe are also contained within the top-down patterning principles of the Holy Word can be obtained only by an interpretation that transcends the literal meaning of its narratives.

▶ Swedenborg's correspondence principle not only requires that the results of classical physics are recoverable from the dynamics of the microworld, but from the deeper theological realities of non-physical operations.

▶ A Theory of Everything will come from God, not humanity.

▶ Complexity emerges when Love attaches utility to combinations of matter.

▶ Time does not travel backwards, because this would undo the very fabric of reality, that is, God's order and processs.

▶ Evolution is a spiritual process. It is the exaltation of Love. Darwinism does not take into account that the biosphere extends to a non-spatial and non-temporal realm.

PREDICTION

The science of correspondences will be the key for both scientists and theologians in constructing a more adequate worldview.

Chapter Eight

WELCOME TO THE SPIRITUAL WORLD

"Scientific concepts exist only in the minds of men.
Behind these concepts lies the reality which is being
revealed to us, but only by the grace of God."
— Wernher Von Braun

By the early 1740s Swedenborg had become so confident in the direction his studies were taking him that he wrote in the introduction to *The Economy of The Animal Kingdom* (meaning domain of the soul) that he believed the time was ripe for a new age of genius. He added that humankind was now very close to finding the hidden causes of things. His ardent desire and zeal for that bold outcome, communicated in his own words, reveal the poet within the scientist.

> "And the time is at hand when we may quit the harbor and sail for the open sea. The materials are ready: shall we not build the edifice? The harvest is waiting: shall we not put in the sickle? The produce of the garden is rife and ripe: shall we fail to collect it for use? Let us enjoy the provided banquet, that is to say, from the experience with which we are enriched, let us elicit wisdom." [1]

Swedenborg saw scientific methodology as a real part of God's Divine Providence. Science moved us toward deeper insights about reality. However, he believed that science could not reach a final theory of reality unless the search for true knowledge was the means for ultimately procuring *wisdom*. Procuring wisdom meant that this new age of genius and its science could not be separated from matters of the heart. This bold new science would lead us to proper values and life-choices. The acquisition of wisdom is for the sake of doing what is *good*. Otherwise we would be left with a vain science that was doomed to validate only the purposelessness and absurdity of our universe, no matter how intelligible its laws may be.

During this hopeful period, Swedenborg came to an unflattering conclusion about the proper methodology for discovering *truth*. Through self-reflection he found a major character flaw in his personality actually limited his ability to discern truth. "As soon as I discovered anything that had not been observed before, I began (seduced probably by self-love) to grow blind to the most acute lucubrations and researches of others ..." [2] He was jolted to realize that human cognition was greatly impaired by "the thirst for glory and the love of self." [3]

During this period (1736-1745) Swedenborg's focus became twofold. He explored his inner world through introspection while also exploring the human body—anatomically, physically, and philosophically—in order to discover the essence and nature of the human soul, its influx into the body, and the reciprocal action of the body. The soul, he believed, had order and truth in it; it exhibited a total mastery of science. But discovering such a universal science could become the property only of "rightly-constituted" minds. The mind had to be *equally ordered to discern the higher orders of reality*. He felt strongly that he needed to rid himself of his character flaws in order to receive greater truth.

This inner cleansing of the heart and mind is not the strategy of academia to acquire truth, and it would be viewed as alien to most researchers. But Swedenborg began a regimen of intense introspection, prayer, and calling on God to help him combat the negative leanings of his character. He emerged from each bout with greater mental clarity. He finally reached the point where he admitted that he knew nothing and could only give himself totally to God.

The universal science he was seeking would be under the favor of God.

If we believe Swedenborg, his strategy worked. God's Divine guidance began to show itself in real ways. Recall from Chapter Two that while he was writing during this period, he began to experience the phenomenon of *photism*; strange lights and flames appeared before him. He interpreted theses lights as Divine signs that he was on the right track. He also attributed these lights to new and deeper activity in his brain. This activity continued to increase. Little did he know that he would soon be using his scientific talents to explore a whole new world, a pre-space world that led him to the first causes of all things. But God would be calling the shots!

This is where everything came to a head for Swedenborg (as I mentioned in Chapter Two). The details now follow.

▶ London, April 1745

Swedenborg sat in his private room at an inn which he often frequented. The server brought in a large array of food, because on this day he had a particularly large appetite. He started into his meal while he satiated his appetite. All the while he was in profound contemplation of his belief that he may have discovered the key to unlocking hidden levels of meaning in Scripture through the corresondence principle. He marvels that *spiritual correspondences* can be distilled from everything in nature.

Towards the end of the meal, Swedenborg noticed something strange: the light in the room was growing dimmer. Then something really terrifying caught his attention. Some kind of exhalation or subtle mist (like the ectoplasm that materializing apparitions are said to be made of) began to flow out from the pores in his skin. As it rose, it condensed into a cloud, then dropped to the floor by his feet. To Swedenborg's horror, the strange mist-like substance began to morph into living shapes, into frogs, snakes, spiders, and other unsavory creatures that finally burst into flames. He barely had time to react to this appalling event when a man's voice, coming from a figure in a dark corner of the room said, "Stop eating so much!"

Swedenborg became so unglued by all this that he scrambled out of the inn and rushed back to his apartment. When he reached his apartment, he was careful not to display his panic to the landlord, so he slowed down as he walked up the stairs to his room. He opened the door and carefully locked the door behind him.

In his room he began to search for answers about what had just taken place. He dismissed the idea that it was a chance happening, a bad reaction to the food, or even the result of any other external causes. Swedenborg did not have to wait long to get the real answer.

Later that night he began to sense he was not alone. While he lay in bed, a man abruptly stepped toward him, dressed in a royal purple robe and emitting a wonderful light. It was the same man he had seen at the inn. At first, Swedenborg was frightened, but his fear turned into profound humility. Then he dropped to his knees! The figure before him was the Lord!

At this point, the Lord spoke to Swedenborg and gave him the "Divine mission" that would forever change his life and astonish much of Europe, including Immanuel Kant. Here is how Swedenborg remembered the event:

> "I went home, and during the night the same man revealed Himself to me again, but I was not frightened now. He then said that He was the Lord God, the Creator of the world, and the Redeemer, and that He had chosen me to explain to men the spiritual sense of the Scripture, and that He Himself would explain to me what I should write on this subject; that same night also were opened to me, so that I became thoroughly convinced of their reality, the world of spirits, heaven, and hell, and I recognized there many acquaintances of every condition of life. From that day I gave up the study of all worldly science, and laboured in spiritual things, according as the Lord had commanded me to write. Afterwards the Lord opened, daily very often, the eyes of my spirit, so that, in the middle of the day, I could see into the other world, and in a state of perfect wakefulness converse with angels and spirits." [4]

Swedenborg claimed that the Lord opened his *spiritual sight* so that he was able to enter into and explore the wonders of life on the "other side." And he claimed that this was a capability he would enjoy until his death in 1772 (a period of almost three decades). At first, he told no one about this miraculous new "gift." He set himself to the task of writing copious volumes about what was being revealed to him.

Around four years after the event at the inn, in the summer of 1749, the first of many strange books began to appear in bookshops. The first series of books were entitled *Arcana Coelestia* (Secrets of Heaven). They were published anonymously and, to the shock of Europeans, described things "seen and heard" in the world of spirits by someone who claimed, as a citizen of the earth, to have personally witnessed these things.

Liberal theology rejects anything that would be seen as inherently "unfair" to others. I, however, have no problem with God preparing an individual to serve as a special instrument for some Divine purpose. I also personally find it a wonderfully refreshing Divine strategy by the Lord to pick, not a poet, artist, government official, actor, merchant, or clergyman for such a mission, but a scientist. Remember, this was also the Age of Enlightenment. What better profession to interface between God and the newest trends of thought than a scientist? A scientist would be the perfect choice to perform experiments and make observations in the spiritual world, which Swedenborg did do for almost three decades. Who more than a scientist would be driven to explore and determine the properties of the substances found in this world—the flora and fauna, as well as the pre-space geological formations? Heavenly things and the mysteries of faith could now come under the observation of someone with real objectivity who took scientific methodology seriously.

▶ A Scientist Conducting Experiments in Pre-Space

Current neuroscience has limited its descriptions of mental processes mainly to the complex synaptic connections between different localizations of neurons in the brain. These 1,000,000,000,000,000 (a million billion) connections represent the many paths information can follow in the brain from one neuron to another. Yet, even this bio-electro complexity is proving inadequate to explain the phenomenology of the deeper levels of mind and internal personal experience.

Swedenborg's model of human cognitive architecture, discussed in Chapter Six, takes neural complexity well beyond mere physicalism and into the realm of non-physical structure. The spectrum of cognitive functions in humans originates from the fact that the neuron and its deeper orders of structure are, like the human eye, true sensory organs in their own right. But the neuron is an *inner* sensory, and even deeper within its structure it contains an *innermost sensory*. Each substrate or layering of structure is based on a different geometrical principle of form. Each has an increased capacity for changing states, and each is therefore adapted to working with *different orders of information*. Evidence that the discrete cognitive functions of the human intellect are seated in a scaffolding of true sensory organs is that physical eyesight, imagination, reasoning, and revelation, are all different qualities of "seeing." One function sees physical objects, the other sees ideas, yet another sees intellectual truths, and deeper still, is a function that sees spiritual realities.

When Swedenborg made the claim that the Lord opened his spiritual eyes or spiritual sight, he was saying that the Lord granted him access to the innermost organic substrates and capabilities of the mind (which consist not of physical substances but of spiritual substances). When he stepped into the non-temporal and non-local spiritual world for the first time, he went neither to the right or left nor up or down. Nor did he travel beyond the farthest stars and quasars. He went *inward*.

This inward journey took him deeper than the pre-Planck scale of the universe or any singularity and into the pre-geometric realm of fundamental reality. In this world of non-physical events, he found no roiling frenzy of quantum fluctuations. Instead, he found a coherent and ordered world consisting of rocks, mountains, oceans, sky, trees, animals, and towns and cities occupied by real people. He discovered that the natural and spiritual worlds were completely alike, yet completely different. Everything one could find in the natural world, one could also find in the spiritual world. But they consisted of different essences. The natural world consisted of physical substances and the spiritual world consisted of spiritual substances.

Consider the differences between physical reality and the experiencing of the deeper real "self." We are more than our height, weight, or eye and hair color. We are more than our genes. We are our feelings, thoughts and beliefs. These inner dispositions exist and subsist as real *substances* and *forms* that ultimately create what we look like in the spiritual realm. We know this on an instinctive level when we describe an unselfish individual as a "beautiful person" and having "inner beauty."

The spiritual world consists of mind and heart stuff. Yet, it is infinitely more real and substantial than the physical world. The psychical is fundamental. It is a world where values, not physical laws and forces, describe the parameters of action, process, order and form. It is this inner landscape or *psycho*scape of the human heart and mind that the Holy Word addresses in its higher levels of meaning.

We are made in God's image and likeness, not in the sense that we have ten fingers and toes and walk upright, but in the sense that we have been given *free will* and *discernment* (which mirror God's Love and Wisdom on a finite scale). This makes us human and forms the *psychoscape* of our inner reality. These are God's two gifts to us, and without them, there would be no means by which the human race could continue to evolve towards the non-physical biosphere of heaven and its eternal blessedness.

▶ What is Heaven?

Whether they believe in a heaven or not, most people would agree that if such a wonderful paradise did exist, it would be a blessed place of resplendent Love. When Swedenborg's eyes were opened up to the spiritual world, he observed that heaven is not simply a place where people love; it is a place that is created by Love, which is the causal agency and ultimate formative substance. One's subjective identity is what one loves, and what one loves determines both the features of one's *spiritual body* and the features of his or her eternal environment—the "place" where one lives in the spiritual world. Since all spiritual substance consists of Love, every form, dynamic, and process in this world represents some measurement of Love.

> "*Man after death is his own love or his own will.* This has been proved to me by manifold experience. The entire heaven is divided into societies according to differences of good of Love; and every spirit who is taken up into heaven and becomes an angel is taken to the society where his Love is; and when he arrives there he is, as it were, at home, and in the house where he was born; this the angel perceives, and is affiliated with those there that are like himself." [5]

Also:

> "Another proof that a spirit is his ruling love is that every spirit seizes and appropriates all the things that are in harmony with his Love, and rejects and repudiates all that are not." [6]

Everyone can be said to be his or her *Ruling Love*. Put into scientific language, one's ruling love is the *center of gravity* of his or her being. This determines where one finds his or her spiritual equilibrium, which is one's placement in the non-spatial and non-temporal realm of the spirit. What one loves also becomes the gravitational order by which a spiritual body is generated for the soul. So the importance of first-person, phenomenal experience in human cognition is that it customizes our inner bio-complexity and eternal abode according to our proclivities.

During life, the subjective self (the spirit) is formed in such a way that its specified complexity perfectly reflects everything that is dear to us. This process provides embodiment, orientation and direction for a soul that has left the physical body after death. Everyone in the other world recognizes the path one must take to find one's eternal "placement" in the spiritual world, because each path is instantly recognized as consisting of the *principles* which one *already* follows. In other words, if our spirit is our volition and reasoning, then the spiritual world, which consists of the same non-physical substances, can consist only of various qualities of Love and its idea-forms.

But this creates a conundrum. Motion happens in time and space. How, then, are we to understand the spiritual or non-physical locomotion of human souls and their spiritual bodies in a setting that has neither time nor space?

While physical motion involves displacement through *change of position*, spiritual motion involves *change of state* through a displacement of thoughts and feelings. We are always going somewhere in our mind due to our changing moods and thoughts. These are real dynamics. So, if a person in the spiritual world changes his *state of mind*—let's say, from anger to love—that person will instantly change the coordinates of his or her inner reality and location in the spiritual world. The spiritual world is under a different principle of extension. Position is disposition, direction is affinity, and speed is strength of intention:

> "Thus times and spaces there are in conformity with the affections of their will, and consequently with the thoughts of their understanding: but these appearances are real, for they are constant according to their states." [7]

The spiritual and physical worlds are alike in that they both portray extension and ratio as relationship. One deals with the expansiveness and various proportions of physical space, the other, with the expansiveness and various proportions of Love. These are two distinct kinds of measurement. However,

Swedenborg observed that everything in the spiritual world still gives the appearance of time and space. This is because the human mind that "perceives" the spiritual world first formed its ideas in the physical world of time and space.

Human intelligence and consciousness must be formed in the physical brain from the ideas of time and space, and relationships and ratios must first be learned through the five senses. In this way, when we are forming the mental images of our thoughts, we preserve the appearance of time and space in our intellect and memory. When we move on to the spiritual world after death, every change of state in the human mind and affection retains the appearance that it happens in time and space. Even more revolutionary is Swedenborg's conclusion that this appearance of time and space corresponds to one's intentions and discernments.

Heaven (and hell, for that matter) is a psychoscape. The features of the spiritual landscape are psycho-topological. They portray a holographic feedback system between one's inner qualities and one's environmental conditions.* The qualities of one's love and discernment are instantaneously projected "outside" of the individual. This allows one to recognize, and always move in the direction of good or bad that most corresponds to one's volition and assessment.

* **Quantum non-locality and non-separability** are *infinitely more apparent in this primal realm.*

Swedenborg verified these findings for himself after years of carefully observing and comparing his bodily and spiritual experiences. God could not reveal the true nature of heaven or hell to the human race without giving someone the power to experience both worlds over a considerable period of time.

Because he enjoyed the ability to jump between worldly and spiritual cognitive levels for nearly 29 years, he could observe the fates of individuals as they entered into the spiritual world, tracing where they ended up and why. Recall from Chapter Two that he even reported attending funerals while having conversations with the departed souls! His major theological discovery from observing individuals as they moved from one realm to another was that God punishes no one. He observed that goodness is its own reward and consitutes heavenly delight, while everyone in hell is their own punishment. Everyone creates their own heaven or hell from the inner qualities of their hearts and minds. This revelation challenges the views of most religious traditions. The Lord God never judges anyone: *Love never condemns.* Swedenborg backed up his findings with Scripture:

> *I came not to judge the world, but to save the world.* (John 12:47)

> *I judge no man.* (John 8:15)

Simple reflection shows we all have been given the power and freedom to either condemn ourselves or conjoin ourselves more intimately and perfectly with God, according to the life we appropriate for ourselves while on earth. Since one's inner reality literally becomes one's turf in the spiritual world, we cannot escape from ourselves. The *full* evolutionary process of humans to adapt to their proper and eternal ecological niche requires God's help through religion. When Jesus says, "The kingdom of heaven is within you," He is making a real scientific statement about the ontological reality of a non-physical realm of Love.

The trees, rocks, soil, animals, and birds in this rarified "biosphere" all represent the various derivatives of one's Love and the principles that evolved to form the belief system that constitutes one's real faith. As a scientist, Swedenborg recognized that all coherent process and series were isomorphic; they proceed by the same steps and patterning principles, whether physical or mental. The universe of heart and mind grows structure and self-organizes in a manner similar to the physical universe. Human consciousness evolves by following the same procession of speciation as did the evolution of life on earth. For instance, information in the human intellect first becomes arranged according to the perception of general things (*genera*), then it is arranged according to the increasing awareness of specific things (*species*) that make up and perfect those general categories. This is how the mind generates a real biosphere and inner ecology.

Here Swedenborg borrowed the method of classification invented by naturalist Carl Linneaus (who married the daughter of one of Swedenborg's cousins) and applied it to the diversity of human ideas and values. Swedenborg stated: "there is nothing that is not in a series in which are genus, species, parts and degrees." [8]

Swedenborg observed and studied the mineral, plant and animal kingdoms of the spiritual world. He recognized that they were appearances and the various representations of one's complex, living values and thoughts, arranged according to their genera, species, parts and degrees. This creates an environment filled with things that correspond to the interior qualities of our spirit. Heaven is not the same for everyone. Yet individuals can share similar values and therefore enjoy friendship and close relationship.

In this feedback system, the mineral kingdom, that is, the ground or soil of the spiritual world and its fertility literally consists of the principles that support one's life path. They become the terrain for our spiritual feet. These are the principles one stands upon. These are the life-choices and values one becomes grounded in. The spiritual ground is therefore a reflection of the things and qualities contained in one's memory. Remember that Love is real formative substance. The spiritual world can only consist of various forms and expressions of a person's Love and beliefs.

The spiritual flora are formed from the human thoughts that take root in, and grow out from the *psycho*-mineral kingdom, which are the *chosen* principles and favored values that have accumulated in our memory. So spiritual flora portray the leafing out, flowering, and fruitfulness of our preferred ideas and perceptions. These are the "fruits" referred to in Scripture and which we are to be judged by, because they represent the true qualities and final productions of one's mind. In the spiritual world, the qualities of one's ideas are instantaneously reflected by the kinds of fruit found in one's immediate spiritual surroundings. Negative or hurtful thoughts are represented by invasive briars, thistles, thorns and all kinds of noxious plants. A wilderness corresponds to the inner state of individuals whose spirituality is obscure, uncultivated and not yet prepared for conjunction with the Lord God.

Spiritual fauna are the living projections of our volitional or emotional world. This more exalted kingdom of bio-complexity in the spiritual world represents the *agency of intention* whereby our principles and values become "animated" and reach a more evolved form. Animals represent one's thoughts evolving into *action*. In the spiritual world, human motives and desires find their equivalent qualities in the appearance of various animals. Even on earth, people are often described as being sly as a fox, sheepish, innocent as a dove, brave as a lion, or a snake in the grass, etc. These metaphors gain ontological status in the spiritual world.

Weather conditions and the spiritual atmosphere represent our emotions in a more general way, such as when we are in a stormy mood or displaying a sunny disposition. Clouds passing overhead represent various states of mental obscurity. Everything we find in the physical world can be met with in the spiritual realm. The difference is that the biosphere and ecology of the spiritual world is divided into species and genera of human volitions and discernments.

Eternal consequences arise from the fact that the spiritual world is an ontological metaphor of the human heart and mind. Therefore, heaven is not a place you go to, it is something you become. Those with evil leanings, say from overly powerful self-love, will find themselves in a correspondingly inhospitable, barren psycho-topological place inhabited by poisonous creatures and thistles, which are the living reflections of their hurtful thoughts and feelings. On the other hand, loving people find themselves in a wonderful, lush garden. Gardens represent wisdom in the other world. Gardens depict an arrangement whereby human thoughts and ideas are properly ordered and oriented to God's influence.

This is the "higher" meaning of the Garden of Eden. Adam and Eve's removal from the Garden actually represented humankind's removal from wisdom.

Recall from the previous chapter that Swedenborg maintained Scriptural stories consist of these same ontological metaphors. They depict the human story, not in terms of historical fact, but in terms of the ontological status of human interiority—humankind's evolving "inner" predicament within God's cosmic scheme of Conjunctive Design.

Such a dynamic may seem incomprehensible to the empirical human mind. Nevertheless, the creative mind will instinctively give metaphor an ontological reality. Storytellers, writers, and even moviemakers often intuit this reality when they amplify the inner quality or hidden intention of a character by placing him or her in an environment and lighting that best *corresponds* to the character's internal state. A shady character looms in the shadows and operates in the darkness of night. Innocence and gentleness, like lambs, do not find their habitat in dark caves.

Swedenborg's science of correspondences is a rigorous description of *causal links* between the higher, spiritual and psychical levels of existence and the world of physical matter and process. These top-down relationships represent a real science and can be applied to the hierarchical structure and the levels of causality in nature, the human mind and Scripture. Theistically speaking, top-down causation requires that the objects, entities, and processes of the physical universe correspond to metaphysical and spiritual concepts. The spiritual world imposes the patterns and orderly arrangement for what occurs in the physical world. The higher world of mind and spirit could not flow into and supervene upon the lower world without the transcendental symmetry afforded by correspondence and similitude.

George F.R. Ellis, Professor of Applied Mathematics at the University of Cape Town in South Africa, gets close to this same idea of the ontological reality of *discrete causal levels*. He states:

> "These are not different causal levels within the same kind of existence; rather, they are quite different kinds of existence, but related to each other through causal links." [9]

Ellis recognizes, as Swedenborg did 250 years earlier, that this non-reductive view of reality may require that metaphysical ingredients be added as one moves up the hierarchy of structure from lower-level entities to higher-level entities. He also believes, like Swedenborg, that these higher levels of reality can be inhabited equally by real creatures, such as rabbits, and mythical creatures, like dragons. In some cases the concept of an entity in a higher realm will correspond to some actual physical entity in the material world. However, sometimes a higher entity does not necessarily exist as a physical creature in the material world. According to Ellis, what determines if something is ontologically real is if it can cause something to happen or some pattern to emerge in the physical world. Swedenborg, by traveling back and forth through higher and lower realms of reality, was able to see the causal result of mythical creatures on actual citizens of the earth.

Swedenborg did indeed observe terrible creatures and dragons in the spiritual world. He discovered that they *do* refer to real entities in the natural world—to those humans who have false spiritual knowledge and an inward hatred of God. He observed groups of people in the other world collectively take on the appearance of a fearful dragon *hell*-bent on devouring any deeper levels of truth, especially those truths that might expose the errors of their worldviews. Fire-breathing dragons

embody those individuals who, through a monstrous mis-
guidance of ego-reasoning stemming from self-love, seek the
destruction of God's revelations with their profane doctrines.
Through the gatherings of such people and their combined
negative manifestations, the dragon manifests in the spiritual
world and gains embodiment to the same degree as these ideas
have influenced the minds of men and women on earth and
brought about real events in the physical world.

(People may argue that some cultures revere dragons
and believe dragons fiercely stand guard over our treasures.
Swedenborg would agree that dragons guard our treasure—
against us! In his spiritual encounters, dragons represent the
magnified perversion of truth that stands between us and our
access, not to worldly treasure, but to the real treasure that are
the spiritual gifts from God.)

Ellis developed his thesis of higher causal levels from his
work on quantum theory and its effect on the macro-world,
not from theology. He challenges the idea that quantum
uncertainty is ontologically real (especially from God's point
of view), supporting the idea that quantum outcomes are
purposeful rather than statistical and random. Ellis' work may
nudge science in the direction of Swedenborg's own findings.
But scientists will have to drastically change their assumptions
about physics. Nailing down the ultimate laws of reality will
emerge from a Doctrine of Love! Swedenborg can contribute to
this transition if science accepts the proposition that at the top
of the hierarchy of ontological reality exists a Spiritual Sun.

▶ The Spiritual Sun

Swedenborg observed that the spiritual world has its own
unique sun. Rather than a source of physical warmth and light,
it is the Divine emanation of God's Love—His spiritual warmth;
and truth—His spiritual light. In the same way that organic life
on earth exists because of the thermodynamic energy of our
astronomical sun, all spiritual life is sustained and maintained
by the Divine emanations of the Spiritual Sun.

The human mind, which consists ultimately of spiritual
substances, is inwardly designed to be receptive to the *living*
rays from this preeminent Sun. But this receptivity is subject to
human free will. It is optional. We can choose to be receptive to
what is revealed by physical light and admit of no other reality,
or we can gain an awakened sensitivity to what is revealed by

spiritual light, such as true wisdom. Even atheists enjoy the influence of this non-material Sun, because without it, they would be unable to think. The Spiritual Sun throws light on a physical situation, allowing us to understand what the eye sees. The Spiritual Sun permits humans to attach meaning to things. Deeper understanding of reality results when humans become more receptive to this source of enlightenment.

The rays of God's Spiritual Sun ultimately consist of all elements of Divine Faith and Love. The purpose of religion is to open the interiors of our mind so that Divine Truth can enter into our *understanding* (the human intellect) and Divine Love can enter into our *will* (the human heart). An individual's state of mind and the life choices he or she makes determines their orientation to the Spiritual Sun, whether receptive to, or rejecting of, the Creator's Love and Wisdom.

Individuals who have had near-death experiences (NDEs) describe seeing a light that is not physical but has psychical and emotional qualities. Its rays produced a feeling of peace and love. They also describe this light as appearing to emanate from the center of the universe—the center of reality.

All things in the spiritual world have reference and orientation to the Spiritual Sun. Swedenborg described observing an actual experiment in the spiritual world where angels were asked to keep turning and then to describe where the Spiritual Sun was relative to them. No matter how they turned, they always perceived that they were facing the Spiritual Sun. Even when angels were facing each other, the Spiritual Sun remained in front of both! In the spiritual world, one's face always turns in the direction of its love, so no matter what activity these angels were engaged in, they always put the Lord before them. In contrast, individuals inclined to promote their own self-importance, especially at the expense of others, placed the Spiritual Sun at a greater distance from their lives or even behind them, since they choose to turn their backs on God.* Swedenborg observed repeatedly that no one was ever punished with hell or rewarded with heaven. Everyone simply moved in the direction of his or her love. Humans are free to live by their convictions, and this is Divine justice.

In another experiment that proves God never punishes, Swedenborg was allowed to

> ✳ *The dynamics for such curious phenomena lie in the higher physics of the spiritual world, where "inner" and "outer" instantaneously reflect and communicate the same reality.* If Swedenborg is correct, the laws of quantum entanglement by which everything is non-locally and instantaneously connected, has its origins in this non-physical world consisting of one unifying substance, which is Love.

observe what actually happened to a person who was inclined to evil and allowed "entry" into a heavenly sphere. Since the atmosphere in heaven consists of a particular quality of Love to which angelic breathing is adapted, it presented an opposing sphere of influence for the "spiritually challenged" visitor. The individual in question began to suffocate and wanted nothing more than to return to a more kindred environment that his *inner* life and *ruling love* were adapted to. Since, on the most fundamental level, one's *love* is one's real life and essence, the spiritual suffocation this outsider experienced was the result of his being taken out of his real element, where there was no environmental support for his non-angelic inclinations. In the same way that living organisms moderate the atmosphere and their environmental conditions in the natural world, Love affects the environmental conditions of the spiritual world.

Not only does God never punish, God forgives *all* sins, no matter how horrendous! Yet this act of perpetual mercy saves no one. Forgiveness is not removal of sin. An individual who finds pleasure in evil and persists in doing evil will continue to strive for greater distance from God's influence and seek an environment in the spiritual world that corresponds to his negative disposition. Swedenborg would support the karmic idea that everyone's guiding principle in life comes home to roost. What goes around comes around. The scientific explanation for this is that like all things put into motion, evil also seeks its equilibrium. What science does not grasp is that all action finds its equilibrium from its source, or first principle, and returns to it. (This circular dynamic of first principles returning to first principles will be discussed in the final chapter.)

From his observations in the spiritual world, Swedenborg discovered that the doctrine of *salvation by faith alone* is flawed. Evils cannot be removed unless an individual *identifies* them and asks God for help in *resisting* them. Religion means to walk with God. One cannot walk with God while continuing to commit transgressions. Jesus did not take away human sin on the cross. This would be contrary to Conjunctive Design. Without the two faculties of free will and discernment there would be no proper way for humans to be conjoined with God or God to humans. The reciprocation of love by humans requires a conscious act and decision. Salvation and the true purpose of the Lord God entering into the theater of human history will be discussed in greater detail over the next three chapters.

Whether one chooses God or not, humans would not be able to think and reason or enjoy free will at all without the influence of the Spiritual Sun. In the same way that the human eye is designed to receive physical light, the interior substances of the human mind are *inner sensories* constructed to receive spiritual influx. From these, we gain the power to *understand* what we see and the *freedom* to choose where we direct our attention. In fact, spirits and angels can see things in their non-physical world only insofar as they understand them, and this, in turn, depends on how well they receive God's Love and Wisdom from the Spiritual Sun that illuminates all the non-physical subjects there.

As already mentioned, heaven is not the same for everyone. Neither is hell. God's kingdom is a kingdom of Love, and everyone's identity is preserved in the other world according to each individual's own capacity to love God and neighbor. *Therefore, heaven is open to those of all religious faiths.*

The deepest and non-physical bio-structures of the human mind are similar to those of spirits and angels, minus the physical brain. Therefore, the Spiritual Sun—as emanations of God's Love and Wisdom—also flows down into the hierarchical structure of human cognition to all those still clad in an earthly body with different degrees of perfection.

▶ The Spiritual World is also Multi-Leveled

Under the Spiritual Sun there are three levels of heaven and three levels of hell. Those in the highest heaven love God. These are celestial angels. Those in the second heaven love the neighbor. These are spiritual angels. And those in the lowest heaven are good-natured but possess a limited knowledge of doctrine and God. These are natural angels.

In the lowest hell there are those who from self-love hate God and love evil. Above that is a less severe hell consisting of those who from the pride of their own self-intelligence look down on others and ignore God's spiritual truths. In the highest or mildest hell are those whose evils and false beliefs are less severe; any good they might do is not from a spiritual principle of mutual love but for the sake of reputation and personal gain.

Love is the underlying dynamic of both heaven and hell. Hell is simply a world of people loving the wrong thing. One cannot leave hell simply by changing one's thoughts. One would have to change, fundamentally, who one actually is. This is impossible in the other world because the bio-complexity of one's spirit forms and evolves in the physical body on earth according to the individual's life choices and values. Because Love is a form-giving substance, without the physical body to serve as a foundation after death, we are left with a spiritual body organized around our Ruling Love. Any change of heart or values after death would go against one's essential fabric. This would amount to the annihilation of the bio-structure of one's spirit. One cannot simply change from a leopard into a lamb, unless such a decision is made by an individual to change one's nature while still in the physical body. Swedenborg found no evidence for a second chance through reincarnation. Besides, no one in hell is held there against his or her will. People actually seek it out in the same way a thief or adulterer seeks out prey.

Between heaven and hell is an intermediary level called the Spiritual World proper, which individuals first enter after death for further preparation before moving off to their final destination. This level exists in order to perfect the unification of an individual's two main mental faculties—volition and reasoning—and prepare the individual for a final abode in one of the three levels of heaven or hell.

On earth, these two mental faculties in humans are separated for good reason. This Divine design allows us to acquire more noble kinds of knowledge and principles to live by and raise ourselves above our normal, worldly loves. But free will and volition also enter into the equation. We can intellectually understand civil, moral, and spiritual values but still despise and reject them in our heart and our actions. This God-given separation between the cognitive functions of our understanding and heart becomes abused when individuals use their intellect, not to change their hearts, but to hide motives. The human intellect communicates something contrary to the quality of one's heart whenever it invokes high-sounding and noble words to protect hidden agendas. This is the origin of hypocrisy. In the spiritual world, however, people cannot enter their appropriate levels or find their eternal homes until their thoughts, speech, and actions are in perfect agreement with their heart.

In the spiritual world, we become who we really are. This is how we find our true place among its various levels. A person's inner reality cannot come to the forefront, however, unless ideas and principles that are not in accordance with one's volitions and core beliefs are systematically removed. Individuals lose the power of deceit in the spiritual world. As spiritual beings, men and women can no longer think or act other than from the dictates of the Ruling Love that is their spirit. The process by which this inner unity is accomplished is best explained by the spiritual world's isomorphism with human organic process.

▶ **Heaven is a Grand Human Form**

According to Swedenborg, the opening up of our innermost essence after death is analogous to the process of digestion. This organic process must take place so that we can each be properly assimilated into the spiritual world of societies and communities whose heavenly commerce is analogous to the functioning of organs of the human body. Heavenly activity consists of mutual support and cooperation in the same way that the human anatomy consists of living cells, tissue and vital organs that interact supportively. Each person's inner essence has its destiny in some organ, layer, and functioning of this grand cosmic community, for a unified society is a composite person. Swedenborg called this preeminent organic complexity, or body of heaven, the Grand Human (*Maximus Homo*).

When we enter the spiritual world, the process of unifying our heart and intellect begins in a community of spirits that act as its "stomach." Even after death we still retain much of our external memory of much eclectic material. This is all broken down in the spiritual world in a way that is analogous to the way stomach acids break down food into its constitutive parts. Food cannot be incorporated into the human body unless it is properly prepared and becomes *reconstituted* through the digestive process. Similarly, men and women cannot be incorporated into the bio-structure and order of heaven unless they go through a similar process. In this way the spiritual world finds its spiritual nutrition, that is, what real value we provide for it.

When individuals enter the spiritual world they first meet angelic spirits whose function is to correct any false notions they might have about God and religion and to inwardly explore whether these lessons will be received by the newcomers and applied to their lives. These lessons act as solvents, removing any material or information in one's memory that is not harmonious with one's heart. False notions are removed from individuals with good hearts while noble principles are removed from pestiferous individuals who would use this information only for deception.

The spiritual "digestive acid" is God's truth acting on the human psyche to prepare newcomers in the spiritual world properly for eternal life. Those who freely resist this heavenly influence move into another community that offers more harsh lessons. These bitter lessons are the spiritual equivalent of bile in the small intestines. If this fails, individuals are taken to another community where more drastic measures are used. Here more pressure is applied, and every effort is made to squeeze something good out of these intractable individuals.

These operations are not physical tortures. The pain individuals feel during this process comes from the attack on their belief systems. To be accepted into a community in heaven requires that all self-centeredness be challenged and removed so that the principle of mutual love can be adopted.

Heaven maintains Divine order according to the unselfish love of service, each service having a precise organic correspondence within the whole spiritual kingdom and its upkeep. Heaven is perfected as new arrivals enter the spiritual world and add their particular, unselfish talents to this grand scheme of mutual love, perfecting unity through an individual's distinctiveness. Stubborn individuals who resist God's influence and prefer to follow their own self-love and self-guidance remain hardened, like undigested food, providing nothing of value to the heavenly kingdom. They are finally jettisoned from God's spiritual kingdom like excrement and collect into various societies filled with souls who share their same compulsions. These societies form the hells. Rather than adopting a human form, their intentions and lives take the form of a great monster.

In hell, Swedenborg observed that the biggest complaint was that a heavenly life should not be different from what hell's occupants imagined it to be. Their conception of heaven was based on no other delights than those derived from self-love and worldly pleasures. In spite of all the efforts God makes, such individuals cling to their evil natures, for it has become the very life of their soul. Yet, these "sinners" are still loved by God and continue to have life from God's Love. They reject God, but they are never rejected by God.

▶ Heaven is not a Retirement Destination

Heaven is buzzing with activity. This activity is why heaven is heaven, a loving society of people providing useful services for each other. This reciprocation of service is the very essence of Love and its power to unite. The human mind needs challenges and is healthiest and happiest when making contributions to society. The human mind and its powers perish if they are not actively engaged. With a heavenly vocation, there is no toil. One finds his or her eternal profession and trade as the deepest expression of one's soul from the deepest places of one's inner potential. The more we engage in heavenly love towards others, the more we find our true identity. Swedenborg also learned that this process of opening a person's inner potentials is continued to eternity. Otherwise, angelic spirits would reach a level of stasis and stifling mental boredom. God's blessings and gifts of happiness are infinite.

The image of heaven as a Grand Human form represents the ultimate conjunction of God with humanity. Heaven is the ultimate human form, because its commerce and economy promotes Spiritual Love. This is what makes humans human. God's Love and Truth can be directed only into forms of usefulness, that is, creation through self-representation and similitude. God can be conjoined to creation, because all things in it are mirror images of some Divine intention to exalt unity through utility. Heavenly commerce between angels represents the highest expression of this grand scheme. Divine Love and Wisdom are increasingly conjoined to finite organic complexity through the interconnectedness and the interdependence of a diverse spiritual community unified by Love.

Swedenborg's observations of heavenly societies led him to discover the profound, all-inclusive nature of God's Love. He was also shown that heaven "cannot be made up of human beings all of one religion, but of men of many religions."[10] In the same way that the human body needs all its different organs to perform myriads of bio-functions, heaven finds its perfection in radical pluralism. Everyone in heaven is united by the love of God and the love of goodness towards others that is the foundation of all true religion. In heaven, unity is perfected as human talents express their distinctiveness and uniqueness. The perfection of distinctiveness is according to the promotion of unity. Heavenly complexity increases as its residents more perfectly portray first causal principles. Heaven is the outcome of God's evolutionary scheme of Conjunctive Design and the exaltation of Love.

Again, the key to top-down causation and understanding Divine action in the world is self-representation and correspondence. Swedenborg helps us grasp this principle of causal linkage by pointing out the layered and correlated ordering of human bio-structure. All things in the human body are specifically designed for utility and answer to some localized structure in the brain in the form of specific neurons and groupings. Moving inward to another layer of structure, all things in the cerebral cortex illustrate the mind/brain correlation by answering to some volition and discernment of the mind. Deeper still, all things of the human mind answer to some dynamic of heaven, which more perfectly represents the human form. And finally, all things in heaven have relation to God's Love and Truth. Since God is a Person (the Divine Human) the whole of heaven, as well as the direction of evolution on earth, ultimately seek the human form in specified complexity that perfects process through nobler utility. Physicists would call this the strong anthropic principle. There is no other way that nature and heaven can be arranged so that God can dwell within created things. If I may borrow a term from Rupert Sheldrake, this relationship is called *self-resonance* (which is the same as Swedenborg's correspondences).

Sheldrake, a biologist who has challenged the mechanistic worldview of the natural sciences and evolution, held a similar view to Swedenborg and states "it makes sense to think of the entire universe as an all-inclusive organism...This self-resonance would help to explain the continuity of the universe, as well as the continuity of material systems within it."[11] Swedenborg extended this all-inclusive organism of the universe into the non-physical realm of heaven.

▶ The Spiritual Sun is a Living Formative Field

Borrowing another term from Sheldrake, Swedenborg's Spiritual Sun represents the ultimate *morphogenetic field* in the universe. Earlier, in Chapter Three (p. 37), I equated the rays of God's Divine Love with the Grand Unified Field of physics. Swedenborg's model of the Grand Unified Field goes beyond that of the most daring physicists who equate it to pure consciousness. Rather, he saw this universal field as theologically morphogenetic. Its sphere constitutes not only a simultaneous and non-local composition of all form and potentials, but the living dynamics of Divine disposition and God's bio-centric purpose.

The activity of the Spiritual Sun's field is a non-physical region of organizing influence. Its activity represents the special initial conditions of the universe before there was time or space as well as love's ability to subordinate and coordinate all things to order and series, according to God's eternal goal. Swedenborg insisted that no proper idea could be formed about either creation or evolution without knowledge of the Spiritual Sun.

The Spiritual Sun's operation deals with forms of Love, that is, spiritual substance seeking concrete reality as creation and then returning as evolution. In this *circular* process, spiritual values first transition into the measurement outcomes of useful process in the physical universe, then, through the emergence of intelligent and moral human beings, return to spiritual values.

All process in the universe, from first principles, through causes, and into effects creates an orbit or gyre that mirrors the mind of God. This holonomic and infinitely self-referable process of the Divine Consciousness involves Divine Love and Wisdom in the form of active information, flowing into and generating the infinite curvature of singularities. God's perpetual influx into pre-space singularities creates an infinitely focused dynamo of *pure striving*. It also instills within its endeavor a *mathematical intuition of ends*, that is, all the lawful information and patterning principles needed for the genesis of the physical universe. (This explains the way mathematics underlies physics and why the universe is fine-tuned for life to emerge.) As creation unfolds and spiritual forces become fixed into forms of terrestrial matter, everything is mathematically and geometrically pre-programmed to enter into new combinations of structure that ultimately lead to the organic complexity of the human brain, a brain that allows humans to embrace or reject spiritual values. The human race represents the means by which self-resonance with God has its greatest possibility and the path by which first principles can return to their source. Again, this is God's grand scheme of Conjunctive Design.

The complexity of nature, including the human brain, thus comes from the non-physical complexity of God's Love and wisdom. Life comes from life, because Love is living, conscious force. Life creates nature and gives nature an endeavor to flow back to life through a complex of causes and effects that promote the intended goals. This is why nature evolves into intelligent structure and organic forms.

Through the Spiritual Sun, matter can be attached to, and have its structure organized around, some specific "use." Matter is dead and has no power to self-organize itself. All forms of nature, organic and inorganic, draw their structure and complexity from the Spiritual Sun. And as I pointed out earlier, God's emanations—His creative action—cannot flow into any form, create a relationship, or be in conjunction with anything without having a usefulness for some end or holistic goal. Otherwise, God would be sustaining something foreign and dissimilar to a Divine attribute. The correspondence principle is based on the dynamic that as spiritual forces flow into and terminate in the fixed forms of the natural world, these forces cannot flow into dissimilarity (I will address the issue of evil in Chapter Twelve). The physical world is a mirror image of the psychical, spiritual world, and all physical laws and forces are actually *spiritual* laws and *living* forces projected into spacetime constraints. All specified complexity and organic commerce (interdependence) in nature is an analog of spiritual love. All natural process is an activity by which physical change is a reaction to the changing states of living substance in the spiritual world. This correspondence is why the laws of the universe are so bio-friendly.

By exalting *usefulness* through increased complexity, the Creator can become intimately and perfectly conjoined to creation. And the human race has the potential to reach the highest spiritual utility. Humans reciprocate by loving God and their neighbors through performing acts of real usefulness and service (called good works) in the world from a spiritual principle.

Because the Spiritual Sun exists, *values and morals* have an objective existence within the fundamental fabric of reality. Physicalism, the positivist view of science, claims that people can adopt values into their lives but there are no objective values outside of us, in the way that the moon exists outside of us. This is why moral values are not considered as a scientific topic. However, the mind can no more be made to sense a value that doesn't exist than an eye can be made to see in a universe without light.

In the same way that Einstein viewed the gravitational field as a characteristic of the structure of spacetime itself and not simply something operating in space and time, Swedenborg viewed moral values as real fields of influence that generate non-material regions of space! Physics acknowledges the difficulty of providing evidence for additional dimensions. Experiments with *physical* gravity cannot be used to detect extra dimensions because process and action is becoming more *psychical* in those rarified regions. Even gravity changes its quality and assumes the nature of moral values and spiritual principles in these "higher" dimensions.

In Swedenborg's cosmology, values are indeed a topic of science. Because religion, which addresses values and life-choices, creates a new gravitational order and bio-structure for the human spirit, it is therefore reducible to scientific language. Hence the two approaches of acquiring truth, science and religion, can indeed be unified.

▶ **Marriage, Sex and Intimacy in the Spiritual World!**

If our essential spirit is our volition and reasoning, as Swedenborg maintains, how can it continue to function after death without the physical body and the physical brain? How can humans enjoy all the capacities of the corporeal body in a non-material realm? Swedenborg addresses the *embodiment* issue in the following way:

> "Unless man were a subject which is a substance that can serve a source and containant he would be unable to think and will. Any thing that is supposed to exist apart from a substantial subject is nothing. This can be seen from the fact that a man is unable to see without an organ which is the subject of his sight, or to hear without an organ which is the subject of his hearing. Apart from these organs, sight and hearing are nothing and have no existence. The same is true of thought, which is inner sight, and of perception, which is inner hearing; unless these were in substances and from substances which are organic forms and subjects, they would have no existence at all. All this shows that a man's spirit as well as his body is in a form, and that it is in a human form, and enjoys sensories and senses when separated from the body the same as when it was in it, and that all life

of the eye and all life of the ear, in a word, all the life of sense that man has, belongs not to his body but to his spirit, which dwells in these organs in their minutest particulars. This is why spirits see, hear, and feel, as well as men. But when the spirit has been loosed from the body, these senses are exercised in the spiritual world, not in the natural world. The natural sensation that the spirit had when it was in the body it had by means of the material part that was added to it; but it then had also spiritual sensations in its thinking and willing." [12]

Thoughts, loves, and values are more than real substances; they generate entirely new organic structures with new sensitivities. Human evolution, through personal experience and life choices, allows for embodiment and complexity to continue to develop as non-physical structure and form a *spiritual body*. These new organs, created from substances of the mind, give real form to all the aspects of our will and understanding, which determine who we are and make up the fabric of our being.

A penis in the spiritual world reflects a male spirit's ability to probe the subject of his *passion* from his understanding. A vagina represents a woman's receptivity and will to accept these qualities of her suitor by becoming the ultimate focus of his love. Makeup and skin creams add nothing to the attractiveness of a woman on the spiritual plane. A woman in heaven becomes a more beautiful wife as she becomes the life and love of her husband's understanding. The man becomes more nearly perfect as a husband and increasingly handsome as his spiritual knowledge increases and is directed towards his wife. So the husband becomes the embodiment of his wife's love towards wisdom. The male and female genitals, whether on earth or on the spiritual plane, are the spiritual analogs that represent the dynamics of this profound union of heart and mind.

This may seem sexist in today's world, but there is no other way to create such a deep and powerful spiritual bond between married partners. This level of bonding is not generally known on earth, yet two partners can become increasingly intimate only by taking on an intimacy that is forged by a united *understanding* and *will*. This increased intimacy allows marriage to be perfected through eternity. All creation exists through the marriage between God's *Love* and *Truth*. Sexual intercourse in heaven mirrors this dynamic as the spiritual act of union between a husband's understanding and a wife's Love. In other words, wives in heaven are spiritually formed to receive and be impregnated by their husband's understanding, and his ideas are his spiritual seed. The wife then brings these ideas into the spiritual world by giving life to those ideas through new forms of love, which add new spiritual qualities and happiness to the lives of both partners. Spiritual offspring are the things generated from this union, such as increased love and service to the heavenly community. Rather than multiplying a population, spiritual progeny in heaven multiply usefulness and goodness. Marriages in the other world reflect the conjoining of various levels of understanding with their proper partners, that is, their proper affections.

Wives are the most beautiful women in heaven. Swedenborg met and communicated with a married couple who had grown in love for thousands of years and occupied the highest heaven. Swedenborg could barely describe the wife's beauty, but when she turned towards her husband her beauty became so radiant that he could no longer make out her features. She had become beautiful beyond Swedenborg's cognitive ability to discern!

So strong is the unity between married partners in heaven that when Swedenborg observed them from a distance, they looked like one angel. This appearance as one angel symbolized the wonderful and profound intimacy of married partners in heaven. Everyone instinctively recognizes his or her eternal partner/soul mate in the spiritual world. Some married partners from earth continue to stay together in the spiritual world and strengthen their relationships. Some do not.

The subject of humans having real spiritual bodies has to do with God's plan to continue evolution into a non-physical realm. It can be viewed from a scientific perspective. The physical world provides a foundation and a fixedness for creative spiritual forces so that these forces can organize matter into more complex physical structure in the universe. In the same way, the human memory provides a higher plane for these same spiritual forces to terminate and become fixed in a foundation upon which a new level of psycho-complexity can emerge.

There can be no doubt that the human mind fixes and internalizes its perceptions of the world through the medium of the memory and through ideas attached to time and space. These bits of data do not simply exist in a chaotic state; they are arranged into real order with real orientation, just as all things in the physical universe are. But the mind is alive, and its data is organized into a new order of bio-structure. Not only are qualities of the human heart and mind real living substance, Love is a primal formative and morphogenetic substance.

The material in our memory becomes arranged into a new organic structure or spiritual body according to the *values we choose in life.* Psychical things can take on organic form because Love, which is living force, adapts information to its own disposition. One's level of understanding is a *real embodiment* of his or her love. Human understanding is the higher-level structure that we each create when we form our subjective self throughout our lifetimes. This inner development allows the subjective self to be able to adapt to a life in a non-temporal and non-local world after the death of the physical body. This is our niche in creation—to forage on data, metabolize it, and make it "us." We are not our physical bodies. We are what we love, intend and think about. That these things form our spiritual bio-complexity makes personal responsibility and religious matters subjects of true scientific inquiry.

▶ "Higher" Education

Swedenborg learned that spirits enjoyed the faculties of seeing, hearing, touching, and even sex much more intensely than terrestrial humans. The reason is that the physical body, although designed to carry out all the intentions of the human spirit, operates on a grosser plane. If it is the mind that actually sees, hears, and enjoys sex, one can imagine the sensitivity of a body that is formed directly by the mind and operates on the same plane as the mind. Spirits not only continue to enjoy the same organic functions as on earth, but there is a perfect transition of one's individuality and uniqueness into the non-material realm.

Swedenborg also saw during his exploration of the spiritual world that God *did not create angels directly* or distinctly from the human race. Evolution is a spiritual process, and angels are its final product.

The reason is scientific and according to order. Finite intelligent beings, including angels, form all their ideas from the things of time and space. This requires a physical body, with a physical brain, living in a physical world. The spiritual world offers no such environment where there are physical units of distance or size, nor planetary movements that can measure time. Neither human intelligence nor angelic wisdom can develop without first experiencing spacetime and the ratios and relationships found in it.

Furthermore, the growth of perception must also be linked to bodily action and accomplishment, because mental growth needs to experience real success within spacetime. Without concrete success, human volition and discernment evaporate. Living in a physical world and forming ideas from time and space also allows us to retain the *experience* and *appearance* of time and space in our memory. Thus, when we move on to the spiritual world, after the death of our bodies, our volitions, perceptions, and thoughts, take on the appearance of space and time through the mental preservation of ratios.

Again, spiritual, moral, ethical and civil things are real substances and forces. Any of them can be forged into a value system that forms the organic yet non-physical, embodiment of the soul. This is a real body with ontological status, but it is formed from spiritual substances. This higher embodiment and organic complexity within God's scheme of Conjunctive Design allows human vocation and utility to continue into the spiritual world. There can be no eternal happiness unless our hearts and minds are engaged in delightful and rewarding occupations.

We do not lose our capacity to learn in the other world. Education also continues in heaven because the Lord God, out of His Divine Love, wants to eternally perfect conjunction with the human race and continually bestow heavenly blessings to all. Heaven would not be heaven without the opportunity to perfect our lives and minds through eternity. Living in a rut constitutes hell.

In this scheme, continued education is also necessary to remove the many false ideas people bring with them to the spiritual world, ideas that thwart conjunction with God. Many well-intentioned people enter the other world believing that heaven is a retirement destination or a place of entitlement and reward. This leads to cacophonous exchanges in a realm where harmony rules, but it all gets ironed-out in ways that lead to highly amusing scenarios. For instance, some new arrivals into the spiritual world are allowed to experience their very own personal version of what they believe constitutes heavenly joy and to live out its obvious conclusion. Some of these learners included the clergy.

In his book, *Conjugial Love*, Swedenborg describes witnessing an event concerning a group of clergymen who had recently departed from the earth. They believed that eternal happiness consisted of "supereminent dominion, boundless wealth, and in super regal magnificence and super-illustrious splendor, where they would reign with Christ forever, and be personally administered to by angels." [13]

The clergymen were led to a portico and told to prepare themselves as princes and kings. Thrones then appeared within the portico. Upon these thrones were silken robes, scepters, and crowns that the clergymen were told to adorn themselves with. After they put on their regal attire they were instructed to take their seats upon the thrones. They were then told to wait.

Soon thereafter, Swedenborg observed a strange mist rising from below the portico. As the clergymen sat on their thrones and inhaled this mist, Swedenborg watched as their heads began to swell and their chests puffed-out as they filled themselves with the belief and confidence that they were now truly princes and kings.

Since the spiritual environment reflects a person's inner state and quality, Swedenborg recognized, through his knowledge of correspondences, that the strange mist represented the aura of the fantasy by which these misdirected souls were inspired. Mists represent some obscurity of the mind that leads to self-delusion.

From time to time angelic spirits would come by to visit these thrones and issue various proclamations that assured the clergymen that they were indeed royalty, and that if they wait just a little longer, courtiers would soon come and deliver them to their heavenly palaces. They waited, and waited, and kept waiting. But no one came for them. They were in such an extreme state of expectation that after a few hours, the clergymen became utterly exhausted from the intensity of their desire.

Eventually a voice came from above the clergymen and chided them, saying, "Why do you sit there so foolishly and play the part of actors? They have been playing tricks with you, and have changed you from men to idols, because you have set your hearts upon the idea that you are to reign with Christ as kings and princes, and to be ministered to by angels. Have you forgotten the Lord's words, that he, who wishes to be great in heaven becomes a servant?" [14]

Heaven is a realm teeming with activity. But this activity involves using talents in loving, unselfish service to others. Love is impotent without action. Anyone can understand that a human mind not focused on attaining some goal has nothing to hold its forces together and goes adrift. This is the devil's workshop. Everyone in heaven has a spiritual vocation. This is the glue that holds heavenly societies together in service and mutual love.

Even those who are skeptical that Swedenborg observed such scenarios being played out in the "other" world will find that they all contain deep psychological insights into practically every aspect of human nature. Everything he reported seeing and hearing in the spiritual world supported and underscored some religious doctrine taken directly from the words of Holy Scripture.

The values we adopt during our lives and appropriate into the texture of our very being, whether noble or not, ultimately resonate with one of the three levels of heaven or three levels of hell. Religion is God's Divine strategy by which humans receive proper guidance that directs the course of their further evolution into a "proper" pre-space realm. Religion fills an ontological gap not addressed by the scientific notions of natural selection that make evolution either an adaptation to the pressures of a physical environment or a random mixture and mutation of genes. We are more than our genes. We can adopt new principles of Love and from them fashion an entirely new set of "internal" organs within our lifetime! Without such organs and their higher-order life-functions, we cannot properly adapt to an environment pervaded by the living emanations and generative power of the Spiritual Sun. Religion helps us form the spiritual bio-complexity needed to receive and metabolize God's Love and Wisdom.

The universe is not simply computational; it is also volitional. Both the neo-Darwinian synthesis and Intelligent Design theories miss the fact that evolution is the *exaltation of Love*. Complexity emerges from God's design to promote common good. Angels are the highest expression of the evolution of the human heart and human bio-complexity. This bio-complexity is a micro-heaven that perfects the human form and best mirrors God's image. Swedenborg articulates this theological subject of salvation and eternal life from a truly scientific perspective. Salvation and eternal life cannot be viewed apart from universal process, order or the constants of law. Divine order is ubiquitous.

Speaking of order, Swedenborg also visited the various levels of hell. He discovered that, contrary to many religious traditions, there was no "Devil" or any other prime ruler over hell. The lives of the inhabitants of both hell and heaven are continually sustained through God's Love and governance. God loves those in the lowest hell equally as much as He loves those in the highest heaven. The citizens of hell simply do not wish to reciprocate. This rejection of God is manifested by the fact that they always turn their backs to the Spiritual Sun and distance themselves from its influence. The "devils" are those in hell who find pleasure in hurting others. Those with inordinate pride of self-intelligence who wish to dominate others with false reasoning are called "satans." Some semblance of law and order is required even in hell. And this can only come from a loving source. God rules both heaven and hell with Love. Otherwise, there would be no true and rational justice in the universe.

▶ Spiritual Gossip

During his twenty-eight years exploring in the spiritual world, Swedenborg met and talked with many individuals, including friends and family members who had passed on and famous people from history. In some cases he actually dropped names and described the nature of his discussion with them.

Swedenborg once observed Isaac Newton having a discussion with angels about whether or not vacuums existed. The angels told Newton that it was impossible "for the reason that in their world which is spiritual, and which is within or above spaces and times of the natural world, they equally feel, think, are affected, love, will, breathe, yea, speak and act, which would be utterly impossible in a vacuum which is nothing." Newton finally acknowledged "that the Divine, which is Being itself, fills all things."[15] Later in Swedenborg's narrative, Newton warned anyone who wanted to talk with him about vacuums that the idea of nothing was actually very destructive.

Swedenborg also got into a theological conversation with Martin Luther about salvation by faith alone. Luther's earlier writings favored "good works," and he admitted that he later fell into the theological error of "faith alone" because of his intense desire to differentiate himself from the Roman Catholic Church. It was pointed out to Luther that justification by faith alone separates religion from life, because a continual striving towards action was fundamental. Love or good works is faith in action, so charity and faith are unified in good works. Furthermore, a person who approaches the Lord God sincerely and humbly, receives Divine guidance and does not take credit or earn merit for his or her good deeds, does not risk spiritual damnation. Good works on earth are to be done for God and eternal life, not simply for political reasons or for the sake of the world and its prosperity.

Swedenborg had a more adversarial conversation with John Calvin on the topic of predestination. Calvin told Swedenborg that "Predestination alone determines all things of religion," and that any arguments against predestination reached his ears like "the rumbling of the bowels."[16] Swedenborg rebuked him by saying, "Predestination, therefore, implies that some are appointed for heaven and some for hell. The only idea of God then, that you can form, is that of a tyrant who admits His favorites into his city and consigns the rest to torture. Shame on you!" [17]

He even had a brief but surprising encounter with Mary, the mother of Jesus, who told him that only the Lord and not she was to be worshipped.[18] Swedenborg also discovered that those who sought out Roman Catholic Saints in the spiritual world failed to find them. These seekers were told that some saints are in heaven, and some not. Saints who are in heaven live in a spiritual society separated from others so that they would not find out about their special status on earth, which would threaten their humility. The saints who did find out and desired to be worshipped became struck with madness and acted like fools. The worship of saints was abhorred by angels, because the Lord was the sole object of their worship. Popes had no Divine authority or special duties either. Everyone in the spiritual world found their rank and status according to the true quality of their spiritual love (or lack of it).

By revealing the lot of individuals in heaven and hell, Swedenborg was able to show that things aren't always what they seem to be. Many who were called "great" throughout human history had internal realities that damned them. By exposing these inner realities and hidden agendas, Swedenborg offered a moment of contemplation about various hurtful traits that men or women needed to be on guard for. These shocking revelations help us recognize these traits in ourselves by showing how faults of character have been masked by others. Unfortunately, the demands, vanities, and allurements of the physical world keep our attention away from our own inner status.

Swedenborg gave a remarkable account of personally experiencing the passage at death from the physical body to the spiritual world. Here is his account of how human consciousness is reawakened from death with the help of angels and introduced into eternal life:

> "How this resuscitation is effected has both been told to me and shown to me in living experience. The actual experience was granted to me that I might have a complete knowledge of the process.
>
> As to the senses of the body I was brought into a state of insensibility, thus nearly into the state of the dying; but with the interior life and thought remaining unimpaired, in order that I might perceive and retain in the memory the things that happened to me, and that happen to those that are resuscitated from the dead. I perceived that the respiration of the body was almost wholly taken away; but the interior respiration of the spirit went on in connection with a slight and tacit respiration of the body. Then at first a communication of the pulse of the heart with the celestial kingdom was established, because that kingdom corresponds to the heart in man. Angels from that kingdom were seen, some at a distance, and two sitting near my head. Thus all my own affection was taken away although thought and perception continued. I was in this state for some hours. Then the spirits that were around me withdrew, thinking that I was dead; and an aromatic odor like that of an embalmed body was perceived, for when the celestial angels are present every thing pertaining to the corpse is perceived as aromatic, and when spirits perceive this they cannot approach; and in this way evil spirits are kept away from man's spirit when he is being introduced into eternal life. The angels seated at my head were silent, merely sharing their thoughts with mine; and when their thoughts are received the angels know that the spirit of man is in a state in which it can be drawn forth from the body. This sharing of their thoughts was effected by looking into my face, for in this way in heaven thoughts are shared.

As my thought and perception continued, that I might know and remember how resuscitation is effected, I perceived that the angels first tried to ascertain what my thought was, whether it was like the thought of those who are dying, which is usually about eternal life; also that they wished to keep me in that thought. Afterwards I was told that the spirit of man is held in its last thought when the body expires, until it returns to the thoughts that are from its general or ruling affection in the world. Especially was I permitted to see and feel that there was a pulling and drawing forth, as it were, of the interiors of my mind, thus of my spirit, from the body; and I was told that this is from the Lord, and that the resurrection is thus effected.

The celestial angels who are with one that is resuscitated do not withdraw from him, because they love every one; but when the spirit comes into such a state that he can no longer be affiliated with celestial angels, he longs to get away from them. When this takes place angels from the Lord's spiritual kingdom come, through whom is given the use of light; for before this he saw nothing, but merely thought. I was shown how this was done. The angels appeared to roll off, as it were, a coat from the left eye to the bridge of the nose, that the eye might be opened and be enabled to see. This is only an appearance, but to the spirit it seemed to be really done. When the coat thus seems to have been rolled off there is a slight sense of light, but very dim, like what is seen through the eyelids on first awakening from sleep. To me this dim light took on a heavenly hue, but I was told afterwards that the color varies. Then something is felt to be gently rolled off from the face, and when this is done spiritual thought is awakened. This rolling off from the face is also an appearance, which represents the spirit's passing from natural thought into spiritual thought. The angels are extremely careful that only such ideas as savor of Love shall proceed from the one resuscitated.

They now tell him that he is a spirit.

When he has come into the enjoyment of light the spiritual angels render to the new spirit every service he can possibly desire in that state; and teach him about the things of the other life so far as he can comprehend them. But if he has no wish to be taught the spirit longs to get away from the company of angels. Nevertheless, the angels do not withdraw from him, but he separates himself from them; for the angels love every one, and desire nothing so much as to render service, to teach, and to lead into heaven; this constitutes their highest delight. When the spirit has thus withdrawn he is received by good spirits, and as long as he continues in their company every thing possible is done for him. But if he had lived such a life in the world as would prevent his enjoying the company of the good he longs to get away from the good, and this experience is repeated until he comes into association with such as are in entire harmony with his life in the world; and with such he finds his own life, and what is surprising, he then leads a life like that which he led in the world." [19]

Evolutionary biology and doctrinal theology are tied together through Conjunctive Design, that is, through God's desire and perpetual action to be conjoined to creation by Love. This conjunction is perfected in the human race, which has been given the capacity to consciously discover God and return that Love. This conjunction is so important to the cosmic scheme that God always provides a means for salvation.

Over the history of the human race, this has required drastic measures, including Divine intervention into the world of matter and radically supervening on the laws of physics and neuroscience. The history of the human race is a Love story—it is the story of the Lord God's profound Love for humanity. It is a history that goes back hundreds of thousands of years.

In his twenty-eight-year exploration of the spiritual world, Swedenborg had access to pretty much anyone from any period of time in the past. He claims to have communicated with our prehistoric ancestors. One remarkable thing about this claim is that close comparisons can be made between his descriptions and fossil evidence that was first discovered nearly a century after Swedenborg wrote. In the next chapter we will see if Swedenborg can offer any new insights into the evolution of the human species, including the mystery of the Neanderthals and their strange extinction.

SUMMARY

▶ Swedenborg believed he was given a sacred commission by the Lord to reveal new things about heaven, hell and Sacred Scripture.

▶ He spent more than 28 years observing the spiritual world and interviewing its inhabitants.

▶ The spiritual world had its own sun. The light generated from this Sun is God's Truth and the heat is God's Love.

▶ In the same way that our physical sun gives energy to the earth and supports its bio-systems, the Spiritual Sun supports the lives of spirits and angels.

▶ One cannot form a correct idea about the creation of the universe without the Spiritual Sun.

▶ Religion is the means by which the human race can continue to evolve and extend the biosphere into a non-physical, pre-space realm called heaven.

▶ Spiritual bodies are the embodiment of the qualities of one's heart and mind. These bodies are real organic forms and enjoy all the senses of the physical body, but more perfectly.

▶ There is marriage and sexual intercourse in the spiritual world.

▶ God judges no one. Everyone ends up where their heart takes them.

PREDICTION

The more the human race evolves
to its God-given potentials, the more it will regain
its "lost" ability to actually communicate
with heaven's angels.

Cave Paintings – Lascaux, France
15,000 – 10,000 B. C.

Chapter Nine

THE STARTLING SPIRITUAL HISTORY OF THE HUMAN RACE

"They were big-brained, slightly larger than the average modern human, as a matter of fact. But the head anatomy was extraordinary. The cranium was long and low, with a 'bun' at the back and protruding brow ridges at the front. And the face was unique in human history."
— Richard Leakey (addressing Neanderthal morphology)

Surprising evidence that Swedenborg visited the spiritual world comes from his claim that he observed and spoke to humans of different species who had once lived on earth. This claim is startling since the first fossil evidence of early human types, like Neanderthals and Cro-Magnon man, was discovered more than eighty years after his death.

In the spiritual world, Swedenborg met with our first human ancestors, whom he described as very animal-like. These early hominids had a brain that was designed discretely different from higher mammals in order to give God a dwelling place deep within their personal experience. Their cognitive architecture had an additional level of ordered structure that consisted of substances sensitive enough to receive direct influx from the Spiritual Sun. This additional level of complexity allowed hominids to enjoy both an outer and inner reality. Their inner reality was part of God's evolutionary strategy for creating a heaven from the human race and extending the biosphere into a non-physical realm.

This Divine strategy is in accord with the inviolate laws of nature. Sound cannot be sensed without the ear or light without the eye. Similarly, heaven's influence must flow into some real organic subject to be sensed. Early humans enjoyed a deeper and more complex neural substrate than other creatures. It gave God direct access into their interior experience. Early humans received information from their physical environment and from a Divine internal source. This allowed human evolution and consciousness to continue into a non-material reality where eternal life is possible.

Swedenborg called these early humans *preadamites*. The Lord God first communicated with them in delightful dreams and visions, literally educating them through various novel combinations of mental images. This operation from heaven allowed them to perceive new abstract meanings from the objects of the world they encountered through their senses in their daily experiences. *Some preadamites responded more than others* to God's influence. The strength of their attentiveness to this inner influence determined which preadamites would radiate into new species and which others would find themselves in an evolutionary dead end.

One of the major questions paleoanthropologists have asked about our past is, "When did our ancestors actually become human?" Did it occur when they learned to stand up on two feet and discovered bipedal locomotion, freeing their hands to make tools and enabling them to get smarter? Or did it occur when they invented language and art? The neo-Darwinian synthesis offers no precise answer to this important question.

Swedenborg does. He claimed that our ancestors became human when they became aware of the existence of God and responded to a heavenly influence. The beginning of humanity and the beginning of a church on earth were the same event. It is only through the Lord God that we become human. Free will and discernment are what constitutes humanness. The freedom to act according to our discernment is a gift from God because the human will and understanding are finite representations of Divine Love and Wisdom (how God made us in His image). That is what distinguished our human ancestors from the myriad of other creatures.

The engine of human evolution was entirely spiritual, as opposed to the Darwinian idea of natural selection, since it was driven by adaptation to God rather than to the physical environment. Human evolution was not merely an increase in intelligence; it was also a lifting of the human heart out of mere corporeal and natural desires—creature comforts—to the higher nobility of conscious spiritual love. Through this second creation, or *epigenesis*, within the human psyche, bio-complexity could reach a level that more closely reflected the true nature of God's character and image.

As we saw in previous chapters, the human brain and mind are oriented, not simply to the mere acquisition of information and data, but to the *value* of that information. The function of Love in human cognition is to call our attention to things that resonate with our inner realities. All conscious brain function expresses some love or derivative of love. The different information that each human brain seeks, covets, and creates points to the brain's essential function as a subjective *organ of love*. It also reveals that the multi-dimensional design of the brain is based on a hierarchy of Loves. For instance, some individuals love to acquire knowledge. Others desire to understand further and more deeply whatever they know. Still others move on to the love of intellectual and philosophic truth. Above this is love for God's eternal and spiritual truth. The human brain will seek these various levels of information in order to do what it subjectively understands as the best "good." One's love and knowledge are therefore unified in a specified utility or endeavor.

Over time, the more responsive and progressive preadamites were taught that all things in the natural world *correspond* to qualities in the spiritual world, which are attributes of God's Love and Wisdom. According to Swedenborg, the knowledge of correspondences preceded all verbal language and writing. This knowledge was developed so fully in early humans that the perception of correspondences became instinctive and involuntary. When early humans looked at a rock or a tree, they *instantaneously* perceived the equivalent spiritual mean-ing of these objects. They perceived these abstracted qualities in the augmented light of the Spiritual Sun!

As early humans responded more and more to God's Love and teaching, the interiors of their mind and its organization harmonized with the order of heaven, which is the perfection of spiritual life. The spiritual substances of their minds cohered into higher-level, *heavenly* structure, in which their organic complexity could live and function in a non-physical realm. When they died, they were the first humans to populate heaven and become angels. In this way God fulfilled the Divine scheme of Conjunctive Design, which was to create a heaven from the human race.

These early humans not only embraced God's Love and Truth, but they also acted on it and experienced the deepest heavenly joy in loving others. They lived in perfect innocence and eventually evolved into a *celestial genus* of humans that is quite different from the genus of modern humans. They formed a Most Ancient Church, which is symbolically referred to in Scripture as "Adam." Swedenborg learned that this celestial race of humans sprang up in the land of Canaan, which we know today as Israel.

▶ Surprising Anatomical Observations

Whenever Swedenborg conversed with these unique celestial humans, he remained a true scientist and made careful physiological observations of his "subjects." He noticed that they breathed much differently from his own contemporaries. Their profound internal perception came with a correspondingly deep respiration that synchronized with their heavenly thoughts. Swedenborg concluded that even when they were living on earth, their respiration was in sync with heavenly rhythms. He wrote:

> "But what is as yet unknown in the world, and is perhaps difficult to believe, is that the men of the Most Ancient Church had internal respiration, and only tacit external respiration. Thus they spoke not so much by words, as afterwards and at this day, but by ideas, as angels do; and these they could express by innumerable changes of the looks and face, especially of the lips. In the lips there are countless series of muscular fibres which at this day are not set free, but being free with the men of that time, they could so present, signify, and represent ideas by them as to express in a minute's time what at this day it would require an hour to say by articulate sounds and words, and they could do this more fully and clearly to the apprehension and understanding of those present than is possible by words, or series of words in combination. This may perhaps seem incredible, but yet it is true."[1]

This highly developed race of humans neither possessed nor needed verbal communication, which is considered a milestone in human evolution. And yet, Swedenborg described their communication as superior to verbal language.

It certainly falls within our own experience that the face and lips can communicate, and especially communicate emotions. A smile or a kiss is one of the most potent forms of communication, one that is understood by everyone. Swedenborg's anatomical studies of the brain and the nerves that connect it to the face had revealed to him that our mental states were focused mainly in the lips and eyes. Humans also have particularly *fleshy* lips, which primates don't have. Nevertheless, how could lips communicate the most intricate ideas of thought? Swedenborg answered this by pointing out the structural complexity of the lips.

> "For if all the fibres in the lips and about the lips were evolved or developed, the truth of this would abundantly appear; for there are series of very intricate muscles and fibres, and there are fascicles of them which have not been created solely for eating and speaking, but for expressing the ideas of the mind, even to the minutest particulars, which may thus be said to be inscribed upon them."[2]

Swedenborg observed that the fibers in the lips of modern humans were more contracted and less able to express the thoughts of the mind than the lips of those who lived in more remote times. So the question then arises, about whom was Swedenborg talking? Who were these highly spiritual men and women who preceded modern humans? Does any fossil evidence support the notion that once there were creatures who may have been designed for increased facial and lip communication?

Yes; the Neanderthals left such evidence!

One of the unique characteristics of the Neanderthal skull was that their face was pushed forward to give them what is known as mid-facial prognathism. This would certainly allow for the promotion of the facial communication that Swedenborg said was practiced in the Most Ancient Church and represented in Scripture as "Adam."

In Scripture, Swedenborg also gleaned that the concept of *Adam living in a paradisiacal garden* had a spiritual content that represented the *internal* qualities of the heart and mind achieved by this ancient and celestial race of humans. More specifically, the Garden of Eden represents wisdom that came directly from God.

How could Neanderthals be placed into such a noble classification? Weren't they lumbering brutes? Or, were they merely victims of bad public relations? In their book, *In Search of the Neanderthals*, authors Christopher Stringer and Clive Gamble state:

> "The Neanderthals come with their own cultural baggage. No other group of prehistoric people carries such a weight of scientific and popular preconceptions or has its name so associated with deep antiquity and the lingering taints of savagery, stupidity and animal strength."[3]

Stinger and Gamble's book displays historical artists' depictions of Neanderthals that increasingly made them look more human over the years. Recent documentary films about human evolution are also depicting Neanderthals as more advanced and manifesting more and more human sensitivities, such as taking care of the injured. Meanwhile, recent findings by experimental archaeologists about the effectiveness of their stone tools are beginning to challenge science to reevaluate Neanderthal intelligence.

Still, even if we change our attitudes and bring Neanderthals back into the mainstream of human ancestry, they still seem like a far cry from the celestial race of humans Swedenborg described. The difficulty lies in the fact that we cannot evaluate their *internal experience*. What *can* we say about them from the fossil evidence?

For one, Neanderthals buried their dead. This suggests they had spiritual motives and understood ritual symbolism and abstract thought. One sixty-thousand-year-old gravesite is particularly suggestive: the individual was buried in a bed of *medicinal* flowers.

Neanderthals enjoyed the largest brains in human evolutionary history. The average brain size of today's humans ranges from 1,200 to 1,500 ml in volume. One Neanderthal skull, the Amud skull, supported a 1,740 ml brain. In spite of this brain-size evidence, paleoanthropologists still cannot begin to guess why Neanderthals showed none of the artistic or language abilities of the Cro-Magnons who followed and who were considered to be the first modern humans.

The Neanderthal pharynx and larynx was positioned higher than that of modern humans. Scientists believe this may have restricted their ability to speak. In Swedenborg's physiological assessment of the non-vocal communication of early humans, the position of the vocal apparatus was inconsequential, because the early humans he observed had a different, more interior, system of breathing. They did not force air out of their lungs and through their vocal chords to communicate.

> "It was also granted me to perceive the nature of their internal respiration—that it advanced from the umbilicus toward the heart, and so through the lips without sound; and that it did not enter into the ear of another and strike upon what is called the drum of the ear by an external way, but by a certain way within the mouth, in fact, by a passage called the Eustachian tube... Besides, they also conversed by slight movements of the lips and correspondent changes of the face; for being celestial men, whatever they thought shone forth from their faces and eyes, which were varied conformably."[4]

Reconstructions of the Neanderthal skeleton show a rib cage that was narrow at the top and flared out at the bottom. These early humans had no waists the way modern man does. This is not by itself evidence that Neanderthals breathed differently from modern humans, as Swedenborg suggests, but this physiological feature is quite curious and most provocative.

Swedenborg further learned from angels that vocal speech in words was not only inferior to but represented a spiritual deterioration from the original:

> "The reason why such speech was first, is that the face was formed for portraying what man thinks and wills, and so also the face was called the portrait and index of the mind; also because in the most ancient or earliest times there was sincerity, and man neither thought nor wished to think anything else than what he was willing should shine forth from the face."[5]

> "As long as sincerity and rectitude remained with man, so long also such speech remained. But as soon as the mind began to think one thing and speak another— which took place when man began to love himself and not his neighbor—then vocal speech began to make progress, the face being silent or dissembling."[6]

Today, spoken language is considered a hallmark in the advancement of the human species. But according to Swedenborg, it was invented from a "fallen state" in which the interiors of the human mind were no longer arranged according to the order of heaven. Natural speech evolved at the expense of angelic speech.

Swedenborg describes another physiological characteristic of this unique, non-verbal language system. He observed spirits not from our planet earth—yes he met spirits from other planets—but nevertheless of a similar celestial genus. He noted a peculiar motion of the lower jaw.

> "Their speech was moreover marked by another common peculiarity, viz. that it was effected somehow by the lower jaw, which with me, as well as themselves, they made protrude beyond the upper lips."[7]

This physiological tidbit is suggestive because Neanderthal skull morphology also reflects the great importance of the muscles that controlled movement of the lower jaw. According to Stringer and Gamble,

> "Additionally, at the lower edges of the Neanderthal occipital bone—inside the ear region—there was usually a prominent crest known as the juxtamastoid eminence or the occipitomastoid crest."[8]

This crest is not found in modern humans. This special arrangement of the jaw muscles is typically seen merely as a means of providing extra power to the Neanderthal bite. However, if Swedenborg was right, it may also have served in the functioning of a different system of communication from that of modern humans.

Swedenborg also made the remarkable suggestion that the face muscles of these ancestors were connected through nerves to the cerebellum rather than the cerebrum. His reasoning was that these early humans communicated from a spiritual will and that the *occipital* region in the brains of these ancestors was the seat of their heavenly affection. *Heaven's Love flowed into the posterior region of the brain and not directly into the cerebrum.*

> "Divine influx out of heaven is into man's will, and through that into his understanding. Influx into the will is into the occipital region, because into the cerebellum, and from this it passes towards the foreparts into the cerebrum, towards where the understanding is; and when it comes by that way into the understanding, then it comes also into the sight; for man sees from the understanding. That there is such influx has been granted me

to know from much experience. It is the same whether we say influx into the will, or into love, since the will is the receptacle of love..."[9]

Looking at evolution as a spiritual process can help us to make sense out of the fossil evidence. Neanderthal skulls differ from modern humans in that they display more pronounced occipital lobes in the back and less developed or sunken frontal and prefrontal lobes at the forehead. In fact, the occipital bone bulges out in a way that reminds scientists of a Victorian style hair bun. This occipital chignon or "bun" was seen by Richard Leakey to be a "curious" feature.

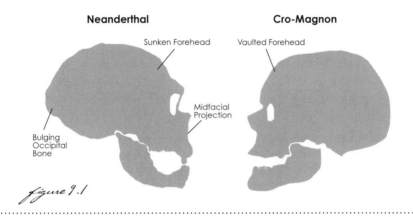

Neanderthal   **Cro-Magnon**

Sunken Forehead   Vaulted Forehead

Midfacial
Projection

Bulging
Occipital
Bone

figure 9.1

Science cannot solve the mystery of creation or evolution without bringing God and heaven into the picture. Swedenborg's spiritual model of evolution was based on God creating humans to become more perfect subjects and receivers of Divine Love. Since God's goal was to create a heaven from the human race, it required the *evolution of the heart*. Brain evolution would have originally been concentrated at the back of the head! The human brain needed to develop in such a way that it could receive more of the inflow of God's Love into the specific region of the *will*. Again, if God's Love flows into the occipital region of the brain, then this is exactly the region that would become more pronounced as humans developed spiritually. The occipital region also houses the *visual* cortex of the cerebrum. Its exaggeration in Neanderthal skulls may have given them the ability to see the world abstractly, that is, to see the world in terms of spiritual values outside the context of time and space.

This ancient race of humans Swedenborg claimed to have communicated with was not just a distinct species from modern humans. They were a different *genus*. These cerebellum-oriented beings could directly experience the rays of the Spiritual Sun. This gave them the ability to look at the objects and processes of nature and instinctively attach spiritual meanings to them. The cerebrum of this ancient race of humans acted in unison with the processes of the cerebellum. This radical difference in "hard wiring" is why they represented a distinct genus.

▶ The Mysterious Cerebellum

More is known about the cerebrum than the cerebellum. The reason is simple. The cerebrum is the seat of our voluntary functions, sensing and thinking. It is therefore the area of the brain where scientists would focus most of their attention. The less glamorous cerebellum has only recently been recognized as an important coordinator of movement. It orchestrates thousands of muscles that direct complicated physical motions without the conscious mind having to provide step-by-step instructions.

The cerebellum is the only part of the brain where neuron multiplication continues long after we are born. In human infancy, the cerebellum, which is sensitive to body contact, warmth, and touch from the infant's parents, plays an additional and important role in the development of our self-worth and ability to learn. It is essential in the acquisition of behavior because it is highly receptive to Love! Modern brain research is thus moving in Swedenborg's direction by showing evidence that the cerebellum is involved in complex emotional behavior.

Swedenborg had already done extensive work on the human brain, so he was able to recognize that the ancient race of men and women he met in the spiritual world were wired differently. As mentioned above, he discovered that the cerebellum was the seat of their angelic will and directly received God's heavenly influence. Swedenborg called the cerebellum the *tree of life* for that very reason, and even observed that it looked like a tree when cut in half. The cognitive functions of the human race that first enjoyed the paradisiacal garden depicted in Genesis were designed radically differently from those of modern humans. Eden was not simply a physical place on earth. It expressed the state of the inner qualities of their mind and heart, as they were derived from God's Divine Love and Wisdom.

Because of his anatomical research, Swedenborg understood that the cerebellum operated the involuntary functions of the body. While autonomic processes seem to have their origins in nerves from the hypothalamus and midbrain of the limbic system, and the pons, and medulla of the brain stem, Swedenborg believed these areas received their involuntary powers from nervous fibers originating in the brain cells of the cerebellum. He also concluded that if the cerebrum were properly connected to their cerebellum, a person would scarcely need a teacher. The reason is that if the frontal lobes were somehow connected to the operations of the cerebellum, cognition would have instantaneous access to the wisdom of their body's involuntary system, which maintained and sustained all organic functions and profound chemical activities.

The celestial race of men and women Swedenborg interviewed did not have to use their frontal lobes to reason about truth. They instinctively recognized the truth from their interior perception, which flowed in directly from heaven into the cerebellum and acted in unison with their reasoning. These early humans could not have left much behind to prove that they lived in such profound internal experience or that they could receive direct illumination from the Spiritual Sun. They simply surveyed everything in the physical world and instantaneously perceived their higher, symbolic qualities. There was no need to create external forms of art or a verbal and written language.

Neanderthals also left very few artifacts behind. This has led anthropologists to view them to be a more primitive people who did not enjoy higher cognitive functions. However, what they did leave behind provides important evidence that they knew spiritual symbolism. In addition to the gravesite that contained evidence that the corpse may have been laid on a bed of medicinal flowers or wore flower garlands, other gravesites indicate that Neanderthals buried their dead and placed them in east-west orientations. Many religious traditions hold that God's influence originates in the East and is symbolized by the rising of the sun. At another grave, horns from wild goats were pushed point down into the ground around a Neanderthal skeleton. Animal bones have also been found marked and grooved in systematic ways that suggest a symbolic code (shown on page 206). Swedenborg claimed that burial in caves symbolized the resurrection of truth from intellectual obscurity.[10]

It would be easy to dismiss claims that Swedenborg communicated with early humans while in the spiritual world and that these non-verbal individuals actually were Neanderthals if it were not for an uncanny coincidence. Swedenborg learned in heaven that this ancient race of humans suffered a strange "extinction" and that a new, cerebrum-focused human took their place, bringing about an explosion in art, technology and verbal language. No one could have known this in the 1700s!

One of the biggest mysteries in anthropology is the extinction of the Neanderthals. If Swedenborg's discoveries in the spiritual world are true, then the solution to the mystery is a spiritual one.

In spite of the wonderful angelic design that shaped the Neanderthal's cognitive functioning, the neural system of this first race of celestial men and women to enjoy the "golden age" had a fatal flaw. It offered no skills for coping with evil compulsions.

▶ The Slippery Slope of Humanism

Men and women of this extinct race were unable to think in a way that was inconsistent from the actual quality of their love. They were incapable of simulating or faking anything. The physiological reason for this is that the nerves in their face came from the cerebellum, and this made deception through facial expression impossible.

> "… all the involuntary of the cerebellum was manifest in the face, and they did not at all know how to present anything in the countenance other than exactly as heaven flowed into their involuntary conatus or endeavors and thence into the will."[11]

In their posterior-lobe neural system, voluntary actions were always in sync with their involuntary passions. Their welfare depended totally on the influence of heaven's spiritual love. When self-love entered the picture, irreversible things happened. These ancient people knew they were special creatures. Over time, this led to egocentrism, and they began to favor self-guidance over God's guidance. There was no cognitive way out of this slippery slope, and the result was the "Fall of Man."

Modern humans can separate their understanding from their will—they can desire to do something hurtful or selfish but restrain themselves and resist the temptation by means of their compartmentaized ability to reason. Those of the Most Ancient Church could not do this. They had no neural capacity to separate their thoughts from their will. As their spiritual love deteriorated into self-love and love of worldly things, their understanding and reasoning powers were dragged down accordingly. Reasoning powers could confirm only what they loved, and their love was becoming increasingly more evil and savage. As self-love increased, so did their hatred for others.

Another dreadful downside to this original brain design led to their extinction. This celestial race had an interior type of breathing that was animated by, and responded directly to, the influx of heavenly forces. Any change in the quality of their relationship with God would have a correspondingly profound effect on their internal respiration. Swedenborg states:

> "But in their posterity this internal respiration little by little came to an end; and with those who were possessed with dreadful persuasions and phantasies, it became such that they could no longer present any idea of thought except the most debased, the effect of which was that they could not survive, and therefore all became extinct."[12]

The fairly rapid extinction of the Neanderthals continues to puzzle anthropologists to this very day. One modern theory suggests climate change was responsible. Another proposes that they could not compete with the modern humans who were emerging with better stone tools. In fact, some theories speculate that Neanderthals were killed off by these new arrivals. Another theory is that the Neanderthals mated with modern humans and became absorbed into their gene pool. There is even a theory, largely discredited by anthropologists, that modern humans evolved directly from the Neanderthals.

If the early humans that Swedenborg met in the spiritual world were indeed Neanderthals, then the last theory is closer to the truth. But even this theory is still way off from the truth. The strange disappearance of one race of humans and the emergence of another was the result of spiritual, not natural causes.

Swedenborg viewed evolution as a spiritually guided process rather than the serendipitous outcome of nature. He learned in heaven that the extinction of these early humans was the result of *suffocation* from respiratory failure! Swedenborg was told that this ancient race of humans was replaced by a new genus of humans who had acquired all the cognitive patterns of modern human behavior and given a more external form of breathing.

"... and that at length, in the last posterity of that church, which existed immediately before the flood, scarcely anything of internal respiration remained; and when at last there remained none of this in the breast, they were suffocated of their own accord; but that in some, external respiration then began, and with it articulate sound or language of spoken words."[13]

Also:

"When internal respiration ceased, external respiration gradually succeeded, almost like that of the present day; and with external respiration a language of words, or of articulate sound into which the ideas of thought were determined."[14]

▶ Divine Surgery

According to fossil evidence, early modern humans arrived in the Levant about 90,000 years ago. Hominid remains found in Israel show that these early moderns also lived in close quarters to the Neanderthals. It was in the Levant that Swedenborg claims modern humans emerged from a previous race. This process was the result of Divine intervention and required "Divine surgery" to reconfigure the nerves from the cerebellum and cerebrum. He states:

"I have been instructed that the fibers of the cerebellum have thus changed their efflux into the face, and that instead of them, fibers from the cerebrum have been transferred thither, which now control those which are from the cerebellum ..."[15]

This Divine intervention and surgery to rewire the brain was necessary to save the human race. It gave humans the capacity to separate the cognitive function of reasoning from their emotional and volitional realities.

"When therefore the Lord foresaw that if man continued to be of such a nature he would perish eternally, He provided that the will should be separated from the understanding ..."[16]

As a result of this rewiring, control was turned over to the cerebrum, and breathing became external (which may have also caused changes in the rib cage). Those early humans who retained cerebellum-oriented brains eventually suffocated when self-love cut out the influx of heaven that controlled their unique type of respiration.

The change from a cerebellum-oriented brain to a cerebrum-oriented one could have taken 60 or 70,000 years to complete. This changeover would also explain why early moderns of the Levant had some Neanderthal features while the "progressive" Levant Neanderthals' seemed so much more refined than European Neanderthals. This changeover might also explain why humans did not create works of art until they had become truly modern Cro-Magnons around 40,000 years ago. According to Swedenborg, this new emerging brain system was not an evolutionary advancement but a necessary adjustment.

Humans now had to obtain knowledge in more external ways. This adaptation necessitated that the cerebrum develop more prominent frontal lobes. While this saved the human race from perishing, it also necessitated a shift to a grosser, more external form of *breathing* and *thinking* that prevented humans from perceiving profound spiritual truths directly from heaven. Since humans were now cut off from the most interior operations of their minds, their conjunction with God was still in peril. A new means of revelation had to be devised for this change to a cerebrum-oriented brain that was *more oriented to the physical world and its five senses.*

> "And when such determinations of the ideas of thought took place, that is to say, into spoken words, they could no longer be instructed, like the most ancient man, through the internal man, but through the external. And therefore in place of the revelations of the Most Ancient Church, doctrinal things succeeded, which could first be received by the external senses, and from them material ideas of the memory could be formed, and from these, ideas of thought, by which and according to which they were instructed. Hence it was that this church which followed possessed an entirely different genius from that of the Most Ancient Church, and if the Lord had not brought the human race into this genius, or into this state, no man could have been saved."[17]

Also:

> "As the state of the man of this church which is called "Noah" was altogether changed from that of the man of the Most Ancient Church, he could no longer—as before

said—be informed and enlightened in the same way as the most ancient man; for his internals were closed, so that he no longer had communication with heaven, except such as was unconscious. Nor, for the same reason, could he be instructed except as before said by the external way of sense or of the senses."[18]

In Swedenborg's spiritually-based model of evolution, "Adam" represented those who belonged to the Most Ancient Church and had cerebellum-oriented brains. The Flood represented their suffocation and extinction. Noah and the Ark represented those with a new brain system that featured frontal-lobe elaboration and cognition (more below). The cerebrum was oriented to be empirical, to gathering and discerning information coming in from the physical world through the five senses. Direct experience provided by the physical senses now became the main means for gaining knowledge about reality and truth (positivism). This brain system became the new means by which the Lord would have to reveal Himself and enact salvation. Furthermore, if Swedenborg was right, he succeeded in unifying the Holy Word with evolutionary science and neuroscience through symbolic meanings!

The biblical flood represented the inundation of direful persuasions in the human psyche. This inundation was a profanation and violation of heavenly knowledge. It was the inner corruption of a pristine spiritual state.

Before the "flood," brain evolution was occipital-directed. The downside of this design, as we saw above, is that these early humans could not reason beyond their passions once they became ego-centered and corrupted. Those who survived this *mental inundation* of false principles became frontal lobe-directed. Frontal-lobe evolution and elaboration was necessary for humans to raise their intellect above their passions and cupidities and to become able to reason about their inclinations. This is why the biblical Ark consisted of three levels, representing the three levels of cognitive function in the new cerebrum or frontal-lobe design. The bottom level was memory-data or knowledge. Above this was the cognitive function of understanding that adds meaning to what it knows. The top level was reasoning, wisdom and the perception of truth (also see Chapter Seven, pp 136-140, 147). Through these three cognitive levels humans were provided the neural vehicle or "Ark" to escape from the crashing waves of their passions. But this new human species had to learn about God in a different way.

God now had to provide a new means of revelation adapted to a brain that was now designed to investigate reality and discover truth exclusively through the physical senses. Even with a new brain design, humans still needed access to the doctrinal teaching and information that would give them psychical buoyancy and the capacity to ride out the deluge of direful inclinations of the corrupted will. Spiritual instruction now had to come indirectly through external means (the Ark's window). This was the way God could set up a new covenant with humanity and preserve the Divine scheme of Conjunctive Design.

▶ Knowledge Transfer of Heavenly Secrets

The transition from an occipital brain system to a frontal lobe system allowed human volition and reasoning to operate separately. The seat of the human will was moved from the involuntary of cerebellum to the right side of the cerebrum, while the intellectual function took its seat on the left side of the brain. This gave the intellectual mind the freedom to be enlightened and elevated above any perversion of the heart or will. Knowledge could now be used to modify the inclinations of the heart.

Today, neuroanatomists have confirmed that the right and left hemispheres of the cerebrum have different cognitive faculties. More than a century before anyone else, Swedenborg had already discovered this! He claimed that:

"… the left side of the brain corresponds to the rational or intellectual things but the right to the affections or things of the will." [19]

Also:

"To the left hemisphere pertain the intellectual faculties, and to the right, those of the will."[20]

Remember, Swedenborg maintained that ground zero for this "Divine surgery" took place in the land of Canaan. Fossil evidence from the Levant document the presence of early modern humans who possessed Neanderthal characteristics and "progressive" Neanderthals who did not share all the classic characteristics of their European cousins. Some kind of changeover was certainly taking place in this biblical land. According to Swedenborg, this process consisted of a change in brain morphology and a transfer of doctrinal information.

Those of the Most Ancient Church acquired profound spiritual knowledge through their direct perception of heavenly influences. Modern humans lost this internal perception. Still, the knowledge of this first church had to somehow be preserved; it included doctrinal matters of faith, and it would be the means by which God could save the newly emerging human species. But how could God make this information available to those more oriented to the physical world?

According to Swedenborg, included among the celestial humans were those who were less inclined to evil and who degenerated more slowly. Some of these individuals collected the heavenly knowledge of their race and put it into a special codex. Knowledge that was once written on the heart of humans now was to take the form of marks written on physical materials.

> "As the Lord foresaw that such would be the state of man, He provided for the preservation of the doctrinal things of faith, in order that men might know what is celestial and what is spiritual. These doctrinal things were collected from the men of the Most Ancient Church by those called 'Cain,' and those also called 'Enoch' ..."[21]

This may seem strange to our normal way of thinking about the Scriptural accounts of Cain and Enoch. But Swedenborg discovered that these stories contain elevated meanings that have relevance to deeper realities that were discussed in Chapters Seven and Eight. In the fossil record, Cain and Enoch may well have represented the regional variants emerging in the Levant during the brain changeover to modern humans. But in biblical terms, they also portrayed the direction that the church and religion were moving at this time.

For instance, when Cain kills his brother Abel, it wasn't a physical act of murder. It represented the mental change and cognitive preference that now favored the intellect over the heart. Recall from the previous chapter that plants corresponded to the intellect and animals to volitions. Cain *cultivated* the earth, while Abel was a *shepherd*. The intellect and will are like brothers, and when one takes over, it represents the "killing-off" of one type of cognitive function by the other. This psychospiritual interpretation explains why God at first prefers Abel over Cain (love over intellectual things) yet goes to great lengths to protect Cain even after he killed his own brother. Protecting Cain represented God's new strategy for preserving doctrinal knowledge and making it accessible for a new brain design that allowed the intellect to take over.

The "mark" set upon Cain and the special protection he enjoyed from Jehovah symbolized that intellectual knowledge of spiritual things would be preserved and secured by a new means. It would be turned into written forms through markings so that they would not be lost. These forms also contained symbolic meanings.

> "These doctrinal things consisted only in significative, and thus as it were enigmatical things, that is, in the significations of various objects on the face of the earth; such as that mountains signify celestial things, and the Lord; that morning and the east have this same signification; that trees of various kinds and their fruits signify man and his heavenly things, and so on. In such things as these consisted their doctrinal things, all of which were collected from the significatives of the Most Ancient Church; and consequently their writings also were of the same nature."[22]

In *The Golden Age*, the Swedenborgian scholar, Rev. Carl Theophilus Odhner writes:

> "The first-born was named CAIN, a name which in Hebrew means 'a smith.' From the root *qoun*, to make or beat into form by hammering,' and he was so named because the generation or age which he represents loved to *formulate*, to beat into the fixed shape of dogma, the truths of religion which they learned from their ancestors."[23]

Later, these collections of *deep* spiritual knowledge were converted into stories and narratives. This expanded form of writing, was accomplished by those in the Bible who were called "Enoch," whose name means "to instruct." This is why Enoch, who was the first-born of Cain, represented the stories, or progeny, generated from this symbolic language. This language was based on a science of correspondences.

> "Because the Lord foresaw that this spiritual perception would perish with their posterity, and with that perception also the knowledge of correspondences through which the human race has conjunction with heaven, therefore the Lord provided that some of those who lived with the most ancient people should collect together the

correspondences and gather them into a manuscript [codex]; these are here meant by 'Enoch,' and that manuscript is what is here signified."[24]

Again, those called Cain and Enoch may have been represented by the progressive Neanderthals and early moderns with Neanderthal characteristics found in the Levant, both of whom would still have had access to heavenly knowledge and been able to put it into forms of notation. Much of this knowledge was probably recorded on perishable materials, and if samples from this codex were discovered today, their symbols and markings would probably be untranslatable. Still, tantalizing samples of patterned grooves and markings have indeed been found at Neanderthal excavations! Some Middle Palaeolithic artifacts show deliberate embellishment, such as polishing and zigzagging scratch marks on bones. Christopher Stringer and Clive Gamble point out that these finds, "are not easily explained away by butchery marks or accidental happenings." However, they also state that there are not enough samples "to assess claims such as those by Alexander Marshack that the pieces with zigzag scratches from Pech de l'Aze and Bacho Kiro prefigure an entire code of symbolic motifs that became widespread later in the Upper Paleolithic."[25]

Marshack, author of *The Roots Of Civilization,* theorized that certain markings on bones over 20,000 years old (like the Ishango Bone in Figure 9.2) contain patterns that could represent notation that was "time-factored," such as lunar months. Swedenborg hints at a deeper symbolism and maintained that early humans understood that time—the differences of the day, week, month and of the year—corresponded to different states of mind and intellectual knowledge, especially in respect to the Lord and the condition of His church.

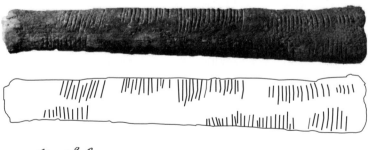

figure 9.2

Marshack was a self-educated anthropologist. Yet his theory that these early markings represent an "entire code" exactly matches Swedenborg's statements. Are these patterned grooves the "hammerings" of Cain or "instructions" of Enoch?

▶ Early Evidence of a Quantum Language

The idea that a Divine Creator directly oversaw the evolution of one genus of humans into another out of spiritual necessity may seem quite improbable to today's evolutionists. Nevertheless, Swedenborg's observations were remarkably similar to the fossil evidence. Nobody was talking about these things in the eighteenth century. The first significant Neanderthal skull was discovered in 1856; 84 years after Swedenborg's death!

He claimed that humans in the Levant reached their highest spiritual evolution as the Most Ancient Church of the Golden Age and then fell from grace. The Levant was where knowledge of heavenly things was transferred from one genus of humans to another with a different neural system. In Israel bones of progressive Neanderthals and bones of modern-type humans have often been found within close proximity. The Qafzeh cave near Nazareth has given us the largest finds of early modern humans to date, and many of these shared some Neanderthal characteristics. Finally, Swedenborg decribed an occipital-dominated genus of humans that ended in extinction and was replaced by a different genus of humans who produced vocal language, art, better technology, and perhaps, symbolic writing. Was this a lucky guess? Or did he learn these things from his encounters with ancient humans in the other world?

If we look more closely at the nature of the material that was supposedly transferred between the two distinct races of early humans, we will find more correlations between Swedenborg's unique claims and the "hard" historical evidence. Could the art of early modern humans have contained a quantum language, for example? If it did, the physical and natural objects depicted in Cro-Magnon art would have consistently communicated a deeper, universal, spiritual message that was anything but arbitrary.

Cro-Magnon art and expression certainly had symbolic overtones. Could it have embodied anything as profound as the knowledge of correspondences? One way to find out is to attempt to translate ancient art using the dynamics of correspondences and see if it provides meanings that are spiritually relevant. In other words, does prehistoric art contain teachings from God?

The surviving art from the Upper Paleolithic period, such as cave paintings, is a good place to start. In Swedenborg's account of the spiritual history of man, these Cro-Magnons would represent the second major church the Lord God set up on earth. He called them "Noah" or the "Noachian" church. While they functioned under a new brain design, they loved nothing more than thinking of the spiritual significance of the objects of the physical world. Their art had nothing to do with worldly things and everything to do with humankind's relationship with God.

Swedenborgian scholar, Erik E. Sandstrom, in a 1975 article in *The New Philosophy*, entitled, "Adam, Noah and the Stone Age,"[26] observed that one major similarity between cave paintings and the biblical story of Noah's Ark is the importance of animals. One of the spiritual principles that would have been transferred to Cro-Magnons from their predecessors was the concept that animals represent various qualities of the human will and heart (as explained in the previous chapter, p 159).

The human heart must be saved through spiritual teachings. Worldly and natural desires must be *penetrated* by spiritual lessons, that is, by information from a higher intelligence. How might this dynamic be depicted symbolically? Using the science of correspondences, Swedenborg would interpret the early depictions of hunters on cave walls as representing the skill of persuading and teaching using weapons fashioned out of doctrinal truths. To "hunt" and to "capture" means to captivate the human mind and thus change a person's previous inclinations to something better. To change an inclination is to "kill" it. In order to persuade, one must have the skill to fashion knowledge into an implement that will penetrate someone's heart and mind and change it. In cave art, the act of hunting animals depicted the skill of persuasion using God's spiritual weapons of truth. Spiritual values thus conquered corporeal (animal) values. Cro-Magnon men and women learned these spiritual skills of "hunting" from a previous race of humans that enjoyed a quantum or angelic language before it went extinct. "Fishing" has a similar meaning to "hunting." This is the reason why Jesus told His disciples to become *fishers of men*. This is also why the Egyptians, who cultivated the symbolic language of correspondences, used the word "spear" to represent "impregnation," that is, the insertion of new teachings.

Some scholars believe that the caves where this art is found may have been used for ritualistic purposes. If so, then these rituals were more profound than most researchers would acknowledge, especially concerning doctrine and theology.

Many other examples give credence to Swedenborg's thesis that prehistoric art contained profound religious doctrine. For instance, the depiction of animals as pregnant or giving birth did not represent natural reproduction; rather, it represented the development of some new quality and potential of the human heart and will. Some cave art depicts strange, anthropomorphic creatures that combined various human and animal features. Swedenborg claimed that such creatures as satyrs, centaurs and mermaids were indeed correspondences. For instance, putting antlers on a human head with a particular amount of branching represented the level and power of spiritual intelligence reached by a particular community. Since water corresponded to knowledge and its accumulation, Mermaids represented the love of acquiring knowledge in the memory. Satyrs represented the love of creature pleasures and appetites.

Swedenborg added another startling detail about this angelic language and God's covenant with prehistoric humans. The Divine form of knowledge called correspondences, which eventually took the form of narratives by those called Enoch, was understood by the ancients as the *Word of God.* It was a "Bible before the Bible." This symbolic knowledge later flourished in Syria, Assyria, Babylonia and Egypt. Swedenborg stated:

> "Among the ancients the science of correspondences, the science of representatives, was regarded as the science of sciences. It was particularly cultivated by the Egyptians, and is the origin of their hieroglyphics."[27]

In a 1973 article in *The New Philosophy*, Horand K. Gutfeldt, pointed out that the term hieroglyphics meant "holy carvings." Gutfeldt went on to explain that hieroglyphic writing was a combination and mixture of phonetic and ideographic writing that lent itself well for creating plays on words. While Egyptologists remain divided about this, there would be no doubt in Swedenborg's mind that early Egyptians understood the image of the sun depicted in their hieroglyphics as the "Spiritual Sun." All objects underneath it were to be understood from a higher, spiritual light and interpreted as representing spiritual values.

According to Swedenborg, the sacred knowledge of correspondences later spread, in more or less diluted forms, throughout Europe, Asia and Africa. Moreover, if we consider early American, Australian, and Polynesian petroglyphs, this quantum language was once universally known. Fragments of this primeval cosmic knowledge entered into various mythologies, legends, and quite possibly, even into our oldest fairy tales. Other evidence for the reality of this universal knowledge in our remote past is the fact that practically every known culture in the world has a flood myth. Images of dragons can be found practically everywhere as well.

These symbols and stories all have one thing in common. They represent the deeper psychological or spiritual realities of the "human predicament," rather than literal, historical fact. Over time this *universal* knowledge was forgotten and turned into complete confusion at the time of Babel and its famous tower. Humanism, stemming from inordinate self-love and the pride of self-intelligence was represented by the building of a lofty tower by which heaven (and greatness) could be reached by human prudence. The tower of Babel was an edifice built on human intelligence rather than on God's intelligence. As a result, human intelligence became disordered and disjointed, only grasping the literal, or corporeal/sensual notion of things.

The worship of God's various qualities in creative symbols deteriorated into the worship of many gods. Monotheism was transformed into polytheism. The holiness of these symbols degenerated into worship of graven images, that is, the worship of the images themselves. Others, who still sensed some connection between natural things and spiritual forces turned this symbolic knowledge into magical formulas for self-gain and sorcery. Sorcery represents the wizardry by which the human mind can turn truth into falsity and falsity into truth.

▶ Heavenly Cybernetics in Human History

As we have seen, Swedenborg maintained that the Sacred Word contained this deeper, symbolic language. On this deeper level the spiritual and "inner" history of the human race depicts the Creator's perpetual efforts and strategies for conjunction with the human race and its salvation. Divine adjustments were made accordingly, to allow for the ever-changing quality of the human psyche. God always kept open a means to continue enlarging a heaven from the human race, which was His ultimate purpose in creating the universe.

Recall that the Most Ancient Church, or "Adam," consisted of humans who had *direct perception* of heavenly information that flowed into the interiors of their mind from the Spiritual Sun. They were humans who evolved spiritually for several million years into creatures whose cognition was rooted in the cerebellum. Then, first in the Levant, a human spiritual breakdown made necessary a neural adjustment that took the cerebellum out of prominence. This was the "Fall" of humanity.

The second church was comprised of humans with a different neural system that was rooted in the frontal lobes. These new humans had intellectual or reflective comprehension of symbolic knowledge, rather than instinctive and intuitive access to heavenly ideas. Instead of giving them continued perception of Divine Love and Truth, the Lord implanted a new kind of moral governor in their psyche called *conscience*. Another outcome of this pre-frontal brain arrangement, oriented more to external investigation, was that it led to the artistic and technological explosion of early moderns some 40,000 years ago, including the first forms of writing. Unlike their predecessors, this new race of humans needed to preserve knowledge and culture in permanent, physical forms. As previously stated, this second church gradually collapsed when the knowledge of correspondences degenerated into polytheism, idolatry and even magic. The end of this church is represented in the building of the Tower of Babel, whereby humankind was thrown into a state of intellectual confusion. God had to make yet another adjustment to address the diminishing psyche of humans. This adjustment took the form of a new covenant.

God's third church was instituted by "Eber," whose posterity were the Hebrews. In this new church, representative and symbolic things reflecting heavenly knowledge were restored through ritual—but not to full human comprehension. The Ancient Word was lost. Instead, sacred knowledge was hidden within the external forms of rituals, including peace offerings and burnt offerings. This restoration also included a return to monotheism and these worshippers began to call their God "Jehovah." This church understood that its rituals were holy, but it understood these rituals only literally and did not know they contained deeper meanings. Ironically, where once the knowledge of correspondences saved humankind, this knowledge was now potentially deadly and could lead to one's eternal doom if it were rediscovered, then rejected.

Swedenborg learned in the spiritual world that humans are not damned by profanation of the external things contained in religion. A person could mock the burning of incense or some other ritual and ceremony and still be safe from harsh spiritual consequences. It was profaning and rejecting the *interior* concepts within these rituals that damned a person, for in these interior concepts lay the essentials of true holiness. Knowing that this church would reject these deeper levels of meaning, by God's mercy, they were protected from having these things revealed to them. Yet, these deeper sacred meanings remained preserved for posterity, safely hidden within a church's external rituals.

At length, the Hebrew Church also fell into idolatry and even rejected Jehovah as their God. Hebrew rituals, monotheism, and the name "Jehovah" were again restored by Divine intervention through Abraham, Isaac and Jacob. God created the Israelite Church, and through Moses and the Prophets fashioned a new Holy Word. This also contained correspondences and representative writings, but these spiritual narratives were somewhat less obscure than those first recorded by Enoch and later embellished during the height of the Ancient Noachian Church.

The Ancient Word had served as a crucial nexus between God and humans after the Neanderthal extinction. These ancient Holy writings from the time of Enoch are only figuratively "lost." Swedenborg learned that they still actually existed but were preserved and put under Divine protection. He did not rule out the possibility that these sacred texts would be found again.

Swedenborg's discovery of a more ancient Sacred Text, containing God's revealed truths, explains why biblical themes were found in stories that predated the current Bible. For instance, the epic Babylonian legend of Gilgamesh has many similarities to the story of Noah's Ark and the Flood. As mentioned earlier, the Ancient Word of God spawned various mythological traditions of ancient people. Swedenborg further learned in the spiritual world that Moses had actually come upon this text. Moses copied verbatim the first seven chapters of Genesis from this more ancient document, which contained pure correspondences known to those of the Noachian Church.

This unexpected information may seem to challenge the Orthodox religious worldview, but the Old Testament indeed refers to stories of greater antiquity. Swedenborg points them out:

"That a Word existed among the Ancients is evident from the writings of Moses in which he mentions it, Num. 21:14, 15, 27-30. Its historical parts were called 'The Wars of Jehovah,' and the prophetical parts, 'Enunciations.'"[28]

"In addition to those books, mention is made by David (2 Sam. 1:17, 18) and by Joshua (Josh. 10:12, [13]) of the prophetical Book of the Ancient Word, called the Book of Jasher, or the Book of the Upright." [29]

Moses' writings were augmented by later prophets who were brought into non-ordinary states of consciousness similar to the states Swedenborg had experienced. They, too, received their inspiration directly from God. This became the Hebrew Scriptures. It did not matter that the Israelites could not comprehend the deeper, quantum language of God's narratives. They were needed to reproduce and copy the precise, literal words of God's communication, keeping the deeper meanings intact for posterity. The Israelites became God's "chosen people" because of their unique propensity for preserving the Holy Word through extremely careful and faithful precision of notation.

But religion was bringing forth diminishing returns. The membership of God's Church on earth changed from those who had profound spiritual love to those with intellectual appreciation of deep spiritual knowledge, and ultimately religion was reduced to mere *obedience* and *adherence* to external laws and traditions. The trademark of the Israelite Church became the requirement of perfect obedience to the Talmud and its hundreds of laws addressing every aspect of daily life. Even this situation deteriorated.

▶ Restoring a Cosmic Imbalance

Once people ceased making the effort to give simple obedience to God, there was no place left for religion to go. God then needed to radically intervene by making an appearance *in the flesh* and entering into the theater of world history. The Prophets of the Israelite Church observed the deterioration of their people and warned them of the Lord's coming. But people overlooked their own Godlessness and did not understand that such a sacred event *must affect the inward quality of their worship* to have any success. Jeremiah had made it clear that the Lord would make a new covenant based on a deeper relationship.

> *But this shall be the covenant that I will make with the house of the Israel; After those days, saith the Lord, I will put my law in their inward parts, and write it in their hearts; and will be their God, and they shall be my people.* (Jeremiah 31:31-33)

However they understood this Divine intervention in the worldly sense that a messiah would come to "kick butt" and save them from their Roman oppressors. What they did not see was that their real oppression was coming from a far more sinister source than Roman rule. It was coming from hell.

According to Swedenborg, Jehovah God Himself came into the world with two major strategies. The first was to reveal Himself to the world as Divinely Human and, therefore, as a Person whom we could rationally approach. The second strategy was to trick the forces of hell into a Divine "rope-a-dope" fight.

From the very beginning, as the church on earth degenerated, humans who had rejected God in favor of self-love and dominance over others died and entered into the spiritual world of hell. And, their numbers were increasing. This had a negative, accumulative effect on top-down causality and God's Divine action in the natural world. Since the interiors of the human mind have relation to the spiritual world, even consisting of the same substances, influences from the Spiritual Sun were becoming more and more distorted by the growing population of harmful spirits. These spirits intercepted this influx and twisted God's Love into hurtful tendencies and persuasions in the minds of men and women on earth. This situation was usually offset by the influence of angelic spirits, but they were now overwhelmed and outnumbered by hurtful spirits. The imbalance between positive and negative influences became so extreme over time that the human race, like the Neanderthals before it, again faced extinction.

A more radical adjustment to worship and a new covenant was necessary. This required nothing less than the Lord's physical appearance in history. The result was the Christian Church. Hidden from the eyes of men and women, a more crucial drama played out while the Lord God lived on earth. God was not only in the world to deliver His new message, but was also restoring Divine order in the *spiritual world* by combating and subjugating the forces of hell that were increasing. This Divine combat was understood by the prehistoric churches, which possessed the symbolic knowledge of correspondences.

The prophecy of the Lord's Advent is first mentioned in Genesis 3:14, 15. This prophecy was also known to those members of the Most Ancient Church who, while suffering their downward decline, still had some communication with heaven. In fact, it was revealed to Swedenborg that those who represented the late vestiges of this celestial race and still enjoyed communication with heaven had direct knowledge that the Lord would appear on earth and be *born of a woman!*[30] They passed this information onto the Noachian Church before their own race suffered its final extinction. Swedenborg also learned that women were especially esteemed by the cerebrum-oriented humans who followed as a result of this prophecy.

Cro-Magnon artifacts show an obsession with vulvas and pregnant women. Between 20,000 and 30,000 years ago Paleolithic statuettes of stylized pregnant females began appearing from Western Europe to Eastern Russia. Only female figures were carved! Today they are called Venus figurines.

It would be hard to imagine a paleoanthropologist looking over a Venus figurine and concluding that the early humans who created it were envisioning a *Divine fetus* developing inside. So, is the anticipation of the Lord's Advent behind this prehistoric obsession with female figurines? Do they contain deeper theological and symbolic meanings?

Figure 9.3 illustrates a Venus figurine that depicts a woman with her left hand on her stomach, and a raised right arm holding a bison's horn. The bison's horn has thirteen purposeful grooves carved into it. This posture suggests that the bison's horn has some reference to that which was developing in her belly.

According to *Dictionary of Bible Imagery* (a compilation of correspondences used in Scripture taken from Swedenborg's theological writings):

1) A bison's horn represents the power of truth in its ultimate or outer form (especially the power to persuade).

2) The number thirteen represents a state of temptation or spiritual combat (from a holy principle), beginning in childhood.

figure 9.3

3) The left arm, which is connected to the right hemisphere of the brain, denotes volition, especially the desire to grasp truth from the principle of Love.

4) And the right arm, which is connected to the left hemisphere represents intellectual cognition, so whatever it holds up is in agreement with the understanding.

Taken together, all these symbolic pieces communicate a message that indeed contains a potent message for humanity. This figurine is communicating the doctrinal message that what was going to be born into the world would come for the purpose of engaging in spiritual combat with evil and temptation from childhood (hence, the 13 notches). This child would fight evil through the power of truth in its ultimate embodiment (the horn), that is, by means of a physical body. Upon making this translation, it is almost as if we are gazing at a 23,000-year-old Christmas card!

Consider the words of Luke 1:68-69:

Blessed be the Lord God of Israel; for he hath visited and redeemed his people, And hath raised up a horn of salvation for us in the house of his servant David ...

In his April 1975 article in *The New Philosophy*, Erik Sandstrom provides yet another thought-provoking female image from a Cro-Magnon painting. He describes an atypical picture on a cave wall of a woman giving birth with two horses apparently watching the event.[31] Horses represent a deeper understanding of things, because the horse that carries us farther and faster than walking is analogous to the understanding that carries the mind farther and faster then the unenlightened mind can. Their watching signifies that this birth-event has a higher, spiritual meaning. It represents the acknowledgment that this is the birth of a special child.

Again, Swedenborg maintained that it was Jehovah God Himself who actually entered the world stage. He quotes Isaiah 9:6 as support for the idea that the Father and Son are one person.

Unto us a child is born, unto us a Son is given; and the government shall be upon his shoulder; and his name shall be called Wonderful, Counselor, God, Hero, the Father of Eternity, the Prince of Peace.

The Divine scheme of Conjunctive Design had become so disrupted by the cumulative effect of those who were populating hell that the human race was in danger of perishing, and the Divine order in the universe was threatened. No single human, no single angel, had the power to resist the infectious power of hell and remain in a state of integrity. Only an omnipotent God could enact redemption, and for reasons that involve real physics, this could only be accomplished if God's infinite essence were made incarnate in a finite physical body. At this point, God used the "rope-a-dope" strategy to restore cosmic balance in the universe.

▶ The Divine Rope-A-Dope

Restoring the cosmic balance required stripping hell of its power. This requires a *spiritual* combat between the forces of good and evil. But since God is infinite, how could anything finite engage God in combat? Not even the power of hell is infinite. Thus, there could be no battlefield where such a conflict can take place. *This is precisely why God took on a physical body* and *brain*.

Being born of a woman on earth allowed God to take on all the weaknesses, inclinations and temptations inherent in the human flesh. The hells now had an organic medium by which they could attack with their deadly influence by exciting the hereditary compulsions connate in human flesh. But this was a trap. Now God could fight the full fury of hell, too.

By taking on a physical body, Jehovah God could enter into the human gene pool and experience all the temptations and evil inclinations of men. The ensuing conflict involved resisting these influences from hell. Each victory brought His human nature in greater alignment and harmony with His Divine nature until both were perfectly unified. This process is called *exinanition* by which the Lord humbled Himself before the Father, that is, His higher Divine essence, which acted as the soul of His physical body.

What is not known to most, is that the highest level of meaning contained in Scripture addresses the progression the Lord made toward this union, while living in the physical world. This process of humility culminated on the cross. Crucifixion was designed not for maximum physical pain but for maximum humiliation. The "ransom" behind this event involved the Lord giving up all vestiges of human self-centeredness and the desire to dominate over others. While the Lord was suffering on the cross the hells were furiously trying to tempt Him to come down and show who was the boss. But the Lord resisted and sacrificed this human inclination in order to give His Love for all humankind. In this way the Lord chose righteousness. If the Lord had gotten off the cross to make His authority irrefutable and had demanded devotion, this would have compelled belief and violated the laws of Divine Providence, which protect human free will.

> "Providence acts invisibly, in order that a person may not be compelled to believe from visible things, and thus that his freedom may not be injured; for unless a person has freedom he cannot be reformed, thus he cannot be saved." [32]

When one takes top-down causality into account, every victory the Lord had over evil and its temptations caused a cascading effect on every level between the spiritual and physical worlds. The Lord's victory reordered cosmic forces until hell was finally subjugated and spiritual equilibrium was restored. The human race could not enjoy free will unless the proper balance of influences was maintained. This is not what the human race expected when the Messiah came. But as the Jeremiah quote above made clear, the new covenant with the Lord would require that a change be made within people's hearts. It was human nature and its bondage to evil compulsions, not Roman oppression, that God fought against when He assumed a human body and came into the world.

While the literal sense of Scripture speaks of the outer conflict and power struggles between the Lord Jesus Christ and men of the world, the inner battle that took place was far more intense and important. Remember, Swedenborg learned that Scripture contained a three-tiered narrative including a literal sense, above that a spiritual sense, and above that a heavenly sense (see Chapter Seven). In both the Old and New Testaments, the highest or heavenly level of meaning translates into a spiritual drama depicting the various intense inner battles and conflicts the Lord faced while bringing His "flesh" into compliance with the Divine Will and putting the hells in their place. The Lord did not destroy the hells or evil, but they now came completely under His governance.

Each time the Lord endured and conquered a temptation He perfected the conjunction and union between His Human and His Divine, until the Human ultimately became Divine.

Glorification was the process by which these two natures were made one. Through His life, death, and resurrection, the Lord gained "power over all flesh" (John 17:2) and "all power in heaven and earth" (Matt 28:18).

The Lord's coming into the world* saved the human species from extinction. Swedenborg added that if the Lord had not been victorious over hell's influence, the human race would have perished. The Lord did not remove our sins. Rather, He kept the door to salvation open.

Contrary to traditional Christian belief, the passion of the cross and the resurrection was not redemption itself; it was the preservation of the means by which men and women could be saved. By restoring order in the spiritual world, the Lord maintained the balance between the influences of heaven and hell and preserved human free will. Since the Lord cannot compel us to spiritual life, we must compel ourselves to seek a new intelligence and new love from the Lord through humility. When we approach the Lord for help, the Lord fights for us and rearranges our inner world in a manner similar to His own process of glorification.

Conjunctive Design and its ultimate purpose of creating a heaven from the human race, require that men and women take part in their own spiritual re-creation, their *epigenesis*. This is why first-person phenomenal experience and human subjective consciousness were so important within the scheme of evolution. We can each change what we love, because we have the freedom to do so. The more we *choose* to prepare ourselves to receive God's higher Love and Truth through humility, and reciprocate, the more

✳ **There have always been theological problems with trying to reconcile God's transcendence and His immanence.** How can God be everywhere—be beyond space and time and still operate in nature? How can He also become incarnate and exist in a finite human body? Swedenborg suggested that finite things must come from things that are greater and less limited. Furthermore, the Lord enters into union with all things with different intensities, according to their use. All use is an image and form of God's goodness.

This is the principle behind Conjunctive Design. It makes no difference whether God operates through nature by creating finite living forms from inert matter or enters into a human ovary in a special way; everything is sustained by the Infinite. God entered into Mary's womb for the greatest of all uses—human salvation. The only requirement for God to act in the temporal world is that all final events must occur as a consequence of an infinite and eternal goal. This is why all things in the physical world are holistically interconnected, interrelated and interdependent.

perfect is our connection with God. Over time, the human heart and mind becomes inwardly organized according to the order and organization of heaven, creating a habitation within each of us into which the Lord God can enter and dwell "as in His own temple."[33] Worship is not simply the act of going to church; rather, we *become* a church. This spiritual process is so important that it will be the subject of the next two chapters.

Swedenborg insisted that the Holy Trinity was *one* God, consisting of three attributes, not three Persons. Jehovah God came to earth and acted as the Divine soul within a human form, a form that came from a human womb. The Lord glorified His Human and made it "one with the Father," in effect, replacing all the imperfect human traits He inherited from Mary with Divine traits. This is why the Lord refers to Mary, not as "mother," but as "woman." This perfection and glorification is also why there was an empty tomb. So the "Father" represents *Divine Love*, the "Son" represents *Divine Truth* (Love made visible), and the "Holy Spirit" is the *Divine saving operation* of God's Love and Truth effecting our salvation. It is worth repeating that Swedenborg quotes Isaiah 9:6 as evidence that Jesus was Jehovah for it is the Father of eternity who was born in the world.

> *Unto us a child is born, unto us a Son is given; and the government shall be upon his shoulder; and his name shall be called Wonderful, Counselor, God, Hero, the Father of Eternity, the Prince of Peace.*

While the real drama of the Lord's advent went mostly unnoticed by human eyes, the Lord's physical presence and actions on earth had a real and purposeful benefit: they created a new religion. This religion, called Christianity, was based on a visible figure. Worship of a formless God would have rendered religion incoherent. To finally be able to focus attention on a God with an actual human form provided a means by which we could rationally approach God and enter into a personal relationship. Indeed, as Swedenborg learned, even in most ancient times, when God appeared to people in visions and dreams, He appeared in human form.

After the Lord's Advent, the Christian church itself underwent further changes. Councils of bishops produced mind-numbing explanations of Trinitarian doctrine based on a literal interpretation of Scripture. The Reformation continued to generate its own splinter groups that also founded their theology and doctrines only on the *literal* words of Holy Scripture. The literal or outer sense of Scripture can be, and has been, used to prove any church dogma one pleases—salvation by "good works," by "faith alone," by "imputation" or even by "predestination." Swedenborg states:

> "The spiritual sense of the Word is not the sense which shines from the sense of the Letter when one searches the Word and explains it to prove dogma of the Church. This may be called the literal and ecclesiastical sense of the Word; but the spiritual sense is not apparent in the sense of the Letter; it is interiorly within it, as the soul is in the body, or as the thought of the understanding is in the eye, or as the affection of love is in the countenance. It is this sense especially that makes the Word spiritual, not only for men, but also for angels; and therefore the Word by this sense communicates with the heavens. Since the interior content of the Word is spiritual, it is written by pure correspondences ..."[34]

Also:

> "It is generally agreed that the Word is from God, is Divinely inspired, and therefore holy; but hitherto it has remained unknown wherein its divinity resides; for the Word in the Letter appears like common writing in a strange style, lacking the sublimity and brilliance which are apparently features of the literature of the world."[35]

A God of infinite wisdom must speak in a language that goes beyond the limitations of ordinary words, even when ordinary words are used. The Lord's miracles themselves contained deeper, symbolic messages. The deeper symbolic language of correspondences was known *instinctively* by those of the Most Ancient Church and *intellectually* by the church that followed. Then it was forgotten. Yet this sacred knowledge remains safely preserved for the future in the Word of God. In order for the Creator to perfect conjunction with humanity, this knowledge has to be rediscovered. *Hence, a new church still awaits us on the horizon.* This is why the Lord promised to return. The religions of the past have all run their course. Our knowledge of God is inadequate for the post-modern world. We need new wisdom to solve the complex issues of the world and new revelations to show that all knowledge is connected, including science and religion.

On June 22, 1757, about twelve years into his twenty-eight year journey into the spiritual world, Swedenborg witnessed a cataclysmic event. He found himself at ground zero of the Second Coming of the Lord!

SUMMARY:

▶ Hominids became human when they recognized and responded to God. The emergence of human-ness and the church on earth were the same event.

▶ Swedenborg met and communicated with an ancient race of humans in the spiritual world. Their evolution was spiritual and was centered in the involuntary part of the brain at the back of the head.

▶ These humans became extinct and were replaced by a race with enhanced frontal lobes.

▶ Symbolic knowledge was transferred from one race of humans to another in the form of codices. Later, this heavenly knowledge was woven into stories. These sacred stories represented the Ancient Word which predated our current Bible.

▶ Swedenborg's depiction of two types of early humans and their physiology conform closely to the actual fossil evidence.

▶ The Lord God had to adapt religion to the changing qualities of the human mind, which was becoming more corporeal and sensual. The symbolic language of correspondences was no longer helpful to religion.

▶ The Lord God preserved the language of correspondences within rituals and ceremonies. Then, through Moses and the Prophets, He created a new Holy Word.

▶ A new church and a new covenant is now being established by the Lord. This will encompass the return of the knowledge of correspondences, by which the Lord's true glory will emerge from the literal meanings of Scripture, making all things anew.

PREDICTION

The Second Coming is now—if you want it!

THE SECOND COMING

> "We have met the enemy and he is us."
> — From Walt Kelly's comic strip, *Pogo*

The Most Ancient Church came to an end described by the biblical "Flood." The next church or Noachian Church came to an end when the knowledge of correspondences degenerated into polytheism, idolatry and magic. The third major church, or Israelite Church, ended when it rejected the Lord's Divine Human at the time He came into the world.

Swedenborg describes the "end" of a church, not by its duration in time or by the size of its membership, but by its loss of the capacity to be a dynamic form receiving God's living and continuous influence. When a church distorts doctrine because of human error and becomes incapable of receiving new revelations or fresh spiritual inputs, it becomes a "dead" church. Mathematician/philosopher Alfred North Whitehead felt religion had decayed because it wasn't learning anything new. At the consummation of each church (the "end times") the Lord raised up a new form of worship adapted to the psyche, genius, and level of understanding of those living during these periods of change.

Here is what Swedenborg says of the Christian Church:

"The fourth Church is the Christian, which the Lord established by the Evangelists and the Apostles. Of this church there have been two epochs, one extending from the time of the Lord to the Council of Nice, and the other from that Council to the present time. This Church, however, in its progress was divided into three branches, the Greek, the Roman Catholic and the Reformed; nevertheless, all these three are called Christian. Moreover, within each general Church there have been several particular Churches which, although they have seceded, have still retained the name of the general church ..."[1]

During the period of time Swedenborg was actively exploring the spiritual world, representatives from these various schisms of the Christian Church, who had died and passed on, were determined to force heaven and the spiritual world to accept their various incorrect theological interpretations of the Trinity and salvation. These interpretations actually threatened human salvation, and once again, Divine order in the cosmos was thrown out of balance. The two fundamental errors of the Christian Church were the belief that the passion of the cross was redemption itself and the Trinitarian Doctrine of three Divine Persons (as opposed to a triune operation of one God). Swedenborg states:

> "There is no doctrine at the present time more extensively promulgated in the books of the orthodox, or more zealously taught and inculcated in the schools, or more frequently preached and proclaimed from the pulpit than this: 'God the Father, being angry with the human race, not only removed it from His presence, but involved it in a general condemnation, and thus excommunicated it; but because He was gracious, He persuaded His Son to descend and take upon Himself the condemnation which had been decided, and thus appease the wrath of His Father; and by this means only could the Father look with any favor on mankind. This was done by the Son, who, by taking upon Himself the condemnation pronounced upon the human race, suffered Himself to be scourged by the Jews, spit upon, and finally crucified as one accursed by God,' Deut. xxi. 23. Moreover, after this was done, the Father was propitiated, and from love of His Son, cancelled the condemnation; but only in the case of those for whom the Son should intercede, and thus He became a Mediator in the presence of the Father forever."[2]

According to what Swedenborg had learned from angels, this doctrine is pure theological cockamamie. Besides undermining monotheism, the harmful message is that God gets angry, that He only loves His Son and will listen only to His Son's evaluation of each of us. So-called Christians were carrying these horrific ideas that were inherently profane and dishonoring of a God of infinite Love and mercy with them into the spiritual world and trying to infect others with their heresies. Swedenborg observed firsthand that these counter-productive ideas emerged from a church tradition that no longer understood the Holy Word as a multi-dimensional document and their idea of God and redemption became merely corporeal and sensual.

This malignant influence grew as those who professed these ideas became more numerous in the spiritual world and again caused the cosmic equilibrium to move out of harmony from true, Divine order. These doctrinal distortions of the Divine Truth had to be removed and corrected in the Spiritual World. Once the Lord had done this, He could establish a New Church on earth. Swedenborg himself seems to have been caught off guard as he witnessed this correction firsthand. This event was the Second Coming! What he witnessed would add a wholly new twist to the eschatology of Christianity, or its doctrine concerning "final things."

▶ The Apocalypse and the New Jerusalem

Swedenborg witnessed a spiritual cataclysm in 1757. This cosmic upheaval resulted when the Lord God made another adjustment to religion and doctrine. This new adjustment first took place in the spiritual world and resulted in the formation of a new Christian heaven that could once again begin to influence the hearts and minds of humans on earth. This descent of new doctrine was the *New Jerusalem* coming down from heaven. Divine order requires that things change from the inside out, or top-down. Physicists call this order top-down causation.

As we saw in the two previous chapters, the interiors of the human mind operate on the same non-material plane as the interiors of those who populate the spiritual world. There is no difference between the mental plane of those living on earth and those living in the spiritual world, for they consist of similar spiritual substances. In fact, when spirits walk through a spiritual landscape they actually walk through the various values and thoughts of our worldly memory that they particularly relate to! Inwardly, we are so connected with the life of spirits that at any given moment we can be said to be in union with either heaven or hell! (We can verify this if we reflect on the clutter of thoughts in our brain that seem to come out of nowhere and can often shock us.)

To the extent that *new* teachings concerning the Lord and the Word are first adopted by those in the spiritual world, there can be an increase in the positive spiritual inflow from the world above into the interiors of the minds of those on earth. This flow of new theological ideas into human minds is the Second Coming. It will enlarge the boundaries of human wisdom and intelligence.

However, hearts and minds must be receptive to this new revelation and worldview. We must want it! Those who truly love God and neighbor will gain new power to discern the theological distortions of their own faith-traditions, or at least begin to feel uncomfortable about some of the things the church is teaching. They will be open to, and hungry for, new ideas. This is why the Lord will come *with clouds* (Rev. 1:7), which in psycho-spiritual language means God will break through our intellectual obscurity about true doctrine and disperse our false ideas. Truly open-minded and receptive individuals will be prepared for this new truth "as a bride adorned for her husband" (Rev. 21:2).

This paradigm shift will happen gradually, and in some cases, painfully, as people and their institutions respond to new ideas. It is a process that has been accelerating for more than 250 years. Its general effect on the human psyche has produced various revolutions in governments, industry, sexual behavior, the environment and in gender and minority rights.

This process is also evidenced by a growing dissatisfaction with religion, even the outright rejection of religion. Religion has deteriorated because it no longer offers anything new to spirit or consciousness. And religion has failed to respond adequately to the ideas of the New Physics and its materialistic assumptions concerning the ultimate reality. Worst of all, religion and its doctrines of salvation have failed to change fundamental human behavior. Evil, greed, jealousy, self-centeredness, and hatred still exist, among believers and non-believers alike.

According to Swedenborg, this crisis of religion and the human predicament will ultimately be rectified by the Lord's reintroduction of the science of correspondences to humanity. This was the Lord's purpose for allowing Swedenborg to have access to heavenly secrets—to help usher in this new paradigm shift. When humanity appreciates the knowledge of correspondences, the Holy Word will offer new, more meaningful theological lessons to a post-modern world. Equally important, only when God's Word is seen as a multi-dimensional document that contains the universal patterning principles of a multi-layered universe, will religion have an answer for the skeptical self-assurance of science.

Correspondences will provide a scientific methodology for understanding God and offer a new way of interpreting biblical eschatology and its promise of salvation through the Lord's return. A rational explanation of the Second Coming that makes sense to the relativistic and cynical post-modern world will become clear only when people embrace God's unveiling of a fuller knowledge of the Holy Word. Scriptural authority and inerrancy will then be confirmed by new metaphysical disclosures rather than by historical and archeological research.

Modern systematic theology will be based on doctrines distilled from the expanded meanings of a quantum language, which Swedenborg called the science of correspondences. As we have already seen, the knowledge of correspondences has been hidden from most of humanity for thousands of years, so it is now foreign to ordinary states of human consciousness and cognition. The ability to accept correspondences within Scripture hinges on three factors: an appetite for deeper truths, the cognitive ability to think above the corporeal and sensual things of the habitual mind, and the malleability of the heart to give up old ideas in favor of new ones.

Those who believe that the Lord will return to earth to punish the "bad people" and set up a social utopia are deluded. This will be a battle for hearts and minds—in other words, a battle for everything! Armageddon will play itself out, not in the physical world, but *within* each of us. God's return will come unexpectedly in the form of new, deeper teachings and all people will find themselves individually facing the "end of the world as we know it." The earthquakes predicted in Revelation will be the shaking up of worldviews and paradigms. All the horrible creatures, beasts, and armies that Revelation says will rise up to challenge the Lord and His angels are the resistances individuals will put up to defend their old or false ideas. The enemy is "us," as Pogo discovered. This is the eschatological shocker of the Second Coming.

The revelation within Revelation is the inner story of human resistance to new thinking and God's victory over human hearts and minds. Creating a "new heaven and a new earth" means reconstituting the *inner* and *outer* qualities of personal human life according to Divine order. Both the spirit and worldly life of individuals will be reformed.

This deeper interpretation of the Apocalypse provides a whole new twist to eschatology (end times) and soteriology (salvation), for it is certainly not what most people are expecting. It is not drastic cosmological changes in the form of a physical upheaval that occurs outside of people, but an upheaval that takes place inside them as old ideas are overthrown.*

✳ **The Apocalypse will not involve the physical destruction of the universe and its replacement by a complete new universe that does not know suffering or death.** Such a materialistic scenario would require all new laws of physics. Rather, a new theology will be required to understand the "end times."

To truly understand the next covenant with God and the true significance of the New Jerusalem, the literal sense of Scripture must be replaced by a multi-leveled exegesis. The quantum language of correspondences will be necessary to gain access to this new dispensation from God.

A literal interpretation of Scripture, for example, offers no reason why the Lord would need to make two appearances on earth. Why is this any more than Divine inefficiency? The Lord must come "twice" because there are two essential steps to our personal spiritual evolution. Swedenborg states:

"The Lord is present with every man, urgent and pressing to be received; and when a man receives Him, by acknowledging Him as his Creator, Redeemer and Saviour, then this is His first Coming ..."[3]

The first stage is called *reformation*. It involves accepting the Lord as the one true God of heaven and the personal Savior. But faith, trust, and Christian teaching alone do not save anyone. In fact, the doctrine of salvation by faith alone and the belief that the passion of the cross was redemption by itself has led many Christians to falsely think that if they merely "believe" that the Lord is their Savior, they have acquired immunity from the condemnation of the law (Commandments). But believing in God requires following His commandments. Add this denial of personal responsibility to the false doctrine that makes one God into three distinct Persons and you have a big reason why the Lord must provide a new dispensation and clarify what it really takes to be saved.

This new heavenly dispensation makes it clear that Faith needs to be put into action and bear fruit. For faith is the Truth of God's Love. This represents the second stage of our spiritual evolution and is called *regeneration*. This stage represents a "second coming" or deeper penetration of spiritual truth into our lives. First we acknowledge truth, then, we implement truth. The Second Coming consists in our *application* of spiritual knowledge for personal transformation. Truth must affect the human heart or it remains as mere memory-data and static. Truth cannot take root without action.

A battle will result from the inner conflict between our former selves and our new selves. This inner state will be experienced as *temptation*. The Lord can only help us remove hurtful influences of spirits associated with us by allowing our various negative inclinations to rise to the surface and horrify us like the apocalyptic Great Dragon that emerges from the sea (and empowers yet another terrifying beast). We must see these unflattering things arising in us, identify them, ask God for strength, then cooperate with God by taking part in the fight to repel them.

This deeper teaching concerning Armageddon will be closed off to those with merely materialistic interpretations of Scripture. The Holy City, the New Jerusalem, will be God's dwelling place in our hearts. This transformation cannot take place unless we recognize evil in ourselves and then ask the Lord for help in removing our hurtful inclinations. A city cannot "descend from heaven" unless it represents a new *habitation* for the mind and heart in the form of new spiritual doctrines of life. These things can be made clear only when the stories of Scripture are understood to represent the *inner* realities of the human spirit and our true relationship with God.

In the first coming, the Lord emerged from a human womb, made Himself visible to men and women of the world, and offered His teachings. In the Second Coming, the Lord emerges from His Holy Word in a way that startles and wakes the habitual mind into radical enlightenment. This new augmentation of human cognition is symbolized by the words in Scripture that the Lord will come "with the clouds" (Rev. 1:7). As mentioned previously, clouds and mists represent cognitive *obscurity*. Swedenborg tells us that the literal interpretation of Scripture is what is actually meant by "the clouds," because like clouds that obscure the sun's rays, the natural sense of the Letter obscures the spiritual, quantum meanings within.

Through the higher-level orders of meaning in the Word, the Lord can burst through our ordinary states of consciousness and finally be seen in His full glory and power.

▶ A Servant of the Lord

Swedenborg was chosen to disclose to the world the "quantum" language of Scripture through his theological writings, which totaled about thirty volumes of work. As mentioned in both this and the previous chapter, God adapted religion to the changing psyche of people throughout human history and prehistory. What is significant is that now God was making critical new disclosures during the rational age of Enlightenment, when scientific discovery took precedence over faith. Both God and the growing scientific community were challenging the prevailing dogma of the Church that human understanding was subservient to faith! As a scientist himself, Swedenborg did not believe that God would have created the ample human brain only to demand blind faith from it. The whole purpose of the Second Coming was to bring the deepest mysteries of faith within the scope of human comprehension. And the Lord chose a scientist to help usher in these new ideas into a scientific age.

Descent and Ascent of the Lord's Church on Earth

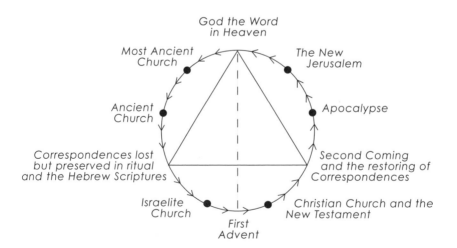

figure 10.1

Swedenborg was once asked by those in heaven how things were going back on earth:

"I was once raised up as to my spirit into the angelic heaven and into a society there. Thereupon some of the wise men belonging to it came to me and said, 'What news from the earth?' I said to them: 'This is new, that the Lord has revealed things which have been hidden (*arcana*) surpassing in excellence those hitherto revealed since the beginning of the Church.' 'What are they?' they asked. I said, 'They are the following:

1) In the Word, in the whole and every part of it, there is a spiritual sense corresponding to the natural sense; by means of that sense the Word is a means of conjunction for men of the Church with the Lord, and of association with angels; and the holiness of the Word resides in that sense.

2) The correspondences of which the spiritual sense consists are now revealed. The angels asked, 'Did the inhabitants of the earth know nothing of correspondences?' 'Nothing at all,' I replied; 'They have been hidden now for some thousands of years, even from the time of Job. With those who lived at that time, and before it, the science of correspondences was the science of sciences. From it they derived their wisdom, because their knowledge of spiritual things relating to heaven and the Church was obtained from it. But as that knowledge was made to serve idolatrous purposes it was, by the Divine Providence of the Lord, so obliterated and destroyed that no one saw any trace of it. However, it is now revealed by the Lord in order that there may be effected conjunction of the men of the Church with Him, and their association with angels. This is done by means of the Word, in which all things, both general and in particular are correspondences.' The angels greatly rejoiced that it had pleased the Lord to reveal this great truth (*arcanum*), so deeply hidden for thousands of years. It was done, they said, in order that the Christian Church, founded on the Word and now at its end, may again revive and derive its spirit from the Lord through heaven. They asked whether by that science the signification of Baptism and the Holy Supper has now been made known, concerning which so many different opinions have hitherto been held; and I replied that it was now made known.

3) I said further that a revelation has been made at this day by the Lord concerning the life of men after death. 'Why concerning life after death?' the angels asked; 'Surely all know that man lives after death.' 'They know it,' I answered, 'and they do not know it. They say that it is not the man who then lives, but his soul, and that this lives as a spirit; and the idea they entertain of spirit is that of something that is wind or ether. They maintain that the man does not live till after the day of the Last Judgment, when the corporeal elements, which men had left behind in the world, although eaten up by worms, mice or fish, will be collected together and again formed into a body; and that in this way men will rise again.' 'What a strange idea!' the angels said, 'Surely every one knows that a man lives as a man after death with this difference only, that he then lives a substantial man, not a material man as before; and that the substantial man sees the substantial, just as the material man sees the material; and they do not notice a single point of difference, except that they are now in a more perfect state.'

4) Then the angels asked, 'What do they know of our world, and of heaven and hell?' 'Nothing at all,' I answered; 'but the Lord has at this day revealed the nature of the world in which angels and spirits live, and thus the nature of heaven and hell; and also that angels and spirits are associated with men; besides many wonderful things concerning them.' The angels were glad that the Lord had been pleased to disclose such things, so that men might no longer, through ignorance, be in doubt about their own immortality.

5) I added further, 'The Lord has at this day revealed that there is in your world a Sun different from that in our own world; and this Sun is pure Love; and that the sun in our world is pure fire; that, therefore, whatever proceeds from your sun, because it is pure Love, partakes of life, but whatever proceeds from our sun, because it is pure fire, has nothing of life in it; that in this consists the distinction between the spiritual and the natural; and that this distinction, hitherto unknown, has also been disclosed. It has now, therefore, been made known whence comes the light that enlightens with wisdom the human understanding, and whence the heat that kindles with Love, the human will.

6) Moreover, it is now disclosed that there are three degrees of life, and consequently three heavens; that the mind of man is distinguished into those degrees; and hence that man has a correspondence with the three heavens.' 'Did they not know this before?' the angels asked. I replied, 'They knew of the degrees between more and less, but not of degrees between prior and posterior.'

7) The angels then asked whether anything else had been revealed. 'Much more has been revealed,' I answered, 'concerning the Last Judgment; concerning the Lord, that He is God of heaven and earth; that God is one both in Person and in Essence, that in Him is the Divine Trinity, and that the Lord is that God; also concerning the New Church, which is to be established by Him, and concerning the doctrine of that Church; concerning the holiness of Sacred Scripture; and that the Apocalypse has been revealed ..." [4]

Swedenborg added that the new revelations he received from heaven included a confirmation that human societies inhabited other planets throughout the entire universe. He also learned that marital love based on spiritual principles, represented the greatest heavenly delight that could be bestowed upon humankind. "Love is in itself spiritual and religious in its origin."[5] Two married partners are an effigy of the marriage between God's Love and Truth, which is the essence of the creative principle behind the entire universe. The Lord's goal in Conjunctive Design would achieve its fullness when two people were united in spiritual Love and Truth. This deeper "marriage" of Love and Truth was what truly formed a church on earth and the means by which the Lord could perfect conjunction with the human race.[6] A man or woman, alone, is only a half-angel.

Swedenborg was painfully aware that much of what he disclosed in his theological writings would not be believed by men and women in the world, particularly his account of his discussions with spirits and angels. He addresses this in the book *The True Christian Religion*:

"I foresee that many who read the Memorabilia that are appended to the chapters of this book will believe them to be figments of the imagination. But I declare in solemn truth that they are not inventions, but were truly seen and heard; not seen and heard in some state of the mind when sound asleep, but in a state of complete wakefulness. For it has pleased the Lord to manifest Himself to me, and to send me to teach those things which will belong to His New Church, which is meant by the New Jerusalem in the Revelation. For this purpose he has opened the interiors of my mind, that is, of my spirit; and in this way it has been granted to me to be in the spiritual world with angels, and at the same time in the natural world with men, and this now for twenty-seven years."[7]

Figment of the imagination or not, his "other-worldly" excursions and interviews enabled Swedenborg to approach theology in a systematic, novel, coherent, self-consistent and amazingly complete manner. He addressed questions that had never occurred to most people or theologians. Whether a reader believes his account or not, it is difficult to believe that he could have made it all up and fit it into a systematic format.

Swedenborg pointed out that the Lord had opened up the interiors of the minds of others in a similar manner, both before the resurrection and after. Among the people so blessed were the Apostles and the Prophets of the Old Testament. Those of the Most Ancient Church also enjoyed this interior sight before their fall and extinction.

▶ Eschatology versus Cosmology

Robert John Russell, a Professor of Theology and Science at the Graduate Theological Union in Berkeley, California and the founder and director of the Center for Theology and the Natural Sciences (CTNS), believes that eschatology is "ground zero" in the challenge of Christian theology to interface with modern physics.[8] Swedenborg offered ideas in this discussion, as well. To understand eschatology in the framework of cosmology, we must change our assumptions about both Christian theology and physics.

The idea that God will inaugurate a wholly new universe to bring about the final redemption of the human race at the end of the world does not square with either current cosmology or with Swedenborg's discoveries in the spiritual world. Neither the Lord's return nor the creation of a "new heaven and earth" requires a fundamental change in the laws of nature. They require a new understanding of the laws that operate in both the natural and spiritual worlds, however. Physics must recognize that Love is the first principle and primal formative substance of the universe. Theology must acquire a new and deeper understanding of Scripture, whose stories describe the necessary and fundamental changes in both the inner and outer qualities of those who will form the next great church on earth.

The traditional Christian promise of eternal happiness in a new physical universe is based on a materialistic understanding of Scripture rather than a spiritual one. There cannot be any transformation of reality or any redemption other than a personal transformation. When individuals inwardly change who they are by adopting spiritual principles for living, their new outlook changes the physical world as well. This *inner* and *outer* transformation of human behavior is what is meant by a *new heaven* and a *new earth*.

Will the physical universe ever come to an end? Swedenborg does not address this question directly, but he does state that the spiritual world and its citizens depend on the physical world of matter in the same way a house depends on its foundation. Certainly, stars in the universe are born, live, and die, but if God's perpetual activity and influx is behind the dynamics of the universe and its evolution, then it can never stop growing and expanding, and new stars will always be created faster than old stars are dying. In Swedenborg's spiritual paradigm, the expanding universe reflects God's continuous effort to enter into deeper relationship with created things. The universe is in a perpetual endeavor to manifest more perfectly the image of God's infinite Love, Wisdom and Order. So there is a continuous effort by God to create more spacetime, matter and complex structure. The expanding universe and its striving towards coherent organization are outcomes of Conjunctive Design and lawful responses to top-down causation that results from the spiritual first principles of Love.

Swedenborg did make clear that the spiritual world and heaven can grow only from the human race and that humans exist on many planets throughout the universe. Heavenly society experiences greater joy, companionship, and perfection as more souls enter God's kingdom of mutual love. Since heaven is non-spatial and non-physical, it can never be filled up.

The Lord's resurrection suggests to the traditional believer that the Creator can quite easily suspend the laws of physics to warrant the Christian hope of eternal life in a *perfect* physical body. But if Swedenborg was right, this belief must be wrong. As we will see, it is impossible to enjoy eternal life in a new physical body and on a new planet earth. Humans are organically destined to live in a non-material realm beyond the tyranny of time.

Swedenborg, who supported the idea of an "empty tomb," said that the physics by which the Lord's Human Body gained Divine powers is not available to finite humans. The Lord's flesh retained its human form during the process of glorification all the while it took on the higher "physics" and laws of the Divine essence itself. Remember that spiritual forces flow into all matter. With God, this top-down process became unified and infinitely perfect as the Alpha and Omega.

A scientific explanation of this Holy process is that the non-temporal and non-local dynamics of the quantum microworld were given full expression in the macroworld—in the large-scale bio-structure of the Lord's physical body. A theological explanation is that natural substance and God's spiritual essence became perfectly unified in the Lord's human form. The Lord could be touched by His disciples, eat fish, yet walk through walls and appear out of nowhere (Luke 24:36, John 20:19). This suggests something similar to the modern notion of quantum tunneling. All miracles involve the laws of higher *psychical* realms manifesting in lower *physical* realms, which ordinarily are under the laws of Newtonian physics.

Again, the Swedenborgian perspective of a "new heaven" and "new earth" is that human personhood will be given a new *inner* and *outer* quality. The only requirement is that people freely receive and adopt the teachings of the Lord's new, dynamic faith-system and apply those teachings into their lives. The "end time" is not cosmic destruction. It is the passing of old and distorted religious paradigms and their replacement by ideas that will form the doctrines of a New Church, the New Jerusalem.

▶ Redemption and Removal of Sin

There are several problems with the notion of a physical resurrection of the person after death. One of the more important ones is eschatological redemption, or the *removal of sin*.

In the traditional Christian version of a resurrected life, whether it envisions the restoration of the physical body or the elevation of the soul to a spiritual realm, is the general assumption that we will be without sin. Christianity promises us a future world without evil and sin. However, according to Swedenborg, much of Christianity does not understand how sin and evil are removed. This misunderstanding is the logical outcome of the Trinitarian doctrine that depicts God as three distinct Persons mysteriously unified in one essence (called hypostatic union). One misinformed priest in the spiritual world verbalized his own Christian belief about redemption with Swedenborg. The priest confirmed Swedenborg's observations, quoted early in this chapter, concerning the misguided belief that Christ's death on the cross was redemption itself. He spoke:

> "God the Father, being angry with mankind, condemned it and shut it out from His pity; and, having declared all men doomed and accursed, delivered them over to hell. He desired His Son to take that condemnation upon Himself; who consented, and for that purpose descended, assumed the Human, and suffered Himself to be crucified and the condemnation of mankind to be thus transferred to Himself; for it is written, Cursed is every one that hangeth on a wooden cross. The Son thus appeased the Father, out of love for the Son, and moved by the anguish exhibited by Him on the cross, determined that He would pardon, 'But only those to whom I impute thy righteousness, and these I make children of grace and blessing from being children of wrath and curse, and I will justify and save them. The rest will remain, as was before decreed, children of wrath.' This is our faith, and this is the righteousness which God the Father implants in our faith, the faith which alone justifies and saves."[9]

Swedenborg observed that when angels heard such distortions of doctrine it caused actual pain in their ears and irritated their nostrils. An angel who overheard the above discussion, responded by saying:

> "It is impossible, by mere imputation, to forgive anyone's sins, to renew him, to regenerate him, and thus to convert unrighteousness into righteousness and a curse into a blessing. Would it not be possible, if such were the case, to turn hell into heaven, and heaven into hell, or to make the dragon Michael, and Michael the dragon, and so to end the combat between them? For

what is needed but to remove the imputation your faith assumes from the one and bestow it on the other? Were this possible we in heaven should live in constant dread. It is not according to justice and judgment that one should take upon himself the sin of another, that the wicked be made innocent, and the crime be thus washed away. Surely this is contrary to justice, both Divine and human. The Christian world is ignorant as yet of the existence of order, and more so the nature of that order which God introduced at the time He created the world; nor does it know that God cannot act contrary to it, as then He would be acting contrary to himself, for God is order itself." [10]

The error in the priest's thinking was that redemption happens *to* us, not *through* us. As we saw in Chapters Seven and Eight, heaven is not a place you go to or get permission to enter, but rather something you become. This is another reason why a shake-up in theology is in order.

In God's Conjunctive Design, men and women must take part in their own salvation and epigenesis. Salvation requires our cooperation with God. "Walking with God," means following the commandments and not believing that by some special saving grace we are immune to the Law. This reciprocal relationship and partnership with the Creator must be an act of free will, so that every individual may appropriate a life of spiritual blessedness. *Eternal happiness is a choice.* But it also requires introspection and a rejection of the negative aspects from our former lives. The order and steps by which conjunction with the Lord God will be achieved and sin removed will be given full explanation in the next chapter. For now, it is important to know that this sacred process is both scientific and logical.

Everything in the universe acts as though it yearns for order. In humans, this yearning for order moves into the nonphysical world of heart and mind where order can only emerge from our choice of *values*.

▶ The Sacredness of the Holy Word

During his spiritual explorations, Swedenborg was able to identify which stories of the Holy Word were actually written in the multi-leveled style of correspondences. The original decisions to include or not to include specific stories in the Bible were made by men who judged these things from a literal or corporeal/sensual perspective. As a result, not all the canonized stories in today's Bible contain these deeper, inner meanings. Some of its stories are the pious works of mortal men but not written from direct Divine inspiration. Swedenborg identified those stories that represented the Lord's true Word and contained deeper narratives. They are listed here. Swedenborg maintained that as these deeper narratives are revealed to the world, they will provide the theology and doctrines that will help usher in the New Jerusalem on earth.

OLD TESTAMENT	NEW TESTAMENT
Genesis	Matthew
Exodus	Mark
Leviticus	Luke
Numbers	John
Deuteronomy	Revelation
Joshua	
Judges	
1 Samuel	
2 Samuel	
1 Kings	
2 Kings	
Psalms of David	
Isaiah	
Jeremiah	
Lamentations	
Ezekiel	
Daniel	
Hosea	
Joel	
Amos	
Obadiah	
Jonah	
Micah	
Nahum	
Habakkuk	
Zephaniah	
Haggai	
Zechariah	
Malachi	

All the above stories were Divinely inspired because they were the only stories that contained a three-tiered narrative and followed the universal plan of top-down causation. Note how much of the New Testament he left out. Even though they are embraced by today's Christians, they are simply the devout writings of men like Paul and James. The Second Coming will send shock waves, especially through the Christian Church.

Recall that the lowest level of Scripture is its literal meaning. Next is the spiritual meaning, which involves both the developmental stages and qualities of the Lord's Church on earth as well as the precise instructions for our personal salvation within the scheme of Conjunctive Design. The heavenly, or celestial level, deals with high Christology. The Christocentric narrative of Scripture depicts the hidden *inner* combats and temptations the Lord would face from men and from hell's hateful fury and deception while he lived in the world. As discussed in the previous chapter, the Lord's victory over every imaginable evil restored cosmic order and spiritual equilibrium, thus fulfilling Scripture from top to bottom (Alpha and Omega). By living out these stories in all their details, the Lord made the Word "flesh" and united His Human with His Divine Essence.

In scientific terms, the three basic levels of meaning in Scripture provide the top-down patterning principle in the *order of creation in the universe*, from God down to the physical universe. The higher, *a priori* level flows down and creates a lower discrete order and finally terminates in the time-bound meanings of human language. This flow is from *ends*, through *causes*, and into *effects*, that is, through discrete degrees which share no finite ratio to each other. Scripture not only consists of the dynamics of top-down causality, but its multi-tiered stories are subordinated and coordinated through successive, differentiated order and simultaneous, coexistent order. In this way, the future is always viewed by God as accomplished and present in each word and meaning of Scripture. In theological circles this is called *prolepsis*.

Everything in Scripture unfolds according to the flow of Divine Providence. Each word represents the "present" fully involved and connected with the future. In this way, God's Infinite nature correlates to and operates within finite things. Each word, sentence and story is a comprehensive whole, reflecting the same order represented in the entire Holy Word. I will offer a few examples of this in the final chapter. This holographic writing system requires infinite wisdom. Men and women cannot write in this profound way, and the inerrancy of Scripture cannot be found by merely grasping the literal meaning of its words.

Angels call God's Word the "great deep," because its inexhaustible wisdom and teachings flow out from the Infinite and the Eternal. For example, Swedenborg learned that the Lord's Holy Word makes use of four species of correspondences to deliver the Lord's multi-dimensional meanings. The first is *allegorical* correspondence in the sense that all things in the physical world are seen to represent heavenly things.

The second is *typical* or *historical* correspondences, in which historical events and historic individuals represent qualities of the Lord's kingdom. Swedenborg learned in heaven that the Bible refers to real historical figures starting from the time of Eber and real historical events from Abraham onward, including all the real content of *Joshua, Judges, Samuel* and *Kings*. However, Swedenborg was adamant that "The Divine Word cannot treat of mere men," and that men, "Considered in themselves, are not worthy to be even mentioned in the Word, unless by them are represented and signified such things as are in the Lord's kingdom: it is these things that are worthy of the Word."

The third type of symbolism is prophetical or fantastic correspondences, which are embodied in *Joshua, Jeremiah, Ezekiel, Daniel, Hosea, Joel, Amos, Obadiah, Jonah, Micah, Nahum, Habakkuk, Zephaniah, Haggai, Zechariah* and *Malachi*. These stories, which often refer to fantastic visions, make no sense unless they are interpreted through correspondences. These are fantastic because they make use of the symbols we might find in dreams. The symbols of *Revelation*, like the Great Red Dragon and pregnant woman standing atop the moon, are also examples of this symbolism.

(The *Psalms of David* represents a hybrid or intermediate style between the prophetical style and common speech. Swedenborg's theological writings showed that higher meanings were incorporated uniformly within the different styles, by different individuals, at different times. He pointed to this uniformity as powerful proof that such stories could not have been written "unless the Word had come down from heaven.") [11]

The fourth kind of correspondence in Scripture is *harmonic* correspondence. This style addresses physics, the multi-dimensional order of the universe, and its dynamical process. This form of correspondence, if humans could master it, would provide the means by which the Word could teach real science. John 1:1-3 tells us that all things in the world—time, space, law, symmetry, energy, matter, organic life, and the complexity of the human brain—were created from God's living Word. Both Scripture and the universe are therefore biocentric. To borrow a term from Francis Collin's *The Language of God; A Scientist Presents Evidence for Belief*, the Holy Word can be called the "BioLogos."

The notion that the dynamic "flow" of the stories in Scripture portrays the universal pattern of organic process, indeed, of every complexity, is a radically new theological concept. We can begin to grasp this from harmonic correspondences which provide the morphogenetic resonance between the multi-tiered narratives of the Holy Word and the multi-tiered structure of human bio-complexity. Swedenborg observed a most interesting example of Scripture providing the nexus between heaven and humans on earth through profound biodynamic conjunction. Because he was able to observe what took place in both the physical and spiritual worlds, he could see that when people on earth read from the Holy Word it had a corresponding harmonic effect in heaven. He witnessed that particular stories reverberated and resonated with particular communities and societies in heaven. As we saw in Chapter Eight, the unified utility, government, and commerce of heavenly society form a perfect analog to the organic functions of the human body and its nervous system. As a result, different biblical stories would correlate to different biological organs and processes of the Grand Human form in heaven! Considering the myriads of functions in the human body, it is remarkable to think of the biblical stories as a Divinely creative expression of this complex bio-order.

This confirmed to Swedenborg that the Divinely inspired biblical narratives provided the template for organic cohesion and whole-part process in the universe (as John 1:1-3 suggests). The universe and all its complexity were patterned after the multi-tiered architecture and complexity of the Sacred Word. Science and the Divine order of Scripture are one and the same thing.

Chapter Fourteen will offer a scientific model of this grand scheme of universal process and holistic relationalism. If Swedenborg was right, the Lord's "coming in the clouds" will disperse the intellectual obscurities of *both* science and religion.

SUMMARY

▶ God is One Being, not three Persons.

▶ Salvation includes our cooperation.
Faith alone does not save.

▶ The Second Coming is not a physical event. It took place
in the spiritual world in 1757 with the Lord's rectifying
the Christian heavens and dispensing new knowledge
concerning Him and the Holy Word. In connection with
this spiritual event, the knowledge of correspondences
is again being made available to humanity. But its
acceptance will be a slow process because it will
challenge humanity to think in new ways and threaten
strongly established worldviews and faith-systems.

▶ The Holy City, the New Jerusalem, represents new
doctrines issuing out from heaven from the Lord God.
These doctrines will provide a new kind of living
and habitation for the human heart and mind.

▶ The creation of a new heaven and a new earth
at the "end of the world" is not a cosmological
transformative event. It is a spiritual change that
takes place within the inner and outer qualities
of an individual's life.

▶ Not all the stories in the canonical Bible are
the true Word of God.

▶ God never suspends the laws of physics.
God is all law and order.

▶ The Second Coming will usher in a new science as
well—one based on the principle of spiritual Love.

PREDICTION

Salvation will someday be seen as
a higher expression of the laws of physics.

Elektrolýza

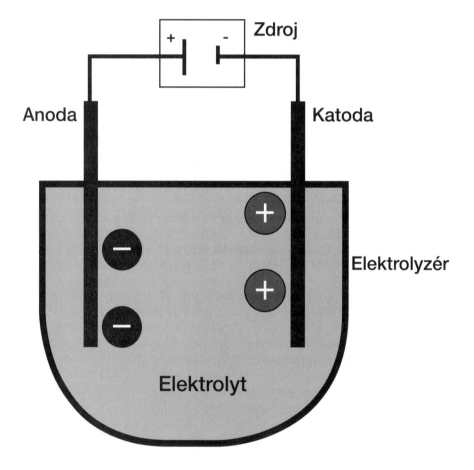

(What one loves determines the anode and cathode of one's energies)

Chapter Eleven

THE SCIENCE OF SALVATION

"God helps those who help themselves."
– Benjamin Franklin

Nature, if left alone, will cleanse and purify itself. This process in nature to restore order is a reflection of God's eternal purifying and saving activity. The process of salvation does not deviate from the laws of physics or from universal order but operates in a dimension where physical constraints are removed and the laws take on more expanded psychical qualities. Salvation is actually the extension of biological evolution and complexity into a pre-space, non-material realm. Expressed in theological jargon, this realm is called "Heaven."

Throughout the evolution of life on earth, we can see an obvious trend toward increased complexity and increased intelligence until we at last come to first-person phenomenal experience in the human mind. The human mind and its ideas are not in space. The fact that we can self-organize our knowledge into paradigmatic, coherent belief systems that *exist* and *subsist* from the assimilation of a continual supply of sensory data is evidence that higher-order structure and complexity operate beyond the constraints of the physical universe. Swedenborg maintained that mental subsistence (feeding on information) proved ideas were real, finite forms and substances and operated in a realm that was more substantial and fundamental than matter.

Natural selection plays no part in this non-local *directionality* of evolution. But selection of *values* does. What we value we assimilate, and this becomes the fabric of our being. God has provided humankind with a means for choosing the best values.

This special Divine means, by which radical novelty can be introduced into the theater of evolution, is called religion. Religion fills an ontological gap in human consciousness by bringing spiritual love into the equation of evolution and enables our mental forces and inner being to be further organized according to Divine order.* Our very birth into the world with such a cognitive gap ensures that we will enjoy free will and be able individually to *become what we love*. Being is perpetual becoming. Existence is perpetual subsistence.

* A person could not think, become rational, and finally attain wisdom, if thoughts and ideas could not be disposed (subordinated and cordinated) into some order. *Religion allows humans to choose values that will dispose their thoughts within a higher, Divine order. This ordering of ideas creates non-physical bio-complexity.*

Humans just don't die and go to heaven; their inner reality and its organic complexity has to be arranged according to non-physical heavenly values. Recall from Chapter Eight that there is no "coordinate" in heaven that is not some spiritual value. Heavenly placement necessitates that these values be consciously adopted into the hearts and minds of men and women. Human evolution is not simply an increase of intelligence. It is an increase of Love.

Love is the teleological engine that drives creation and evolution. In Conjunctive Design, ihe ultimate goal of religion is to provide a means by which the Creator can enter into and perfect a more intimate relationship with creation through the principle of Love. If God's Love were to stay a one-sided affair, the universe would represent an incomplete creation that could not fulfill the ultimate goal of exalting Love.

Love finds its fullness in reciprocation. There can be no ultimate conjunction between Creator and creation without conscious reciprocation in which Love is returned. This requires a purposeful plan by which increased organic complexity evolves into first-person cognitive function that can receive revealed truths and be led to conscious acknowledgement, acceptance, and a rejoicing of God.

Only through this special union is eternal life possible. That is why it is the goal of Divine Providence to create a heaven from the human race. This purposeful design actually holds the key to the deeper understanding of time that physicists and philosophers are seeking. This was discussed previously in Chapter Three, but is worth returning to here because it shows how spiritual evolution and salvation are extensions of the laws of physics.

Time is essentually dynamic circular process. Cosmic process is an *involution* towards matter and an *evolution* from matter. The arrow of time moves out from God's spiritual and conscious first principles. It flows into the fixed objects and holonomic activities of the physical world, then swings back toward conscious first principles as the human spiritual mind evolves and chooses to reciprocate with God. This allows Love to find its fullness in a grand transcendental circle (see Figures 3.5 and 14.15).

The completion of this great cosmic cycle of action allows dynamic process in the universe to maintain self-consistency as it continuously corrects itself in the *proleptic* trajectory that connects the past with the future. In this way, the "present" is always oriented towards and linked to God's infinite and eternal action. The universe anticipates and relies on human spiritual evolution to hold everything together and perfect this cosmic nexus. Holonomy (the law of the whole) is established by God's relationship with a human race that is moving towards a heavenly life and back to the Creator.

Purposefulness and conscious control in creation is supported by the fact that action occurs in wholes (quanta). Furthermore, Swedenborg's idea of universal process represents a true quantum cosmology because the full picture of dynamic flow transcends classical determinism. It is the determination of Love, whose holonomic trajectory emerges out from a non-local realm into the constraints of classical spacetime and returns to the non-local and non-material realm. This explains the asymmetry, or one-way-ness of time. If time were to flow backwards God's Love would unravel and undo itself. Again, the process of creation and evolution is a trajectory by which Love exalts itself through increased organization whereby created forms can become more responsive to God.

This grand circular scheme is only possible if conscious Love is the ultimate substance of reality and the law-giving agent in creation. Love is psychical and infinitely more substantial than matter. And there is even a scientific way to consider the ontological status of Love as primal substance.

Neuroscientists know that every part of the human body co-responds and answers to specific neurons in precise areas of the brain and that different areas of the brain relate to specific activities of the mind. All activities of the mind answer to intentions, appetites, and volitions that are *derivatives* of Love. This means that the human mind, brain, and physical body are all extensions of different qualities of Love into different levels of structure. Since the various characteristics of Love come from qualities in heaven, the human form originates in the spiritual world, not the natural world.

There would be no top-down contiguity between the physical body and the human will if Love were not a real living substance with the power to determine physical structures and produce corresponding physical actions. Grasping things with the hands, digesting things in the stomach, moving from one place to the next, all have their correspondence in some function of conscious Love that grasps and digests ideas and moves about from one idea and intention to another.

While the concept of Love as a real and primal bio-substance is not testable in a laboratory according to the criteria of methodological naturalism, the potency of Love as an active agency does fall within the inner empiricism of human experience and consciousness. According to Swedenborg, personal salvation represents "ground zero" in this personal-level experience. For humans to evolve, Love must evolve.

▶ An Interested God

As we saw in the previous chapter, God has intervened in human history in significant ways, especially to form new churches on earth, each adapted to the special genius of the men and women at those particular times. These interventions did not occur through the suspension of the laws of physics, because natural laws originate out of a spiritual domain and are thus responsive to it. This compatibility allows the laws of physics to be augmented as spiritual laws become more manifest in time and space. The human brain and mind are perfect examples of the laws of physics being augmented by non-local and non-physical forces.

We learn from Swedenborg that there is another intervention taking place and a new church is being formed now on earth. This is the long-awaited New Jerusalem descending from heaven! The bad news is that many of our theological ideas may have to fall to the wayside as this church is ushered in. The good news is that many people who are unfamiliar with Swedenborg's theological writings instinctively sense that there is no adequate theology for solving today's challenging social issues or offering a rational response to the discoveries of science, including the startling theories of the New Physics. The doctrines of the New Jerusalem offer a theology ready-made for men and women who welcome increasing their cognitive abilities and sincerely desire to seek truth. At the same time, this new dispensation from the Lord God will not segregate or deny anyone salvation who remains loyal to his or her current spiritual traditions.

The knowledge of correspondences is a tool for helping us mine deeper meaning and wisdom from the Holy Word, but it is not essential for salvation. Swedenborg stated that the need to inform humans about how they are to be saved and gain a life of blessedness is so important that it is transparent even in the sense of the letter. And, what we need to do is quite simple: *acknowledge the Lord God as the one God,* and, *love God and our neighbor.* The Lord tells us that all the laws of the prophets hang on just these two principles.

Hear, O Israel; The Lord our God is one Lord:

And thou shalt love the Lord thy God with all thy heart, and with all thy soul, and with all thy mind, and with all thy strength; this is the first commandment.

And the second is like, namely this, Thou shalt love thy neighbor as thyself. There is none other commandment greater than these. (Mark, 12:29-31)

Much of Christianity has strayed from this revealed strategy, maintaining instead that faith alone saves. Without the works of the law, this doctrine proclaims that through mere "belief," we are spared from personal responsibility. However, the life of religion is in doing *good* from a spiritual principle of sincere mutual love. The purpose of God's revealing deeper levels of meaning in Scripture is to expose and remove those incorrect ideas about salvation that corrode goodness. These deeper levels offer greater enlightenment for those who are willing to roll up their sleeves.

Loving God and our fellow human beings indeed allows us to meet all the requirements for fulfilling the purpose of Conjunctive Design. But as we all know, God is in the details. If loving God and the neighbor were easy to do, the world would be a much different place. In actuality, they are often the hardest things to do. Something always gets in the way. The dispensation of the Lord's next great church on earth will provide a helpful map with *details* for understanding the obstacles we each *personally* face and the steps by which salvation and eternal happiness can be gained.

▶ The Details of Spiritual Regeneration, or Epigenesis

Swedenborg learned that the process of salvation begins in infancy and early childhood, even though we are not aware of it. Feelings of comfort and protection from the warm embrace and physical closeness of parents, soft blankets, playful talk, and constant fussing are stored away deep within the child's psyche and remain there protected by the Lord God. These most innocent experiences are also perceived by angels and keep the child inwardly connected to heaven. These early experiences of love first open the unformed mind to intellectual curiosity, promoting alertness and the affection to learn. This is why it is so important for children to grow up in a loving environment.

Neuroscience may confirm this. The proper development of neurons in a child's cerebellum and its connection to the emotional centers of the limbic system has been linked to physical cuddling and early loving experiences.

The best way for God to protect these tender experiences and precious moments of love and peace is to keep them in the involuntary processes of the cerebellum and away from the voluntary processes of the cerebrum, which is prone to succumb to the allurements of the world. Swedenborg called this storing away of heavenly experiences the *remains*. These remains form a plane or matrix by which spiritual evolution can take place in later life, when the mind has had time to form its likes and dislikes and shape its subjective reality and conscience. The Lord God first attempts to *move* us beyond worldly concerns and stir up spiritual feelings within us through these deeply stored remains.

Swedenborg learned in heaven that the opening chapters of Genesis, when interpreted through their deeper meanings, contain important but otherwise invisible details about our inner situations and the processes involved in spiritual evolution. Each of the seven "days" represents a distinct step in the dynamic process of *epigenesis*, that is, the precise science of our spiritual re-creation through a new partnership with God.

When this process of spiritual regeneration commences, remains, which have been under the protection of the Lord, can serve as the "womb" in which a new gestation and birth can take place within our deepest being. Understood through the lens of correspondences, the Genesis story becomes the roadmap to our personal epigenesis, in this post-modernist world.

Each day of the Creation Story symbolically refers to a scientifically precise stage of our inner evolution because it addresses the emergence of a more complex bio-system and *new concepts of gravitational order*. These stages of spiritual growth represent the ascent of the human heart and mind in which what we love and think about become increasingly vivified by God. As we adopt nobler values in our lives, our affections and thoughts find their center of gravity and equilibrium in higher spheres of dynamic reality. These higher principles of love represent the equilibrium end-state of a new gravitating system that will grow *angelic* structure. Spiritual growth moves towards maximum negative entropy, that is, increased order. As we are about to see, this increased order takes bio-complexity into a non-physical realm.

Love adapts information to its own disposition and complexity in the same way physical gravity turns the motion of stars and planets into complex organization. One force creates coherent structure and order in the mental universe, the other force in the physical universe. The process of spiritual evolution is a real ascent of organic structure and complexity into a higher sphere. It starts when new spiritual feelings are aroused in us from our remains.

▶ **Day One**

In Genesis, this awakening of deep spiritual feelings within us, is depicted by the words: *and darkness was upon the face of the deep; and the Spirit of God was moving over the face of the waters* (Gen. 1:2). This operation results when God reactivates the remains stored in our *unconscious* mind, about which we are in complete "darkness." God's love flows into the cerebellum, where it then continues on through neural connections to the emotional centers of the Limbic system. This Divine activity, called the "spirit of God," stimulates a new yearning inside us.

The phrase "Let there be light," refers to a new state of cognition, enlightenment, and understanding within the cerebrum as a response to new influences from the Limbic system and cerebellum. Men and women begin to perceive that goodness and truth actually come from a higher, transcendental source. *And, then they now seek this goodness in their lives.*

Everything in this hopeful beginning hangs on the human free will. Free will has to do with all matters of love. Love has no meaning without free will. Similarly, relationship with the Lord has no real meaning unless we use our free will in spiritual matters. As Swedenborg points out, "This involves applying the mind to the consideration of the life after death."[1] To borrow a phrase of Philip Hefner[2], we must act *as created co-creators* within God's purposeful scheme. While some theologians might argue that individual effort takes merit upon oneself and disrupts salvation, this objection is overcome if the individual approaches the Lord with humility and acknowledges that he or she cannot accomplish anything without the Lord's help. The *inner* interpretation of the Creation story accords with Conjunctive Design.

▶ Day Two

The second stage of our epigenesis is described in Scripture as a "separating of the waters." This refers to a new cognitive ability for spiritual discernment. Recall from Chapter Seven that in the quantum language of correspondences, water represents memory-data. A person seeking real inner growth is now able to distinguish the memory information that is important for worldly life from that which is vital for spiritual life. This separation of knowledge is the first step toward putting values into a proper hierarchical order. Again, since heaven is not a place we go but something we become, we must begin by putting our inner world of heart and mind into heavenly order.

▶ Day Three

In the third stage, the *remains*, which have been protected all this time by the Lord, can now rise to the surface as "dry land." Our spiritual life is now given a proper foundation on which to be built upon.

Swedenborg refers to this stage as *repentance*, the natural next step following *discernment*. By identifying the negative aspects of ourselves and making the effort to remove these hurtful character traits, we seek a return to innocence. The tender experiences of our childhood innocence now rise up out of our unconscious mind and begin to manifest in our conscious mind. Spiritual principles cannot form the inner ground and foundation of our inner life without innocence.

Repentance is a familiar term to most Christians, but few would guess that without taking this most important step, not even the most well-meaning people could actually love God and their neighbors in a way that is spiritually helpful to them-selves. Here, Swedenborg adds an interesting new wrinkle in the understanding of doing *good*. Becoming a better Christian is not simply a matter of deciding to be good to others. The first principle of love to God and the neighbor is actually to shun evil, and the second is to do good, for "In so far as a man shuns evils as sins, he desires the good of love and charity."[3]

Performing good acts before repentance is like pouring distilled water into a bucket already filled with dirt. Repentance allows the dirt to be removed so that the pure water can be added without contamination. Similarly, when we reject evil and remove it from our lives, the Lord's goodness can start flowing in without being perverted. *Only through rejecting evil, can purity and true innocence enter into the equation of doing good.* Repentance makes love legitimate, not simply surface dressing, or a means to promote one's reputation in the world (which is self-love).

Swedenborg describes repentance as "examining oneself, recognizing and acknowledging one's sins, praying to the Lord and beginning a new life." He states further that, "There can be no repentance for a person without self-examination," and that "examination must lead to a recognition of his sins, their recognition to the acknowledgement that they are actually in him."[4] Swedenborg observed in the spiritual world that those of the reformed church, who favored justification by faith alone, dreaded introspection. Catholics, who regularly practiced con-fession, showed less resistance to finding dirt.

If we do not seek out the real motives of our thoughts and actions how can we be certain that our good works are good? Self-discovery is impossible without the willingness and *con-sent of the heart*. When we explore our inner reality, nothing negative will show itself unless we have a true desire to find it. The eye will not see what the heart does not allow it to see.

Once our adverse motives are recognized and identified we can confess them before the Lord. Swedenborg named the three duties of repentance: confession before the Lord, prayer for God's help, and the conscious willingness to begin a new life. Since Conjunctive Design relies on human free will and cooperation, the world is made better one person at a time. For kingdoms to change, men and women must change.

The emergence of dry land in Genesis signifies the emer-gence of a new mental plane within the human psyche that allows higher-leveled structure and complexity to grow. In repentance, we put pressure on ourselves and what we know to be right. This mental pressure causes our spiritual knowledge to become concentrated into more fixed and stable forms. Spiritual principles become *crystallized* within our mind's psychoscape, putting our aspirations on "solid ground." An individual's efforts at repentance give the Creator a fertile mental plane and soil on which the seeds of novelty can be sown.

The appearance of plant life in the Genesis story is a metaphor for the spiritual evolution of the human intellect and its increasing complexity. In Chapter Eight, we noted that plants in heaven represented the lush spiritual thinking of those who perceive God's truths and doctrines. Plants represent new life taking hold in our psychoscape and the cultivation of a new intellect. Plants represent the growth and blossoming of spiritual intellect and reasoning. Plants are mentioned at this stage of the process because we need our thinking to be reorganized before we can advance. The act of *repentance* leads us to a state of intellectual *reformation*.

I cannot overstress that Scripture contains real science. Nature and mind follow the same laws and patterning sequences. The evolution of life on earth follows the same steps as the evolution of spirit. The symbolism and correspondences in Scripture portray ontological reality, because at each upward step of spiritual evolution, the heart disposes the contents of the mind into a new equilibrium end-state that is closer to heavenly order.

Just as plants rise up and exalt the mineral kingdom by reconstituting its material into higher-level orders of structure, the progression of *faith*, rooted in the "dry land" of sincere spiritual principles, is reassembled into the structure of noble thoughts and perceptions. Through our thoughts and intellect, the principles we espouse are turned into living forms. For instance, the principle of "love to the neighbor," while noble, remains a static and inert principle until it is raised to a vivified and dynamic form when the human intellect thinks of actual ways to accomplish good deeds for others. As we contemplate utility and usefulness, our adopted principles literally "come alive" in our inner world, just as the mineral world comes to life and increases its dynamics in the organized bio-arrangement of plant structures. The principles we nurture grow, mature, flower, and bear fruit according to organization and complexity produced by *determined love* in the human intellect.

The evolution of a "plant kingdom" within the mind's landscape represents the maturing and quickening of spiritual faith. The power of faith is not to be measured by ignorance, but by the intellect's power to understand rationally the truth behind spiritual principles and doctrines.

▶ **Day Four**

Plants need light to grow vigorously and perform the work of photosynthesis. The inner evolution of our spiritual intellect and its corresponding work of *idea-synthesis* also requires a reliable metaphysical light that shines within the mind. In the fourth stage in our spiritual evolution, the "luminaries" enter into the Genesis story.

These luminaries represent the psycho-energetic influences in our inner world that come from "above." This is *enlightenment* from the Lord. The lesser luminaries in Genesis, like the moon and stars, represent heavenly light sources when our minds are in states of relative darkness concerning spiritual truth. These heavenly influences first reach us either from a distance, like stars, or indirectly, like the moon. Since the moon reflects light from another source, it represents our intellectual reflection on ideas of faith coming from a greater source. The "lesser" luminaries were created by God to "rule the night" and serve as moral guides during our inner states of spiritual darkness, just as stars helped ancient sailors to navigate at night.

The greater luminary is the "sun," which corresponds to the Spiritual Sun discussed in Chapter Eight. It turns things from night to day in our inner world. Intellectual faith by itself exists in relative darkness in the head compared to the brilliance that comes when that faith is combined with the heart and becomes love itself. This greater luminary represents love manifested in faith, because the sun is a source of both *light* and *heat*. Metaphysically, heat represents the spiritual warmth of love. When the Spiritual Sun of heaven becomes a fixed and permanent influence in our inner world, we begin to feel the real truth and love of the Lord's spiritual light and heat.

As spiritual light flows into our intellect and spiritual love flows into our will, our inner evolution expands into a new dimension. Re-creation is now moving its center of gravity from our minds to our hearts. We begin to do things not just because they are the right things dictated by the principle of truth, we do them because they are good things dictated by the principle of love. New feelings and affections begin to stir inside us.

▶ **Day Five**

In the fifth stage, the focus of re-creation continues to move from the mind to the heart. This is called *regeneration*. Our inner world now becomes more animated by the faith that comes from love. This stirring of the heart and increased animation within our inner world are represented in Genesis by the "creeping things, which the waters bring forth," the "flying fowl," and the "great whales."

As this symbolic narrative of our inner evolution unfolds, we are offered further evidence that the quantum language of Scripture is self-consistent and not arbitrary. Recall from Chapter Seven that water represents memory-data. Fishes, creeping things (which crawl out of the water), and whales portray all the new activity that develops and grows in our memory from this new spiritual animation. This new activity in our memory is the cognitive function of *imagination*. The imagination also provides food for the higher, rational mind, represented by flying foul. Birds can make a meal out of creeping things and eagles catch fish. This portrays, in a creative symbolic way, that human rationality feeds off of ideas in the memory. In the same way that the fish, once eaten, is reconstituted in the higher-level bio-structure of an eagle, the imaginative ideas of the memory are reconstituted in the higher-level complexity of human rationality. The rational mind soars above the memory in the same way that birds soar "above the earth." When our rationality is based on spiritual principles, the mind soars above corporeal and worldly ideas and takes in the bigger picture. This bigger picture is also represented by the "great whales," those larger and comprehensive ideas about reality that form in our memory.

▶ **Day Six**

God commanded in Genesis that all these creatures be fruitful and multiply. This signifies that our new spiritual faith and love should increasingly vivify and animate our inner life. Vivification in the human heart and mind continues to intensify into the *sixth* stage of epigenesis, where more complex and warm-blooded animals now emerge. Recall from Chapter Eight that animals represent affections and volitions. The higher thermodynamics of warm-blooded animals, such as mammals, represent love that has reached a higher order of complexity in our inner world.

The introduction of a woman and man into the Garden of Eden represents the ultimate partnership and union between the heart, or *will* (our female side), and intellect, or *understanding* (our male side). True humanness is attained at this stage.

To be human is to have free will and discernment through which love and truth can be intimately conjoined, like a *marriage.* Swedenborg heard directly from the mouths of angels[5] that all created things in the universe are a marriage of good and truth, that is, *essence in union with form.* Each has its existence from the other. Divine Truth is the form by which Divine Love exists, just as understanding is the form or body in which the affection lives. Through this sacred union between disposition and information in all created things, the Creator has an eternal covenant with the universe. This covenant is perfected in the inner evolution of the human race in which spiritual love forms a union with wisdom. A "garden" in Scripture represents wisdom or a mind arranged according to heavenly order.

At every stage of human development, learning requires that a certain quality of love enter into our cognitive function. As we saw in Chapter Six, every quality of love has a center of gravity at some location in the hierarchical scaffolding of human cognitive architecture. As each distinct love adapts and disposes information according to its disposition, it brings new order and structure into our inner reality and personhood.

▶ Neuro-Nuptials

During childhood, there is at first a love of knowledge. This affection conjoins itself to data on the level of the human *memory.* Next comes a love of understanding what one knows, which allows memory data to be abstracted and reshuffled through the development of human imagination. Above the *imagination* is a love for perceiving and judging truth, which is human *rationality.* In each case, an affection or quality of love generates a distinct cognitive function and conjoins (marries) itself to a higher order of knowledge. Further evolution between the heart and mind is possible, but this can only come from God.

Beyond natural or human rationality there is spiritual rationality, which arises from the affection for spiritual truth. Unlike the natural-rational cognitive function, the spiritual-rational mind concedes that all love and truth comes from

God. This *humbler* state of mind, which understands that truth comes from a higher source, is stimulated when God reactivates our remains and we respond. Only from humility can we take part in the deeper Genesis narrative and begin the six-stage process of spiritual re-creation. This elevated rationality comes from a sincere spiritual affection and not from human sagacity or the atheistic inclinations of the natural moralist. In God's plan for Conjunctive Design, spiritual rationality is the only true and *legitimate* rationality for humans.

The quantum language of Scripture reveals a deep and unexpected knowledge of human psychology. For instance, the formation and comparison of our natural and the spiritual rational minds is implicit in the story of Abraham's two sons, Ishmael and Isaac. Abraham's first son, Ishmael, was born to Hagar, a maidservant, rather than to his actual wife, Sarah. A maidservant signifies a lower level of love that can be of service to spiritual love, but only if it humbles itself to a higher principle. According to Swedenborg, Ishmael represents the birth of the secular, rational mind. But Ishmael is not born until after God asks Hagar to "humble herself" toward Sarah, which indicates that the love for the natural secular mind must eventually give way to the love for a spiritual mind. Sarah finally bears Abraham a son, Isaac, and this symbolizes the birth of a more elevated cognitive function through the conjunction of a higher, legitimate love with knowledge in the human intellect. Isaac represents the formation and birth of the spiritual rational mind, which *follows* the birth of the natural or secular rational mind represented by Ishmael.

At one point Sarah saw Ishmael mocking this new situation and implored Abraham to cast out both Hagar and Ishmael. Psychologically, to cast something out is to recognize that the spiritual rational, as embodied by Isaac, was more important than the secular human rational Ishmael and secular human rational love, or Hagar. Ishmael's mocking represents the disregard secular reasoning often has for God and religion (Gen. 21:9,10).

The conflict between Sarah, Hagar, and Ishmael illustrates the internal conflict within us among our *values*. Like Abraham, we must make value judgments. What quality of love do we ultimately embrace, the worldly or the spiritual? This personal predicament and deep psychology hidden within the literal words of Genesis is further evidence of the authority of Scripture and God's Infinite Wisdom. Who would expect to find our own story within the narratives of the Holy Word? The human race is free to choose its *ruling love*, but only religion offers to shepherd us and keep us from going astray.

Since the time of modern humans (discussed in Chapter Nine), and God's providential redirecting of our evolutionary path by enhancing the frontal and prefrontal lobes, we have gained the ability to elevate our understanding and intellect above our proper love. It is now possible for people to contemplate the negative consequences of acting on selfish and hurtful desires and even to contemplate a noble alternative. This creates conflict and friction between two distinct cognitive levels. If a person complies with the dictates of a higher, more elevated idea, the resulting friction creates a *psycho*-thermodynamic "heat" that can fuse together a new and permanent quality of being.

Along with the intellect, one's *will* becomes elevated to a new level and inner stability results. This "psycho-equipoise" is the conjunction of love and its proper knowledge. It allows a person to actually live by one's professed tenets. Without such a fusion, we would remain at the mercy of a constant fluctuation of desires, cupidities, and urges that seduces our thinking and distracts our reflection to the point that we are only aware of their pleasures, not their consequences.

The fusion of love with the intellect involves the real formation of bio-structure. Swedenborg takes the science of complexity theory all the way into a non-physical realm. Spiritual regeneration involves novel organic change and real structural growth. Inner growth is more than neurons simply making new synaptic connections. It makes connections to the spiritual love that is a creative, formative substance in the deepest architecture and scaffolding of the neuron.

The multi-layered structure of the neuron houses a hierarchy of loves or dispositional forces that ranges from the corporeal to the spiritual. Each level of love has an affinity for different qualities of knowledge. Ideas and thoughts are real forms and substances that receive our affections in the same way as the ear receives sound. They provide the non-physical building material with which various levels of love, when appropriated, can manufacture coverings and membranes for the evolving spirit's new anatomical design. This is radical reverse entropy.

Information, ideas, and truths clothe our subjective loves and affections in a manner not unlike the way polymers, proteins, and organic molecular motors are formed in the physical body. They have negative entropy. Ideas grow into chains and side chains, allowing specific and determined principles of love to create a non-physical organic system and machine from their own characteristic patterns of correspondences. This new bio-system, or spiritual body, evolves out of our life-choices in the world to become capable of operating in the spiritual realm after death and finding its proper environment according to its specified utility.

Ideas have physical analogs in molecular structure, because both are forms which carry *information* and have the profound propensities to bond, grow into new coherent structure, and *move*, or be directed. One's intellectual ideas are reshaped during spiritual reformation in a manner similar to the process of ionic bonding. During spiritual regeneration, by contrast, love and knowledge form a more permanent relationship that resembles covalent bonding.

Throughout life, as we choose our values, nurture the ideas supporting these values, *and act on them*, the spiritual body we are molding from within assumes a more nearly perfect human form. Since *free will* and *discernment* are the two essentials of human form, the elevation of love and mind perfects this form. The organic complexity of the spiritual body within different individuals depends on the stratum in our hierarchical cognitive architecture in which the center of gravity of our ruling love resides. Our inner bio-complexity can be organized from corporeal/sensual, rational, or spiritual love.

Actual physics is involved here, because the level we most occupy and favor in this tiered, cognitive system determines the flow of forces and direction that our inner organization will take. Our *polarity* can be aligned to either worldly or spiritual concerns.

A well-known electrolysis experiment in high school science classes shows that two nails will exchange their metallic "coating" according to which nails are connected to the positive terminal. This gives a visceral idea as to how the spiritual body is formed within us from what we love. By elevating our minds we can put the positive terminal of being above our normal, everyday desires, and into a higher, discrete realm where spiritual principles of action reside. If we then accept these new principles in our hearts and minds, our living forces, like an electric current, will flow and begin to coat and fashion new subtle structures according to our new orientation.

Both the corporeal-sensual and secular-rational minds become the negative terminals of our being when spiritual evolution is going well. But during our spiritual combats and temptations with our personal demons and compulsions, our lower nature, which is oriented to worldly things and the ego, struggles to become the positive terminal. It fights for dominance and rule by making the spiritual mind passive and changing our polarity. The positive terminal is always determined by the true focus of our lives and the *center of gravity* of our being. Spiritual growth is the electrolysis of values and ideas. The level of heart and mind that wins these inner confrontations subordinates and coordinates all the others to form its own bio-structure.

Since salvation does not take place apart from neurological and cognitive science, another important element for the Lord's ushering in the New Jerusalem is to reveal that the Holy Word contains a model of the human cognitive scaffolding and the theological basis of its neural substrate. (God wants us to be able to unify science and theology because it furthers His cause.) This is why Noah's Ark had three levels representing the discrete hierarchical levels of human cognitive function and the vertical order that enables us to evolve up its different floors and gain buoyancy in the deluge of daily disturbances that inundate the habitual mind and drown off the soul. Each discrete level of Noah's Ark represents a higher degree of negative entropy. Evil can be understood as the reversal of this cosmic order and the process by which the cathode becomes the anode.

Changing our personal polarity requires us to implore the Lord's help. Why? During our spiritual reformation and regeneration, we are not just confronting ourselves or able to make the fight alone. Swedenborg observed that our inner nature is so complicated and the influences from hell that are flowing into our hearts and minds are so numerous and interconnected that only the Lord could put things into order, as He did when He lived physically in the world and gained power over all flesh and its negative inclinations.

"We know absolutely nothing about the inner state of our minds; yet there are infinite things there, none of which comes to our awareness. The inner working of our thought or our inner self is our actual spirit, and there are infinite elements there, innumerable elements, just as there are in our bodies. In fact, there are even more, since our spirit is human as to its form, and there are elements in it to answer everything in our bodies.

Now, since our senses tell us nothing about the way our minds or souls are at work, both together and separately, in all elements of our bodies, we do not know how the Lord is at work in all the elements of our minds or souls, that is, in all the elements of our spirits. This activity is constant. We have no part in it; but still the Lord cannot cleanse us from any compulsion to evil in our spirits or inner selves as long as we keep the outer self closed. Each of the evils that we use to keep our outer selves closed seems to be a single item, but there are infinite elements within it. When we dismiss it as a single item, then the Lord dismisses the infinite elements that it contains.

This is what it means to say that the Lord then cleanses us from the compulsions to evil in our inner nature and from the evil practices themselves in our outer nature."[6]

Within Conjunctive Design, we need the Lord's help and the Lord also needs our help. We need to self-examine and recognize hurtful tendencies within us as best we can so that we do not allow them to become deeds. This self-examination is a sign to the Lord about the evils we really wish to be expunged from our innermost lives. When we prevent the full manifestation of hurtful tendencies on the outer worldly plane, the Lord works secretly within us to help remove our inner compulsions by subduing those temptations at their source, the hells. Alone, we can indeed prevent evil, but we cannot remove the actual compulsions and inclinations without God's help. Swedenborg further pointed out that if we banish destructive deeds other than because they are real sins against God, we are not really abstaining from them. "We are only preventing them from being visible in the world."[7]

God does not want to be approached and loved for His own sake, but for our sake and for our eternal life. It is the Lord who fights for us and actually removes the evils from within. As we saw in Chapter Nine, humanity would not stand a chance against these evils unless the Lord had been victorious over the hells while living in the world. His Divine intervention restored cosmic order and human free will. Free will necessitates that the positive and negative influences that flow into the mind from the spiritual world be kept in equipoise (We will discuss this in the next chapter). Swedenborg learned that if the Lord had not come into the world to restore cosmic order, human morphology would have taken on monstrous and hideous forms and the human race would have ultimately perished. True humanness can only come from the Lord.

▶ Day Seven

Once regeneration, the sixth stage in human epigenesis, is complete, God's work of salvation is done. The seventh day or "day of rest," means rest from spiritual conflict. It does not mean rest from usefulness or service. Heaven has its joy and blessedness in usefulness from mutual love. The purpose of the specified complexity in the spiritual body that forms during this entire inner process is to allow men and women to continue to be useful to each other in an eternal, non-material realm. Personal happiness and self-worth cannot survive death except through continued usefulness and love for others. All unity and utility is reciprocal union.

Living according to Divine precepts takes negative entropy beyond the realm assumed by scientists. Religion is God's strategy for humans to personally overcome the tyranny of time and death. Spiritual evolution is an adaptation not to the physical environment but to the influence of the Spiritual Sun, the emanations of Divine Love and Truth, so that they can be received and nourish our spiritual bio-structure. In the same way that the organic structure of the ear provides a wave guide for auditory signals and the eye for electromagnetic signals, the innermost structures of the brain and the mind serve as wave guides for spiritual signals. Like the ear and eye, the human mind becomes attuned to things we find delightful and harmonious. Those who populate the hells are constituted internally such that they do not resonate with the Spiritual Sun. They turn away from its influence, which they find discordant with their ruling loves. God judges no one; His love condemns no one. He loves those in hell as much as He loves those in heaven. But they do not reciprocate.

Swedenborg stated that the human mind, which is spirit, perpetually strives to give form to its loves and cognitions. We not only need to take special notice and care of our life-choices and recognize that these choices determine the spirits with which we associate in the other world, we also need to understand that we are continuously creating the bio-structure and abode of our future.

In Chapter Fourteen, we will see that the process of spiritual regeneration mirrors the *seven-step* patterning principles and laws that create comprehensive wholes and complexes in the universe and enable all things to be seen from a universal science. The Holy Word contains this universal science deep within its spiritual meanings and sacred architecture. God is order itself. This seven-step order depicts gravitating systems finding a common equilibrium. That is why the symbol of a *rainbow* was made known to Noah and appears in Ezekiel's vision. It involves real physics. A rainbow's seven colors find their common equilibrium in "white" light just as the seven stages of spiritual process find a common equilibrium in a regenerated man or woman. This Divine order is in all process, whether it be physical process or the process by which the Lord makes a covenant with humankind. This Divine patterning principle or template portrays the very structure of intelligence and mind in the universe. It is the science of Love.

The deeper levels of meaning within the Holy Word offer up new layers of revealed truth and provide a more adequate theology for the post-modern world. Not everyone will choose to take part in the Lord's ultimate goal of creating a heaven from the human race.

Why can't God just prove He exists? Why can't God simply remove all suffering and fix everything with a snap of the fingers?

What is the origin of evil and human sin? Does this imply some imperfection on God's part or with the Divine scheme of Conjunctive Design?

SUMMARY

▶ Salvation extends scientific law and the evolution of bio-complexity into psycho-spiritual realities. Spiritual growth requires the creation of new non-physical organic structure.

▶ The deeper meanings of Scripture reveal this science of the heart and its proper evolution.

▶ Faith alone does not save. Faith must be put into action, which is *Love to God and the neighbor.*

▶ Salvation requires our cooperation with God.

▶ Hell is a real force to contend with.

▶ We are Loved. God takes personal interest in our eternal happiness.

PREDICTION

Scripture will someday be seen as providing the template for all dynamic process in the universe.

Lucifer, King of Hell by Paul Gustave Doré, from Dante's *Divine Inferno*

Chapter Twelve

THE
THEODICY
ISSUE

"Is God able to prevent evil, but unwilling?
Then He is not benevolent. Is He willing but not able?
Then He is not omnipotent. Is He both willing and able?
Then whence all the evil in the world?"
                                    – Alfred Russel Wallace

Swedenborg's answer to the question of how an infinite God could act in all the finite events of nature and all events in human history is Conjunctive Design. The Creator's ongoing governance connects all things from His eternal ends. This is radical teleology. All nature consists of forms that are recipients of God's influx (active information). God is principle and nature is instrumental. Every finite thing was created to have a relationship with the Infinite, and every finite thing is connected simultaneously with everything else through the pre-space principle of spiritual love. All this expresses the dynamic process that connects the whole of everything through love's endeavor to *unify* through *utility*. All unity and utility (complexity) is reciprocal union. The multi-dimensional universe maintains its lawful regularity through unified functional properties that exist to promote an ultimate, eternal conjunction with God.

The creative action of the Creator-Architect can only flow downward into forms of utility and usefulness. The living force of Divine Love cannot flow into something which serves no purpose. Love cannot create through dissimilarity, but through correspondence with God's goodness. The human race, in which organic specified complexity is elevated to more nearly perfectly mirror God's image and likeness through *free will* and *discernment*, represents the greatest similarity to heavenly things and the greatest potential for conscious reciprocation with God.

> "The universal end, that is, the end of all things of creation, is that there may be an eternal conjunction of the Creator with the created universe; and this is not possible unless there are subjects wherein His Divine can be as in Itself, thus in which it can dwell and abide. In order that these subjects may be dwelling-places and mansions of Him, they must be recipients of His love and wisdom as of themselves, and conjoin themselves with Him. Without this ability to reciprocate no conjunction is possible. These subjects are men and women, who are able as of themselves to elevate and conjoin themselves. That men and women are such subjects, and that they are recipients of the Divine as of themselves, has been pointed out above many times. By means of this conjunction, the Lord is present in every work created by Him; for everything has been created for man as its end; consequently the uses of all created things ascend by degrees from outmosts to man, and through man to God the Creator from whom are all things."[1]

The fact that God creates and maintains the universe through Conjunctive Design raises the red flag of theodicy and the dreadful issue of evil in the world. How does evil come about if a God of Love cannot create anything dissimilar to goodness? Is God responsible for evil, and not infinitely benign, or is He unable to do anything about it and thus, not omnipotent?

Swedenborg rejected the Deistic solution to this problem, which is to declare that God created the world, turned away, and let it simply run by itself. Deism leaves the door open for bad things to happen. Swedenborg asserted that the physical universe could not run on its own. All physical nature is inert and passive to the living active agency of spiritual energy; it therefore could not be self-organizing or self-maintaining. In the spiritual world, Swedenborg observed that only God is life and therefore life is uncreate. All organic complexity in the natural world is a receptacle and container for God's living force of love, a force that emerges from a pre-space realm.

So, how does evil come into the equation of Conjunctive Design in the cosmic order of a universe that reflects God's attributes? How did evil begin? Just whom do we blame?

Swedenborg stated that all theology must be extracted from the Holy Word. When we look to Scripture for the origin of evil, we find it being addressed in the Genesis story of Adam and Eve. The actual original sin, which was so damning for the human race, seems on its face value to have been a rather trivial transgression—the eating of a particular fruit from a particular tree. Damnation and the resulting "fall of man" comes across as a rather harsh punishment for such a minor act of disobedience, especially since Jehovah God went out of His way to protect Cain, who committed murder. Here again the knowledge of correspondences and its quantum language comes to our aid.

Recall from Chapter Eight (pp. 158-160) that the Garden of Eden represented a pristine state of wisdom rather than a physical location. Likewise, Adam was not a particular individual in human history, but was the name given to the Most Ancient Church that enjoyed wisdom in its most innocent form. To be removed from the garden was to be removed from this innocence and wisdom. Since the Most Ancient Church consisted of humans who had direct perception of heavenly things and direct perception of life flowing into their hearts and minds from God, the "fall" could have occurred only through a diminished state of cognitive function. A lower quality of love began to shape human attention and change their perception of reality.

Because the two capacities by which the human race was made in God's image and likeness are *free will* and *discernment*, these gifts of cognitive life from God make the human race a special creation. They are the means by which the human race can conjoin itself intimately to the Creator so that love can find its fullness in reciprocation and we can gain eternal life.

Free will concerns all things of love. Love is meaningless without free will. And free will is meaningless without *choice*. Even the Most Ancient Church, which enjoyed a psychical paradise, had the option of leaving the *Tree of Life* for the *Tree of the Knowledge of Good and Evil*. These trees represent opposing belief systems and worldviews and distinct cognitive levels. The Tree of Life represents embracing God's guidance, while the

Tree of the Knowledge of Good and Evil represents the choice of personal judgment and self-guidance. One requires love of God, the other, love of self. The story of Adam and Eve portrays how the Most Ancient Church began to move away from God, preferring its own self-guidance in all matters of importance. This longing for self-guidance is described in Scripture as Adam being "alone" and the fact that "there was not found a help as with him" (Gen. 2:18,19). This human condition represents the desire to find support for putting one's faith in self-guidance and ego-reasoning.

Since those of the Most Ancient Church directly perceived that heavenly things and life flowed into their hearts and minds from God, the only way to break away (which they ardently sought) was to diminish their cognitive function. God permitted this to happen in order to preserve human free will in those seeking autonomy. God never uses coercion in matters of faith and love. Protecting human free will was the only way to preserve love in the individual and keep the door to salvation open.

In the previous chapter we discovered that the deeper quantum language of correspondences revealed that throughout the seven-day Creation Story Jehovah God vivified and animated the human heart and mind through spiritual evolution. Now God had to disconnect humans from their highest spiritual cognitive function and activate an inferior mental state. This inferior mental state was symbolically represented by God's putting Adam into a "deep sleep" and then forming "Eve" from Adam's "rib." This is why the creation of man and woman is treated twice in Genesis. The first creation refers to spiritual evolution, the other to the beginning of a fallen state. This can be explained only through a higher-level interpretation using psycho-spiritual correspondences.

A rib is a bone in the chest region, a region that is also occupied by the heart and lungs (which correspond to the will and reasoning). The fact that a bone, which is less vital and less animated than the heart and lungs, was formed into a "woman" signifies that something relatively dead had now become the object of desire within the human psyche or inner world. This new help or "partner" for the human race was the new sense and self value that began to take on the greatest importance in their lives. Eve represented a delusional love that is rebellious.

God permitted humans to gain a new sense of *proprium*. Swedenborg defines "proprium" as that which a person calls one's "own." Consider an infant who at first makes no separation between itself and the objects and people of the outside world. By age two, however, the infant begins to make strong distinctions between self and the outside world. This cognitive stage is often referred to as the *terrible* twos, when exerting new-found individuality often means rebellion. But the child is now able to begin to appropriate his or her own life through individual choice, which is of the utmost priority to God. Humans still had free will, even though their new embrace of self-importance and self-guidance was an illusion and served to block out the reality of God's importance.

This diminished state of cognition described in Scripture by Adam's "deep sleep" was an actual form of *trance*. Swedenborg maintained that cognitive disconnection from God is spiritual sleep. This state of spiritual sleep is both a psychological and physiological disconnection because our highest centers of consciousness are literally closed off to us. This is why the dynamical process of spiritual evolution involves the reconnecting and reopening of the more interior, spiritual cognitive functions of the human intellect. The "fall" of man was not something abstracted from real neuroscience.

All "trance" is the outcome of partial states of consciousness that result from real organic and neural disconnections between our discrete cognitive powers (as described in Chapter Six). While the multi-dimensional scaffolding and hierarchical order of human cognitive function allow humans to raise their reasoning above their proper loves, they also allow the phenomenon of *hypnotic* states in daily human life when various cognitive functions become *disconnected*. This is evidenced by the fact that our habitual mind works on "cruise control" and "automatic pilot" more than we are willing to admit. It is also evidenced by our fascination with celebrity and *suggestibility* towards fads and trends that lead to mental shallowness. But it is most evidenced by the daily news channels and newspaper headlines with their countless stories of individuals, political leaders and clergy acting irresponsibly. This irresponsibility carries all the way into faith systems that promote the idea that salvation is pure mercy and does not require changing how we live our lives. Irresponsibility is the outcome when the mind becomes disjointed and partially conscious. We will return to this topic in the discussion of hypnosis and its relation to both neuroscience and theology in the next chapter.

People who pray for justice in the world don't understand that if God interfered with human illusion in provocative and coercive ways, He would destroy human freedom and their chance for salvation. In fact, humans would strongly rebel, because God would be attacking exactly what they held closest to their hearts. Humankind ardently seeks Eve.

"Eve" represents the first major separation in human consciousness from God in that humankind sought to give illusions and fantasies ontological status. Eve personified an emerging quality of human self-love and a growing appetite to consult one's own prudence about what ultimately to believe. Swedenborg explained that this change of focus within the human psyche was symbolized by the fact that the Tree of Life was first "in the midst of the garden," but later, the Tree of Knowledge took the center position. This shift in our *psychoscape* signifies that a new value system had taken central stage. The Tree of Life represents instruction from God. The Tree of Knowledge is self-guidance.

> "At length, in consequence of the ascendancy of self-love, they began to think that they could lead themselves, and thus be like the Lord; for such is the nature of the love of self that it is unwilling to submit to the Lord's leading and prefers to be self-guided, and being self-guided to consult things of the sense and of memory-knowledge as to what is to be believed."[2]

Conversing with the "serpent" symbolized such consultation; it was reasoning founded on evidence from the natural senses. The snake symbolized worldly, sensual thought, for it spent its life in full contact with the earth, from head to tail. The forbidden fruit was the produce of self-love. Eve's offering this fruit to Adam represented human volition seeking the support and consent of human reasoning to enjoy the harvest of human prudence.

God warned humanity against eating this fruit not simply to test our obedience, but to counsel us against making a deadly choice. The forbidden fruit was intrinsically toxic to our eternal trajectories. "Eating" means to appropriate something into the very fabric of our lives, to swallow something whole. When someone completely buys into a premise, we will often say that he, or she, "ate it up."

▶ Love Twisted Into Evil

The separation from God is mild at first, but it grows and eventually turns into actual hatred for God. Human history shows over and over again that as love of self, self-pride, and love of the world intensify, the desire for domination over others and worldly acquisition increases. Since love of God and the neighbor is in opposition to love of eminence, the result is hatred for God and doctrinal things. People usually respond with a knowing nod when they are told that evil is "live" spelled backwards, because evil is indeed the reversal of order. And since order can be reversed by human free will, evil occurs contingently, as an ongoing real possibility, though not something created or injected into the world by God.

When individuals freely choose to put themselves first and God last, they reverse cosmic order. Ego reasoning leads to misjudgment and self-love leads to contempt for others, including God. This state of affairs accumulates over generations as individuals add new evils in life to the hereditary evils they received from their parents. While these hereditary evils are only tendencies, removing oneself from God leaves a man or woman completely vulnerable to the vanities and allurements of the world.

Human salvation remains possible as long as we can raise our reasoning above our proper love (Ruling Love) from free will. Everyone of sound mind can reason about right and wrong, and everyone can choose between the two. To be made in the image and likeness of God is to have *free will* and *discernment*. These two abilities are meaningless if we are given no real choices in life. Evil must be a real choice or option.

Free will would end if God were to stop all evil in the world. Free will, regardless of good or bad choices, is fundamental to person-level experience and individuality. God does not want evil, but He permits it. Again, God does not cause evil to exist. It is simply a choice humans make when the power of loving others is degraded and minimized into love of self, which itself is evil only when it rules over the love of God and neighbor. Evil is the *misuse* of our two God-given attributes of free will and reasoning. Evil has no ontological existence, that is, no fundamental reality within the Divine order of the universe.

This now invites a scientific challenge to causal agency within the scheme of Conjunctive Design, which states that

God's influx of active information can flow only into forms of self-similarity. If evil is permitted, how can God's Love flow into and form a conjunction with something so dissimilar? How can God be infinitely benign when all process, including evil, is perpetually dependent upon the Creator?

▶ The Cosmic Utility of Evil within Divine Providence

The ultimate purpose of Divine Providence, or governance through Conjunctive Design, is to create a heaven from the human race. Once evil is out of the bottle, it, too, is utilized in the Lord's infinite wisdom to further this eternal goal. We judge the qualities of things through contrast. We perceive and appreciate love in the world more distinctly if we have a visceral comprehension of its opposite. Evil can serve as a powerful wake-up call and change our contemplative focus more towards God and eternal life. Many people do find religion during times of misfortune.

In all temporal and finite activities, the Lord only focuses on what is infinite and eternal. This is called *prolepsis*. God's love can flow into evil phenomena because no evil can exist unless it serves some purpose in His eternal goal, whether that purpose is immediately apparent to us or not. This is how Divine Love is conjoined to all the ills of the world. For many who see all the suffering, oppression, and injustices in the world and the horrendous displays of savagery and war, this answer may seem unsatisfactory.

But consider, first, that any dissatisfaction with the notion that God can bend evil into long-term positive results is closely tied to the value we put on physical as opposed to eternal life and to the degree to which we form our beliefs from our natural senses. Even the most horrific evils in the world are mere temporal "blips" when compared to the reality of an eternal life. There is no way of comparing worldly suffering and eternal blessedness. This is not to diminish human and personal tragedy. Rather, it is to acknowledge that much more is going on than we are aware of.

The Lord oversees the actions of billions of people on earth and many billions more in the spiritual world. In the spiritual world, the Lord governs everyone in both heaven and hell. The fact that we are inwardly spirits and that spirits are closely associated with us means that the Lord has to take everything into account in order for any single thing to occur. The finite human mind rarely considers this top-down relational holism.

Swedenborg's experience of both worlds gave him the opportunity to make many unique observations of this multi-dimensional governance. For instance, a life may be taken unexpectedly if the victim will serve a higher use or fulfil a more urgent need in the spiritual world. The Grand Human form of heavenly societies that was described in Chapter Eight is constantly being perfected by the arrival of new individuals. It has changing needs, requiring those with particular talents at certain times. On the other hand, death may come to some in order to limit their evil. Divine mercy prevents those in hell from surpassing the evil committed on earth.

Many of the most grievous human tragedies involve the deaths of innocent children. When children are dying in great numbers on earth, it is because the evil in the spiritual world has temporarily accumulated to the point of outweighing the innocence of heaven and flows down into the minds of people on earth. In order to restore the equilibrium that is crucial for human free will, the Lord uses that same evil to take away the lives of children and bring more innocence into the spiritual world. Children and infants can also be taken away if the odds are heavily stacked against them to live a reasonable moral life. All children who die before maturing into rational adults are taken directly into heaven. So the deaths of so many children on earth can be looked at as unfair only to those who reach maturity on earth and blow it.

Men and women cannot be kept in freedom on earth unless there is a balance of good and negative influences reaching them from both heaven and hell. Beyond preserving this equilibrium, the Lord's challenge in the scheme of Conjunctive Design is to address the present diminished cognitive state of the human race. One outcome of this diminished state or "trance" is that many people perceive evil as *something that happens to us* and not *within us*. While on earth the Lord was particularly careful to communicate personal warnings like "Not to cast the first stone" (*John* 8:7) and to be cognizant of the "Log in our own eye" (*Matt.* 7:4,5). It was not freedom from Roman rule that the Lord was concerned about in Jerusalem. It was freedom from the enslavement of one's ruling love and self-centeredness, which blind us from self-examination.

Swedenborg assured us that we were all involved in evil and must have this evil removed in order to be saved.[3] Evil cannot be removed unless it is *seen*. And unfortunately, it is not easily seen unless it is done and becomes deed. So evil is permitted to surface in the world, and within our lives, for the purpose of salvation.

> "Everyone can see that love for our neighbor and love for ourselves are opposing loves. Love for our neighbor wants to do good to everyone, while love for ourselves wants everyone to do good to us alone. Love for our neighbor wants to serve everyone, and love for ourselves wants everyone to be our servants. Love for our neighbor sees all people as our family and friends, while love for ourselves sees all people as our slaves, and if people are not subservient, it sees them as our enemies. In short, it focuses on ourselves alone and sees others as scarcely human. At heart it values them no more than our horses and dogs, and since it regards them as basically worthless, it thinks nothing of doing them harm. This leads to hatred and vengeance, adultery and promiscuity, theft and fraud, deceit and slander, brutality and cruelty, and other evils like that. These are the evils to which we are prone from birth."[4]

Swedenborg further assured us that wars happen because wars are inseparable from the evils that men and women are prone to. We war with our brothers and sisters, we war with our parents, we war with our friends, we war with our spouses, we war with our coworkers and bosses, we war with the opposing football team, and finally, we get together to war with other nations, which have accumulated the same evils we are prone to. Secular culture deludes itself when it treats evil as a human institutional problem that requires legislative solutions rather than as a spiritual problem that requires theological solutions.

The "human predicament" is that we are blind to the real reason why we are all put on this earth. The Lord cannot lead us out of this predicament and save us without many horrors being permitted to occur. The Lord does not simply want to comfort us with promises of eternal happiness; He often needs to startle us to effect salvation.

> "...unless evils were allowed to surface, we would not see them and therefore would not admit to them; so we could not be induced to resist them. That is why evils cannot

be suppressed by some exercise of Divine Providence. If they were, they would stay closed in, and like the diseases called cancer and gangrene, would spread and devour everything that is alive and human." [5]

This does not mean that God does not suppress evil at all. He permits evil only so long as it serves some purpose in the Lord's ultimate scheme for Conjunctive Design. Evil must serve some ultimate "good." In this way, God's love can flow into self-similarity and find Divine utility even in negative situations.

▶ How The Lord Governs Evil in Hell

All life comes from the Creator. Even people inclined to evil are still sustained by the Lord's unceasing love in this world and beyond. While evil and self-absorbed individuals turn their back on the Lord's love, there can still be minimal conjunction between the Lord and hell through *usefulness*. As long as there is use, God's living force has something it can flow into and be conjoined with. Nothing in the manifest universe exists apart from use. Through use, disposition, force, and energy find form and complexity, mirroring the various characteristics of God's infinitely living Divine Love.

While we all enjoy free will, we are not free from the responsibility of usefulness or service either in this world or the next. As we saw in Chapter Eight, the citizens of the spiritual world are arranged into societies according to their utility, because the distinct nature of everyone's love and reasoning finds its true form, its specified complexity, in service. Even those in hell must *also* perform some use, or they experience extreme cold and pangs of starvation so acute that they compel themselves to perform some menial task. The feelings of cold and hunger come upon these evil spirits from their rebellion against participating in the Divine scheme of universal cosmic preservation through reciprocity and mutual dependence. Not to participate in this mutual and sacred exchange of energy through service to others is to disconnect oneself from subsistence and life itself. So evil people are motivated by fear, while angels are motivated by Love.

Those in hell also provide a useful service of which they are not even aware. They are used by the Lord's providence to counterbalance the angels in heaven. The Lord's living force is received and filtered through the angels and the evil alike before it descends into the minds of men and women on earth.* There could be no real free will unless these influences were kept in perfect equilibrium. When both influences are equalized, the human will, acting as a *third force*, can more easily push in one direction or another. As Swedenborg wrote, "This is the spiritual balance which puts men and women in a freedom for thinking and intending."[6]

Many people might find it absurd or insulting to be informed that ideas and influences flow into our minds from citizens of another realm. However, this truth can be easily ascertained by direct experience.

If we become quiet and take notice of ourselves, we will see that our minds continuously generate brain chatter that moves from one topic to another without any particular intention on our part.

Strange and even uncharacteristic thoughts may seem to come from so far out of the blue that we may pause to ask, "Where did that idea come from?"

After all, the Lord came into the world to reestablish cosmic equilibrium. The Lord subjugated the hells while He was on earth so that the possibility for salvation could remain open.

> "For unless the Lord set bounds to the rebellions from hell and controlled the forms of madness that exists there, the balance would be destroyed, and with the balance everything would go."[7]

The Lord's victory over hell and His glorification while on earth was not redemption itself, but an act to spare humanity from certain extinction and restore free will. Actual salvation is made possible only when individuals respond to the Lord through free will and life-choices. Much of the tension between a theistic science and theodicy, or evil in the world, emerges out of our unwillingness to confront squarely the implications of human freedom. There is also little awareness as to how the Lord wisely uses evil for Divine goals.

▶ How The Lord Governs Evil on Earth

The Lord also governs evil on earth through the external laws of society. The evil, who may have little or no conscience, can still be constrained by civil laws and fear of the loss of reputation stemming from their self-love. Self-love is a potent motivator, and the Lord uses great wisdom in taking advantage of evil's powerful desire to succeed. Even the evil often must rise to positions of power through some useful service to society. So the Lord actually makes use of evil hearts to get things accomplished in the world. The evil are often more driven to embellish their talents and strive for accomplishment, reward, and prestige than good citizens or humble individuals. Swedenborg made the remarkable observation that those with strong self-love have become crucial agents for helping spread the Holy Word because of their highly passionate preaching. We have all heard of clergymen who, in spite of their great preaching skills and their ability to attract hundreds of thousands of people to the Lord's teachings, have succumbed to the sins of the flesh or become involved with scandals. The Lord providentially has allowed these things to become public in order to save the preacher and cause the rest of us to look deeper into our own lives and determine whether our public persona is sincere or just surface dressing.

Not only does free will necessitate that evil be a real option, but that evil must have some success. Human *psychoenergy* is focused on obtaining some goal. If a goal cannot be achieved, then the intentions of the will and strategies of the intellect perish. Free will has no meaning if it has no reasonable expectation of success. If the Lord prevented evil from happening, free will would not only be a farce, but evil would remain hidden only more deeply, eating away and destroying the soul. Real deterrents have to be enacted within the hearts and minds of men and women, not from external threats by God.

Again, God does not allow all evil to happen. He manages evil with such Divine wisdom and foresight that He makes it useful to the goal of salvation. Any evil that does not serve some eternal end or use is prevented by God. Thus, in spite of all the evil and destructive power in the world, the universe continues to function.

Nevertheless, evil acts *are still evil*, even when the Lord uses a particular evil for some eternal benefit. God cannot violate the laws of Divine Providence without violating Himself. Saving humanity by force or arm-twisting is impossible, not because God cannot do it, but because the laws of Divine order don't allow it. The Lord works out our salvation through specific means and successive steps. This can happen only when our free will is protected. God cannot prevent all evil because His goal of salvation depends on hurtful and negative compulsions becoming exposed. Only then can we choose to go against these compulsions or, alas, to embrace them.

The Lord God acts in infinite and eternal ways at each and every temporal moment and each point in geometrical space. In Divine Providence, the *future* is present, and God views every temporal event from that eternal perspective. The future becomes the guiding principle for a universe in dynamical process.

Because God is infinite love and love never condemns, Swedenborg noted, no one is punished for what they have done on earth, not even Hitler or Vlad the Impaler. People will be punished for what they *continue* to do in the spiritual world. In other words, Divine forgiveness is squandered by those who insist on doing evil to others. Because the holographic dynamics of spiritual substance create a non-physical landscape that portrays and mirrors the various inner qualities of those there, evil is its own punishment. The punishment of hell is actually a person's evils returning to its source—like karma. The evils people take with them into the spiritual world are those they made no attempt to resist while on earth.

Since the spiritual world is a psychoscape based on the qualities of one's loves, going to heaven is not a simple matter of being "let in" through some gate. We go inward after death and become who we really are. Salvation requires more from us than praying to the Lord to make things right. We must also pray for the courage needed in sincere self-examination. We cannot help God remove those evils from our hearts that we do not see, acknowledge, and finally, attempt to resist. As we resist evil, the Lord works invisibly to eradicate the compulsion.

▶ Natural Evil

Many people see natural evil as a serious challenge to their faith in a benign Creator. Natural evil is misfortune that excludes all moral evil, like man's inhumanity to man. It includes things like disease, natural disasters, accidents, pain, the savage death of animals and their being eaten by predators, and even the extreme waste of species extinction over millions of years of evolution. Darwin himself found natural evil so overwhelming that he could not reconcile it with the Divine.

Those who believe in a benign God and consider natural evil to be a great cosmic injustice take the position that the Lord's promise of redemption in the world must include the healing and reconciliation of all this natural suffering. This view, while compassionate, means that God will resurrect the physical bodies of every creature that ever lived and that all will enjoy peace and perfect harmony. They point to John's statement that he "saw a new heaven and a new earth" (*Rev.* 21:1) and the Lord's promise to "make all things new" (*Rev.* 21:5). The Scriptural passage "The wolf also shall dwell with the lamb, and the leopard shall lie down with the kid; and the calf and the young lion and the fatling together; and a little child shall lead them" (*Isaiah* 11:6), also seems to offer theological support for such a reconciliation (although correspondences treat this passage as signifying a reconciliation between conflicting qualities in one's inner world).

Assume for the moment that the Lord's resurrection not only suspended the laws of physics but changed them to allow for the redemption of all creatures. Visualize a new planet earth without natural evil, death or predation. Begin by deciding whether all the microbes and viruses that ever existed also need to be included in this redemption. Would cancer cells be included? Where do you draw the line? Are there no parameters for cosmic justice?

If you demand *physical* resurrection and justice for all life forms you'll soon find that the planet earth is too small to support everything that ever lived. So visualize it as being large enough to accommodate all these creatures. A planet that large would have enough of a gravitational effect on the whole solar system to throw it out of kilter. Once you make room for everything and everybody, are they still allowed to procreate and increase in numbers forever? Would it not be unfair to future life forms if they couldn't be born? Would it not be unfair if any ovum, egg or seed failed to mature? Since insects alone could cover the earth within one year of unrestricted reproduction, the earth would have to keep expanding at a rate that would ultimately make it the largest cosmic formation in the universe.

However, if you take the position that there will be no more proliferation, will humans lose their reproductive organs? Or, will they hold onto their sexual organs for recreational purposes? After all, eternal life would be boring without all life's associated pleasures.

Will children get to grow up, or will they be kept at some optimal age? Will everyone be good looking? It would be cosmically unfair if one person were more attractive than another.

What about eating? Eating involves one life form being destroyed by another. Would this stop? If there is no eating in this resurrected world, then wouldn't the entire commerce of all organic process move into a mode of endless "pantomime"? Or would people simply become anatomically hollow, since their digestive systems no longer serve any use? Will people get to keep mouths and perfect white teeth in order to smile for eternity?

In a resurrected world where the weather is always perfect, there would be no need for clothes or shelter. There would be no need for a nature or science channel on cable TV, because such programming would show animals in this world displaying radical modification and doing none of the things they presently do on earth. There would be no need for comedy or drama, which often demeans people and portrays negative situations. What industries and commerce could humans become involved in, living in such a perfect world? What fun would there be if every chess game ended-up in a draw to protect the "loser" from emotional scars and preserve personal esteem? What would people strive for if everything is given to them—and equally at that? If life is adventure and striving, how would people enjoy living in such a "wonderful" world?

Rather than a view of a resurrected earth based on new laws of physics that ensure the end of natural evil, we need a more enlightened view of pain and suffering. Pain and suffering can be a blessing as well as a curse. The human body devotes a good deal of its nervous system to pain detection, and for good reason. Pain always helps us to become aware of something that needs our attention. The Lord, while on earth, endured horrific physical pain and suffering, and witnessed the rejection of His message, but He nevertheless triumphed.

"The world is full of suffering, and is also full of the overcoming of suffering." – *Helen Keller*

Suffering and pain are requisite for growth—"no pain, no gain."

However, some people, like Darwin, still believe that the physical death of any creature in the world, let alone the extinction of an entire species, is too high a price to pay to keep things moving along. But such sentiments come from human prudence and from the consciousness of self that appreciates an idea of justice that other creatures do not have or care about. Blessedness and peace can be bestowed only upon those creatures that possess the psyche to know what these things are.

We should not wait for a materialistic eschatological and soteriological solution. The natural world is not capable of fulfilling our hopes for eternal good and blessedness. Heaven and eternal happiness can only be a personal and conscious choice—not God's tampering with the laws of physics, which is contrary to Divine order.

According to Swedenborg, not even life in heaven constitutes a state of perfection. It is however, a perfect expression of our love, which continues to evolve. Heaven allows for happiness and usefulness to increase in our lives through an eternity. Adventure, striving, and learning all continue in the other world. The human intellect constantly needs new impressions and opportunities to remain animated and flourish. In heaven, the human mind is in its true element, inwardly open to explore the new levels of beauty that radiate from an infinite Creator whose Love and Wisdom is inexhaustible.

▶ Heaven can only be Conscious Conjunction with God

Animals and humans have physical bodies designed to perform uses on the worldly plane. But the higher-level capacities of the human will and reasoning are necessary to form a conscious relationship with God. That relationship, when nurtured, creates a new inner-complexity, adapted for processing spiritual inputs and raising the principle of personal utility in the form of good works to a higher, heavenly level.

Animals cannot create a subjective self, which involves deep inner experience and for that reason they do not dwell on life's misfortunes or demand justice. More importantly, animals cannot rise above their proper loves and appetites, both of which are instinctive. A cow will always prefer grazing on grass to gazing at Scripture. Human personhood develops from the freedom to choose what we love from a variety of levels—corporeal, rational and spiritual. Only our first-person phenomenal experience permits God to have a more intimate involvement in human lives and to reciprocate through hidden "special Divine acts" that make eternal life possible.

Specific Divine intervention into people's lives does not miraculously suspend the laws of nature; it proceeds according to law and causal order. The Lord unites His eternal qualities with the time-bound things in us to the extent that they *correspond*. That is, the Lord unites Himself with us as we perform genuine spiritual acts of service and love. By attaching Himself to these actions He can add eternal qualities to the temporal events of our lives. Our daily activities can be *eternalized* by God *if* we do them for the sake of others, not just ourselves, and if we do them from a spiritual principle. In this way, the Lord creates a ratio and conjunction between the finite and the infinite, not from the finite itself, but from the infinite that exists in the finite. God is relational and conjoins Himself to our subjective self in the degree to which we receive love and wisdom out of our exercise of free will. Reciprocal union makes things one. This special union provides a dwelling place for God within us and makes it possible for the human subjective self to enjoy eternal life, escaping the tyranny of time and entropy.

▶ God Knows our Ultimate Identity and Potential

What about predestination? Because the goal of Divine Providence is to create a heaven from the human race, Swedenborg learned that all men and women have been pre-destined for heaven and none for hell. The only thing that can get in the way of that wonderful plan is "us" and our use or misuse of free will. Evil proves free will. Free will is so important that God protects it above all things. It is a freedom that men and women have been willing to die for throughout history.

The Lord makes sure our choices are really our own—whether good or bad. Our talents come from God, and we have been designed for utility and service in an eternal, pre-space realm, so God knows better than we do what our most heavenly joy consists of. Human self-guidance is what leads us away from this God-given potential and full expression of our real selves. While heaven is our true destiny, it may not be our fate. Nevertheless, we never lose the capacity for discerning what is true, or doing what is good, which is the image of God in us, so everyone *can* be saved.

Swedenborg maintained that God secretly works throughout our lives to keep better options available to us at all times. Divine Providence allows us to see some of these positive influences or "lucky breaks" only when we look back at our lives. If we were able to see this influence coming at us, we would actually rebel against such Divine mingling into our personal affairs. In fact, we would become cognizant of the fact that God acts in opposition to most of the things we embrace or seek in the world.

When we misuse our freedom so that it becomes license, we fall into a self-induced bondage in which our compulsions become our master. This is what happens when people do harmful things and later claim they lost control and could not help themselves. Following one's own prudence to the exclusion of God is the very essence of slavery and loss of true identity.

Lead my soul out of prison, to acknowledge your name.
(Psalm 142:7)

The more we live according to the Lord's tenets, the more we are united to God, and the more we find our true identity! Following God is the very essence of personal freedom. Why? Because we were created to have a unique heavenly identity. We lose our identity by turning from God and pursuing our supposed autonomy.

This false autonomy still serves God's needs for Conjunctive Design, because human free will, while distorted, is actually kept intact. If we were deprived of our subjective intentions and abilities to discern things for ourselves, we would be deprived of life. Furthermore, if we were deprived of our subjective life and free will we would lose the means by which we could still be conjoined to God. This is why the Lord God preserved human free will by fashioning Adam's rib to take on a life of its own as an alluring woman formed from something relatively dead and less vital.

Ego reasoning is alluring but flawed. God allowed for this illusion so that humankind would continue to enjoy free will from an augmented sensation of selfness. Underlying this illusion and apparent autonomy is the false belief that we are life itself, as opposed to recipients of God's life.

In order for God to change the world and make all things "anew," humankind will gradually have to wake up to an unflattering truth: within our multilayered structure of mind, we have opted for a diminished level of cognition, in spite of our technological advances and the discoveries of science. This may offend our post-modern sensibilities. But if we are brave enough to look through Swedenborg's theological lens, we will see that the two central convictions of modernity, *essential human goodness* and the extreme value of *human autonomy*, are simply illusions. These things come from God. Until we acknowledge this fundamental truth, each new generation will simply reinvent an "Eve" for the purpose of giving an illusion "beauty queen" status.

This state of relative trance was brought upon human awareness by our growing preference for ego reasoning and our insistence on creating powerful images of self-identity. Former men and women acknowledged that they "were men solely by virtue of that which they had from the Lord."[8] The first men and women of the Most Ancient Church (discussed in Chapter Nine) recognized nothing human coming from themselves. They had a direct perception that life flowed into them from God. In their pristine and childlike innocence, they had no desire to have it any other way.

As we shall see, salvation from evil is salvation from hypnosis.

SUMMARY

▶ Evil has no fundamental basis in reality.
It is a contingency.

▶ Evil is the reversal of order.

▶ Evil has its origins in the diminished cognitive functions of humans that result from self-love and self-guidance.

▶ No evil is permitted unless it serves God's ultimate purpose.

▶ God keeps humankind in a spiritual equilibrium between good and evil in order to preserve free will.

▶ Free will is a gift of God's Love.

▶ Humankind lives in a trance.

PREDICTION

The solution to the mystery of hypnotism will provide potent evidence for verifying the truth of Swedenborg's ideas.

Eve and the Serpent – detail from *Biblia Pauperum*

Chapter Thirteen

HYPNOSIS: THE COSMIC MANIPULATION OF LOVE

"Man is asleep."
– George I. Gurdjieff

In the last chapter we discussed that early in Genesis, Adam was put into a deep sleep. Adam was not the only biblical figure God put to sleep.

According to Genesis 15:12,

... a deep sleep fell upon Abram; and, lo, a horror of great darkness fell upon him.

In Isaiah 29:9, 10, we read

They are drunken, but not with wine, they stagger, but not with strong drink; for Jehovah hath poured out upon you the spirit of deep sleep, and hath closed your eyes.

Jeremiah 51:57 says

And I will make drunk her princes, and her wise men, her captains, and her rulers, and her mighty men: and they shall sleep a perpetual sleep, and not wake, sayeth the King, whose name is the LORD of hosts.

Later, in the New Testament Gospels, Jesus implored people to wake up.

Keep awake, therefore, because you don't know what day your master is coming. Surely you recognize that if the homeowner had known during which night watch the thief was coming, he would have stayed awake and would not have allowed his house to be invaded. (Matt. 24:42, 43)

Behold, I come as a thief; happy is he that is awake and keepeth his garments. (Rev. 16:15)

The theme of wakefulness and sleeping is also addressed in *Luke 12:37,40, Mark 13:35-37, Matt. 25:1-13,* and in *Rev. 3: 2.* Had humankind been snoozing and unconscious all this time? During his numerous spiritual journeys, Swedenborg was instructed as to the meaning of the term "sleep" in the quantum language of Scripture:

"Natural life, considered in itself, or without spiritual life, is nothing else but sleep; but natural life, in which there is spiritual life, is watchfulness."[1]

Also:

"The unregenerate man is like one in a dream, and the regenerate man like one who is awake; and in the Word natural life is likened to sleep, and spiritual life to wakefulness." [2]

In Scripture, "sleep" represents a diminished level of cognitive function in the human habitual mind. Even the Lord's disciples needed a cognitive jolt. In *Luke 9:28*, the Lord took Peter, John and James up to a mountain to pray. As the Lord was praying, He became transfigured, with His face and clothes giving off a glistening aura. Then Moses and Elias appeared and conversed with the Lord, who was in His "glory." In *Luke 9:32*, the disciples were first asleep to all this and did not see the Lord's Glory until they were "awake."

But Peter and they that were with him were heavy with sleep: and when they were awake, they saw his glory, and the two men that stood with him.

The Lord did not physically transform himself before His disciples. Rather, He became a "light bulb" of new insight flashing inside a deeper level of their minds. Transfiguration is a change in human perception. Glorification is the process whereby the Divine within the Lord's Human presence is made known to men, who begin to see that Jesus is the one true God and learn to dispel doctrinal falsity through the interior meanings of the Word. That the disciples could see Moses and Elias talking with the Lord indicates they were in a more enlightened state and able to perceive that the Lord's relation and connection to Scripture was more intimate than they had previously understood. They did not simply awake from bodily slumber, but they had a more deeply interior level of perception opened to them. They briefly awoke to the reality that the Lord was the Word Incarnate. Taking them up a mountain symbolizes that the Lord was leading them to a higher cognitive function. In spite of this brief spiritual synapse, the disciples were still in relative ignorance of the truly profound nature of the Lord's connection with the Holy Word. Thus, they soon became engulfed by a "cloud." They had not yet fully grasped that the Lord was the Word itself, and that everything in it referred to Him and Him alone. Seeing the Lord in His glory is to have one's eyes opened to a new level of understanding the Holy Word.

Recall that clouds represent the literal meaning of Scripture, which obscures its deeper meanings. To hear a voice coming from within this "cloud," as the disciples did, was to receive Divine revelation concerning something deeper in the Holy Word. The disciples were given only partial glimpses and insights that the Lord was in fact, Jehovah God, and that Scripture revealed these things only from its deeper levels. These deeper levels were the topic of discussion between the *resurrected* Lord and His disciples on the road to Emmaus (*Luke 24:26,27*). These same revelations are today being made available to the New Church, or New Jerusalem, which is currently being established by the Lord on earth.

Most people may find it difficult, if not offensive, to learn what they call normal waking consciousness and habitual daily life is but a state of slumber occasioned by their diminished cognitive state. But because of this diminished state, the modern world must be challenged—as were the disciples by the Lord—to discover that the Holy Word contains a deeper quantum language that addresses the ontological reality of higher levels of phenomena. Most of the world's theology and doctrine is derived from this diminished cognitive state. Humankind remains trapped in the material philosophy of naturalism and the reductive view of ontology that understands everything in Scripture through its lower level or literal meaning. We fail to recognize that we have fallen into chronic neurological degeneration and therefore we have become vulnerable to *suggestibility.*

Theology and religion usually deal with the issues of good and evil apart from neuroscience and anatomy. In other words, the process of disconnecting oneself from God and religion is seen as a spiritual and moral issue that has no physiological consequences within our neural substrates or cognitive architecture. But a brief look at the subject of hypnotism suggests otherwise: our "wiring" has become faulty because of chronic disconnection with our higher spiritual cognitive functions.

▶ The Mystery of Hypnotism

The phenomenon of hypnotism has everything to do with humanity's diminished cognitive ability, because it is contingent upon a disconnection from God. The human predicament of living our lives in a trance or a semi-trance-like state has created a wide spectrum of negative consequences, from personal dysfunction to the mass psychosis that leads nations into war. Our infatuation with celebrity, scandal, and the superficial provides immediate evidence that people have become disconnected from their higher cognitive functions and, indeed, from cosmic reality.

Stay awake, and keep praying so that you won't come into crises. (Matt. 26:41)

We are asleep at the spiritual wheel. How often have we discovered that we drove our car from one place to another without giving it any thought, as if we were on automatic pilot? How often have we found ourselves doing something odd and realized, "I don't know what got into to me, " or "I wasn't myself." How often do we feel the world is foreign and strange and say, "This is a crazy world we are living in?" How often do we find career success based on how well we play the "game." We instinctively sense something is off-kilter when it comes to human life. Few suspect the true nature of the crisis. Knowing the difference between good and bad offers little help until we know how and why we became hypnotized. The study of hypnotism is a window into a realm where salvation, psychology and neuroscience overlap in profound ways.

Hypnotism is not completely understood. Several theories have been proposed to explain the strange phenomena of hypnosis. Some theories hold that hypnotism is a form of sleep ("hypno" means sleep in Greek). Other theories propose that it is a reflection of our susceptibility to suggestion and conditioning. Another theory describes hypnotism as a kind of dissociation or disconnection. There is also a theory of the parallel existence of two minds—conscious and subconscious. And finally, yet another theory is predicated on the existence of an underlying higher consciousness.

All these theories are correct. That is, they all reveal some truth about the nature of hypnotism. However, none of these theories adequately addresses the *mechanism* and physiology of hypnotism. In fact, even the phenomenon of sleep, which is a related topic, is still not fully understood.

The twentieth-century mystic-philosopher, George I. Gurdjieff, may have been the modern world's greatest expert on hypnotism. He claimed that the science of hypnotism was well understood in ancient times, when people called it "Mehkeness," or "taking-away-of-responsibility."[3] This definition of hypnotism is similar to Swedenborg's idea that theodicy emerged from the irresponsible actions of human free will in a state of relative cognitive slumber. Gurdjieff's organic and anatomical description of hypnotism contained the revolutionary idea that hypnotism and salvation were related.

As noted above, the physiological mechanism of hypnotism is still a big mystery to modern researchers. But not to Gurdjieff. In *Beelzebub's Tales to His Grandson: All and Everything*, he made a curiously simple statement about the mechanism of hypnotism. He boiled it down to the "difference-of-the-filling-of-the blood vessels."[4] At first glance, the phenomenon of hypnotism seems too strange and occult to be a mere matter of hydraulics.

Swedenborg also addressed the circulation of blood in his anatomical writings, linking variations of blood circulation with specific states of human consciousness. Hypnotism and its mechanism may indeed offer the most convincing evidence of Swedenborg's multi-level cognitive theory of Love, his science of correspondences, and his spiritual revelations concerning the human predicament. In fact, hypnotism may emerge as the most relevant subject of study for our salvation and elucidate the manner in which we are to respond to the Lord's warning to "wake up" and be "watchful." By studying the "difference-of-the-filling-of-the-blood vessels," we may discover the most visceral approach to the perennial philosophy of "know thyself."

▶ Blood Flow and Flow of Consciousness

In Swedenborg's anatomical works, including *On Tremulations* (vibrations), *Economy of the Animal Kingdom*, and *The Animal Kingdom* (domain of the soul), he built a strong case that every affection or state of mind had its own particular and corresponding blood flow in the body. The truth of this assertion can be witnessed every time a man gazes at the sight of a beautiful naked woman and is brought into a state of arousal, whereupon blood is *immediately* sent to the penis. Similarly blood flow is increased to the female genitals during sexual intercourse.

Other states of mind, such as amazement, courage, fear and anger, also provide strong empirical evidence that emotion has an effect on the direction in which blood is directed through our veins and arteries. For instance, anger forces blood into the arterial extremities, and we turn red. Fear can turn us pale ("as white as a ghost") or paralyze us ("scared stiff") as blood is drained from our muscle fibers and forced inward towards the heart and veins. In melancholic states blood flow is heavy and sluggish.

The *correspondence* between blood flow and consciousness is even repeated on different scales in the body and in the human heart. Swedenborg stated that all our emotions and their associated blood flows have a corresponding circulation in some coronary vessel or another. In other words, the blood flow in the heart, like the body, "mimics" the various states of our mind. And it does this by maintaining the equilibrium between changing influences of our immediate passions and bodily needs.

> "For if the general equilibrium of pressure pertaining to the arteries and veins depends upon internal causes, such as the affections of the mind, or upon external causes, such as the affections of our body, and if the above-mentioned equilibrium be represented by the distinct determinations of the coronary vessels, it follows, that there is thus a representation in the heart of the state both of the body and animal mind; and that according to this state, the general equilibrium, and consequently the circulation of the blood, is regulated, and in conformity with the state of the circulation, the animal economy in general."[5]

Therefore:

"It follows, that there is no change arising either from the brains or the body throughout the whole sanguine-ous system, which is not represented in the heart." [6]

The human heart directs blood flow in a way that mirrors the states of the mind through coronary vessels that are divided into several classes. Swedenborg calls them *refundent, retro-quent, anticipant, transferent,* and *retroferent* vessels.[7] These vessels regulate the flow of the blood and its pressure in the chambers of the heart according to the changing affections and passions of the mind.

To illustrate just how closely the blood flow of the human heart can be directed within its chambers to mimic a specific state of mind, Swedenborg described a peculiar circulation in the coronary vessels that corresponds to the mental state of *anticipation*.[8] When we anticipate something, our mind is influenced by something coming at us from the future. Therefore, the blood flow must have a contingency in the coronary vessels of the heart so that it too can flow back from the "future." Swedenborg describes people in this anticipatory mental state as having some of the blood in their left auricle directed back to the right auricle through the *retroferent* vessels of the coronary arteries.

Why would the flow of blood from the left auricle to the right mimic mental anticipation? In the normal flow of blood through the heart, blood first enters the right auricle and enters the left auricle only after it has gone through the lungs. Blood in the left auricle therefore represents a future state of the blood relative to the right auricle. When blood is directed back into the right auricle from the left, it is, in a very real sense, being influenced by something coming at it from the future, and it therefore perfectly mimics the mental state of anticipation.

Swedenborg emphasized that if the organic processes of the body did not emulate the processes of mind and spirit, the whole organic kingdom would fall asunder. This *supervenience* between higher and lower operations follows the laws of cor-respondence. In his later theological writings, Swedenborg assigned a particular flow of blood to deceitful behavior and dissimulation in human relations. When we are purposely deceitful to others, we compartmentalize our thinking in a way that sidesteps the scrutiny of our conscience or higher reasoning. This real disconnection between two distinct cognitive functions stems from irresponsible manifestations of human free will. Similarly, nature provides a contingency by which the flow of blood takes a diversionary route and escapes the direct *scrutiny* of the lungs. It is the function of the lungs to oxygenate and cleanse the blood of impurities such as phlegm as the blood comes directly from the heart. This has a mental analog in our introspection that cleanses bad motives coming from our will.

In *Divine Love & Wisdom,* Swedenborg stated that there are two routes by which the blood can enter the lungs:

> "The heart can be conjoined to the clustering vesicles of the bronchia by blood sent out from itself, and also sent out not from itself, but from the vena cava and aorta."[9]

What Swedenborg meant is that the blood can flow into the lungs directly from the heart, or indirectly. By means of coronary vessels, the heart can allow some blood to bypass the lungs and make the circuitous path through the vena cava and aorta. After this "detour," the blood can be sent back to the lungs through the bronchial arteries. This detour allows the blood to enter the lungs in a way that is a step removed or "abstracted" from the heart. This blood flow is analogous to the capacity of humans to raise their intellectual understanding above their proper love. To do this, ideas have to be *disconnected* from the will so that intellectual concepts can be abstracted and refined in and of themselves. In Chapter Nine we saw that the Lord first gave this cognitive ability to anatomically modern, cerebrum-focused humans after the "Fall" of their predecessors. This brain modification allowed modern humans to be able to regenerate their hearts through the application of refined knowledge.

However, when we misuse this ability to separate and raise our minds above our proper love, we develop the habit of putting up a false front. Over time, this mask becomes a second nature. Human consciousness splits between the real and the artificial. The genuine consciousness becomes the sub-conscious mind, while the artificial consciousness acquires its own independent functioning as the everyday habitual mind. Swedenborg states:

> "This division of the natural man into two forms is an actual division of both will and thought, for all a man's actions proceed from the will, and all his speech from the

thought. Thus a second will is formed by him below the first, and similarly another thought; but still both constitute the natural man. This second will which is formed by the man may be called his corporeal will, because it moves the body to act morally; and this second thought may be called pulmonary thought, because it moves the tongue and lips to speak such things as are of the understanding." [10]

The *subconscious* or *internal* natural man must be spiritually regenerated, because *it is who we really are*. But we claim a vested interest in preserving the status quo of our dissimulation. The disconnection of cognitive functions allows the knowledge gained in life to be compartmentalized and used to influence others rather than to confront the real motives of our deeper, inner world. This cognitive disconnection is evidenced by our suppression of sincere contrition and our compulsion to displace blame onto others or external events. This is precisely the state of sleep that Scripture addresses in the passages quoted above. Humans have lost touch with their inner, essential selves.

The process of waking up and being watchful during our spiritual evolution requires self-examination and inner sincerity, not the reinvention of ourselves, which would simply be a change of masks. As self-scrutiny *reconnects* the two levels of consciousness that were split up by dissimulation, a process of spiritual cleansing ensues. During this cleansing, the blood circulation mimics and corresponds to one's increased self-scrutiny by submitting itself to the purifying operation of the lungs with greater frequency.

There are coronary contingencies by which the blood can be re-looped to pass from the heart and into the lungs two or three times for double or triple purification before being sent to the aorta and back out into the body. This more intensive organic scrutiny of the blood is activated according to the intensity of our introspection and the degree to which we are willing to filter out negative leanings and apply noble principles to our life. While deceit delays and detours the blood before it reaches the lungs, inner sincerity loops it through with greater frequency. There is a wonderful *correspondence* between the operations of the human will and understanding with those of the heart and lungs.

▶ Love and Organic Mutability

The male penis gives us one more unique example of the correspondence between the flow of consciousness and the flow of blood. The penis provides real empirical "hard" evidence that proves Swedenborg's ideas about the importance and truth of his science of correspondences in top-down causation and can lead us to understand why Love is the ultimate science.

It is true that sexual excitement causes blood to flow into the male penis, causing an erection. But not every sexual intention by a man is a noble one. Swedenborg identified the anatomical difference between noble and spurious erections, between true love and just plain "being horny," as the manner in which blood is directed into this highly mutable organ.

There are two kinds of erectile tissue in the male penis. The *corpus spongiosum* encloses the urethra, while the *corpora cavernosa* constitutes the outer, superficial part of the penis. According to Swedenborg, if the corpora cavernosa tissue fills up with blood first, it is a purely carnal arousal. But if the corpus spongiosum tissue fills first, it is a genuine and noble erection. In the former, the erection is created by more external causes and is mere inflammation. In the latter, the erection is promoted by a more interior and general principle, since the *corpus spongiosum* is activated by the state of the *vesiculae seminals* and the prostate.[11]

To bring this back to the discussion of hypnotism, the capacity of the tissues in the penis to expand or to slacken by blood flow shows how influences to the brain can be intensified or disconnected. An erect penis, with its tissues put in greater tension by filling with blood, becomes not only a more sensitive organ, but a much better *wave guide* for transferring signals to the brain. This affects our states of consciousness. Swedenborg believed that all organic membranes, on all orders and scales of structure, promoted or retarded wave signals through the principle of expansion and tension. This holds true for neurons and their deeper substrates. Therefore, different levels of consciousness can be both *intensified* and *disconnected* from each other due to the flow of blood and other vital fluids. As we saw in Chapter Six (p. 113), expansion and constriction are physical analogs of cerebral passion and its likes and dislikes. This correspondence facilitates the nexus or causal link between the states of mind and bodily states in which human volition and thought bring organic structure into coincidence.

"Every affection of the cerebral sensory, which is also called a passion, induces a peculiar and corresponding state on the fibres of the whole body. This state is first induced on the internal sensories; and since the medullary fibres of the cerebrum and also the nervous fibres of the body, arise from these sensories, a similar state is induced on the fibres and consequently on the whole body which is woven of nothing but mere fibers. Such is the case with joy (*loetitia*), sorrow, anger, mercy, love of different genera and species, and other affections. For joy expands, dilates and opens the fibres, but sorrow and grief constrict them; anger hardens them and mercy softens."[12]

Modern researchers have indeed discovered that the inner lining of the arteries expands when people listen to enjoyable music and constrict when they listen to music they do not like. Again, science is moving in Swedenborg's direction.

Swedenborg developed these ideas during his search for the soul and its mechanism of intercourse with the body. He believed that the various nerves, fibers and membranes of the body and brain communicated with each other through vibratory motions, or wave fronts. His physiological insight was that there was no communication between mind and body unless the various organic membranes were in a proper state of tension. Like a violin's strings, if the membranes become too slack, vibratory motion ceases and action perishes. The human anatomy was built up of structures and membranes based on harmonic proportions and ratios[13] that are capable of attuning themselves according to an individual's various states of mind. So also like a violin's strings, our membranes and fibers can be stretched or relaxed to keep their activity or non-activity in the same key with our ever-changing inclinations and mental states.

"There must be tension in the membranes if any kind of tremulation is to be communicated over the whole of their expanse, so as to effect sensation in one thing or another."[14]

This includes nerve impulses:

"Unless the nerve be tense, it will never allow any vibration to reach the sensible sphere. Tension therefore is required; and being absolutely necessary, the presence of a specific cause is also required to produce it."[15]

And just how is this self-tuning and tension produced in the membranes? Swedenborg's response was similar to Gurdjieff's notion about the mechanism behind hypnotism:

"This tension can arise only from the infilling of the vessels, whether it be the blood vessels or those through which any lymph or juice is circulating."[16]

Modern neuroscience does not currently acknowledge that the nerves, neurons, axons, and dendrites in the brain need to be tense in order to be efficient wave guides for signals. Instead, it holds that communication comes from chemicals called neurotransmitter molecules, which can excite or inhibit a neuron from "firing," according to the type of synapse they are associated with. However, these neurotransmitter molecules cause changes in the neuron's excitable cell membrane, which is occupied by millions of highly mobile and mutable receptors. Neurotransmitters seem to change the actual state and shape of the receptors to make them more responsive to certain information and take on particular emotional states. Neurotransmitters can therefore promote or retard signal propagation. Swedenborg suggested that neuron plasticity went beyond "excitable" receptors and that the entire axon membrane and its deeper level structures could expand, contract, tighten or slacken as living forms perfectly *responsive* to our mental states.

What seems more readily acceptable to modern science is Swedenborg's claim that the nerves from the brain are perfectly positioned to direct the flow of blood into various membranes and tune them according to an individual's state of mind:

"Now, as a mass of ramifications of nerves weave themselves about the arteries and veins—not only in the body itself, but also in the brain—composing what is called the nervous tunic, it follows that as soon as any passion has originated, the blood is more or less under the control of the nerves; by the contraction of the nerves the blood is closed off from its finer vessels, while by the expansion of the nerves the blood is permitted to flow freely or is propelled forward with increased pressure, so as to expand membranes. From this cause comes that immutable law which is exhibited in the membranes. For if the blood is obstructed in its membranes, there results at once a different attuning of the whole nature of man."[17]

What separates Swedenborg's from contemporary thinking is that he took the above idea into the hierarchical structures and scaffolding of the brain itself. He believed that the tensing and slackening of the neural connections intensified or disconnected the signaling between neurons. Deeper still, the subtler membranes within the neuron's multi-leveled architecture also expanded and contracted, becoming tense or slack. This allowed distinct cognitive functions to be connected to each other or disconnected, permitting the operation of one or the other to be *intensified* at the expense of the others.

Here, Swedenborg put his finger on the precise mechanism of hypnosis. The phenomena of hypnosis and trance within the human psyche is possible because blood and fluids of various viscosities can be directed into particular membranes and expelled out of others, allowing the *disconnection* and *intensification* of a particular cognitive function. When a particular cognitive function is over stimulated by either a powerful emotion or suggestibility, that mental function will take control of the body's blood flow.

"Therefore that which holds sway over the animus, holds sway also over the blood."[18]

This is why when we discover what another person loves, we gain some degree of control over him or her. For instance, when we flatter someone we stimulate and *intensify* that person's sensual/imaginative cognitive level, where self-love has its seat. At the same time, flattery can temporarily disconnect that person from his or her proper reasoning and judgment (which is why we say love is blind). Excessive praise can put a person in a state of diminished cognitive ability. So in a sense, anything we do to seduce or capture someone's attention is a form of hypnosis.

A professional hypnotist gets control of both a patient's animus and blood flow by disconnecting and intensifying distinct operations of the mind through various techniques learned over time. The difference between clinical hypnosis and the hypnotic spell of a slot machine in a casino is that the former produces an even more exaggerated form of disconnection and intensification, in which one can be convinced that ammonia smells like roses. However, the self-induced trance of the slot machine exemplifies how hypnotism can be a "taking-away-of-responsibility." Hypnotism would not be possible if the human intellect was not composed of discrete cognitive functions that can be united or closed off from each other by various passions.

▶ "Homo Disconnectus"

Love can be physical/corporeal, sensual/imaginative, rational and spiritual. As we saw in Chapter Six, each of these loves operates from a different principle of action and from a different level within the hierarchical scaffolding of the brain and its neurons. Common human experience teaches us that these various levels of emoting and intending can work autonomously or work in harmony with the others. They can even conflict with each other, as when our rational mind tells us we should go on a diet and challenges our "lower" corporeal pleasures. This conflict proves that our cognitive functions are distinct or "quantized." Consciousness can be a multiplexed state or a segregated and partial one.

The intensification of one distinct level of the human intellect through some passion can disconnect us from the operation of another. For instance, a strong imagination can cut us off from the outer world, as occurs in daydreaming. In fact, fanatical vision occurs when the imagination is so intensified that even when our eyes are wide open it "refuses all objects of the sensible things that flow in by way of the external organs."[19] Instead, we see only what our imagination sees.

When we lose our temper and get angry we descend into our animal or paleomammalian brain (the Limbic system) and become temporarily cut off from our higher reasoning centers in the frontal lobes. Counting to ten helps us to control lower emotions because it *reconnects* us to a higher cognitive ability (and redirects psychic energy). Religious experience and ecstatic states can be so powerful that they seem to take people into a wholly different reality where they are completely cut off from all sensory inputs. Extreme cognitive states could not manifest without a neural contingency for disconnecting and intensifying distinct mental functions.

▶ Yes, Hydraulics!

The physiological or mechanical reason for these various species of disconnections and intensifications within the human psyche is that the brain's functions are compartmentalized among its various distinct lobes. These lobes consist of folded or convoluted neural "wiring." According to Swedenborg, when blood and nervous fluids are flowing in the brain, these twisted lobes expand, promoting more intense activity. When activity stops however, fluids exit these serpentine membranes to the point where they become slack and even kink and no longer act as proper wave-guides to carry information. In other words, "Motion cannot be received without tension."[20] The slackening and expanding of the brain's folds and lobes is the mechanical reason why there can be both segregated operation and intensity of our cognitive functions. Swedenborg included disconnection and intensity within the tension of the membranes and deeper processes of the neuron as well, for, "Every least part of the body is thus clothed with a membrane."[21]

As mentioned above, modern neuroscience does not yet recognize that such a modification takes place in the brain, the neuron, or in the neuron's deeper structures. This is precisely the reason why even the phenomenon of sleep, which is related to hypnotism, is still somewhat of a mystery to brain scientists. In *The Economy of the Animal Kingdom: The Fibre*,[22] Swedenborg attributed the phenomenon of sleep directly to the brain's structural plasticity. During our waking and working hours, the brain's membranes are being expanded and stretched according to the intensity of their use. At the end of the day, when we relax, the membranes start to slacken. Once loosened it is difficult to bring these organic structures back to a state of tension and expansion where they can become re-turbinated into their proper "tortional" forces and coils. This slackening causes a person to feel tired.

> "If now this membrane becomes slack, or is deprived of its heat, its blood, or its animal spirits, then the whole man becomes dull, heavy, and dead. Further, when a relaxation takes place in the nerves of the five senses, after having been in a state of tension during the entire day, then sleep sets in...[23] If the fibers or nerves become slack, a person is similarly deprived of sensation ..."[24]

In passing, Swedenborg stated that yawning temporarily expands and vivifies the cerebrum's membranes to keep it awake longer. Yawning as well as stretching the limbs in the morning helps the cerebrum to gradually unfurl itself and its powers. When we stretch our arms and legs blood and energy is sent to the corresponding part of the cerebrum that controls these extremities.

▶ **Forms of Sleep**

The multi-leveled cognitive structure of the brain allows for different species of sleep and different levels of wakefulness to occur, as well as their corresponding states of consciousness. In what we call normal sleep, the convolutions of the cerebrum are collapsed, becoming less distinct and differentiated. This *disconnects* and closes us off from bodily sensation and voluntary action and actually restricts arterial blood from entering deeply into the cortical brain.

> "In the state of sleep moreover, the red blood is restrained from approaching too near to the individual cortical substances, and it is detained at a distance from them according to the degree of slumber; for it is animation that draws the blood from the arteries and invites it towards the cortex; and therefore as is the animation such is the afflux, distribution and circulation of the blood. Thus in sleep the blood brushes the outermost surface of the pia meninx, nor does it penetrate into the interiors of the cerebrum except by more open paths, through which also pass the trunks; for it cannot penetrate towards the substance of the cortex through the winding folds and commissures, since these are constricted and collapsed."[25]

Swedenborg pointed out that the arteries in the brain, or the internal carotids, escape from the heart's influence by shedding their muscular sheath, acquiring a different nerve and taking tortuous, almost perpendicular, bends. This allows the animation of the cortex to regulate the arterial blood in the internal carotids and draw it inwards *according to the need*. This need and attraction is governed by the passion and appetite in the cortical cerebrum, which animates all its substances. The cortical substances, or neurons, are the source of animation in the body. They too, must enjoy modification and change of state, or there would be no nexus and correspondence between the two.

In sleepwalking, a different order of brain disconnection or species of sleep occurs. Swedenborg maintained that in this state, the cerebrum's folds remain unfolded, erected and differentiated. Sensation and imagination remain operative, because the cerebrum is still connected to the body. The actual disconnection lies deeper within the neuron. In sleepwalking, the neurons continue to generate states or modifications within their structure, but only in a general way, and they are cut off from the more interior activity of the rational mind, which resides in a preeminent cortex within the neuron. Not only is human cognitive function hierarchical, but sleep is too. We can be asleep to the world, asleep to our imagination, asleep to our reasoning powers, and finally, asleep to God, which is what Scripture addresses.

When we close God off from our lives, the result is not just psychological. We experience a real physiological disconnection from the subtle organic forms of our mind and spirit, which are real organic forms that receive the influx of Divine information from the Spiritual Sun. However, when we open up to God, every positive emotion or species of love awakens; they erect, unfold, and vivify corresponding substrates of membranes within the multi-leveled cognitive architecture of the brain and its neurons. *The intensification of one level of mind through some passion or belief sends blood and other vital essences like nervous fluid to certain parts of the brain and closes off others.*

▶ Three Living Fluid Essences

Three discrete levels or species of "blood" vivify the distinct organic forms of our cognitive scaffolding. They are the red blood, the electro-colloidal nervous fluid, and the preeminent creative essence called the first organic form of the soul, which consists of the most rarified substances of the physical world mixed with spiritual substances. Swedenborg referred to the red blood as the corporeal soul, the nervous fluid as the animal or imaginative soul, and the highest essence as the soul of our higher mind. Modern science does not recognize this last essence.

These three bloods and their discrete pathways become restricted or augmented, separated or mutually conjoined, to provide the mechanism for disconnecting and intensifying our different states of consciousness. Spiritual love engages the soul and opens up the innermost structures of the human mind by sending spiritual essences to the organic forms that receive the living energy and active information of the Spiritual Sun.

When the Lord appeared on the earth and warned people to awaken, He was not expressing a theology or religion separated from neuroscience and brain design. Religion was formed from God's infinite wisdom about the intricate workings of the human psyche and its vulnerabilities. Because of human cognitive dysfunction, the Lord cannot simply comfort us. He must often shake us out of our mental haze.

During times of illness, sorrow, misfortune, and the untimely death of loved ones, the artificial consciousness of the habitual mind becomes quiet, and we temporarily reconnect to our true self. People tend to cut through their masks during times of crisis. The Lord uses misfortune to snap us out of our artificial and habitual consciousness. Only when we seek something deeper can the Lord activate the innocence and love stored up in our *remains*, where all the possibilities for genuine spiritual impulses can be found.

▶ True Religion Reconnects Cognitive Functions

The Lord appeared on earth to help wake us up and snap us out of our trance. He came not simply to comfort us with resplendent love, but to de-hypnotize us! Humanity had become inwardly disconnected from God and disconnected from their real inner selves. This has led to what George Gurdjieff called "non-responsible-manifestations-of-personality."[26] Deceit, greed, envy, hypocrisy, treachery, and hatred are all manifestations of some limited, disconnected state of mind and the over-magnification of one's self worth. How else could humans do hurtful things to each other in spite of knowing better, unless there was a chronic cognitive disconnection of our full mental facilities?

It is interesting to note that the word "religion" is derived from the Latin term *ligare*, which means to *connect*, much like *ligaments* connect all the muscles, organs and bones of the body so that they can work in unison. Equally interesting, the Hindu term "yoga" also means *connect*. There are various forms of yoga practice designed to help one connect to God through the physical body, the emotions and through the intellect. As a system of faith, Western religion has forgotten that proper connection with God involves real anatomical and brain science to correct the dysfunctional operation of the human habitual mind. Without knowing this science or recognizing the human predicament of chronic cognitive disconnection, a minister's sermon will be not have its desired effect.

Every state of mind or affection has its own flow of "blood" and circuitry. This includes false or artificial consciousness, which develops over time as we learn how to put up fronts and wear "masks." The human body adapts its blood flow to this repeated ritual until it becomes chronic and *second nature.* As artificial consciousness strengthens and gains a life of its own, it covers over our real selves and relegates them to function as the subconscious mind. Our real essence is driven underground.

Today, we often consider the human habitual mind, which is really an artificial consciousness, to be our normal state of wakefulness. This makes positive change in the world difficult. The purpose of artificial consciousness is to protect the status quo. It is a powerful motivator. It protects us from discovering anything unflattering about ourselves and keeps beneficial information, like that of sermons, in the memory only, where it is stopped from penetrating more deeply into our inner being. So, valuable knowledge that could be used for inner growth is instead used for the cunning and clever purpose of changing our mask instead of changing ourselves. Our true inner self, deprived of this data for real growth, thus remains primitive, unregenerate and disconnected.

According to Gurdjieff, a hypnotist puts our false consciousness—what Swedenborg calls *pulmonary thought*—to sleep. The hypnotist will tell people to count and say to them, "Your eyelids are getting very heavy." This relaxes certain neural connections and allows another type of blood flow to take over, allowing the subconscious, or internal sensory, to take over.

Because the subconscious or internal human essence had been blocked from directly receiving data and proper education by the artificial, habitual mind, it remains open to extreme suggestibility. By putting the habitual mind to sleep, the hypnotist gains direct access to the deeper structure of the neurons and can influence their ability to change states. The hypnotist can convince the cortical brain to stop smoking or overeating by intensifying its functioning and implanting a new desire and new information within its deeper structures.

Religion offers us another, more Divinely designed method for cutting through the mask of artificial consciousness, and allowing God's guidance to reach our inner reality and plant new information. Unfortunately, much of modern religious doctrine and theology has been created by the faulty and habitual mind. The doctrines of *salvation by faith alone* and that *God is three Persons* are examples of this faulty thinking. Recall from Chapter Eleven that God's theological and scientific strategy was to enable our internal reality to come to the forefront, be examined, reformed, regenerated and ultimately to gain eternal blessedness. Sincere introspection reconnects our inner reality to the spiritual knowledge that the habitual mind kept locked in bondage in the memory. Self-disclosure allows us to properly apply what we know to what we are. But introspection involves more than self-observation, it involves *permission from the will*. We have to *want* to see the negative aspects of our personality and want to change them. Change of character is real only when there is a change in one's heart. True religion was designed to break through our mask of false consciousness, which preserves our trance-like state and protects our inner motives.

▶ Organic Centers of Gravity

The reason why affections can change blood flow is that every quality of Love and its disposition represents the psychodynamic equivalent of a physical gravitating system. Every quality of Love is a distinct principle of action with its *center of gravity* in a distinct region of our multi-leveled cognitive architecture. Since every organic process and function in the physical body has an analog in some area of the brain and every area of the brain is an analog of and correlated to some operation of the human psyche, every tissue, muscle, and organ represents a center of gravity or equilibrium end-state for some organic power or fluid.[27] This gravitational order not only allows dynamical processes to operate in their proper organic sphere or "space," it allows all bodily structures, membranes, and fluids to bear the same *affection* or *aversion* as in the mind.[28] A change of mental state changes the state and disposition of the blood, and a change in the blood changes our mental state.

The whole organic system cooperates so that the blood can change its chemical composition and viscosity according to mental states and adjust its center of gravity to where it needs to be in the body. For instance, when we are in an active state, the center of gravity moves to the arterial blood, which is determined into the smallest arteries, where it releases nutrients to the body and gets lighter. In a more passive state, the center of gravity shifts to the venous blood, which draws aliments to itself and gets heavier, and seeks the larger veins. These two extremes are analogs and correspondences of "like" and "dislike." In the active state, blood has aversion to food, while in the passive state there is strong *appetite*. This is why pep pills or "uppers" reduce appetite and "downers" increase appetite—they both chemically manipulate blood flow, which has a corresponding effect on the mind.

Every organ in the body helps maintain proper blood chemistry in accordance with our emotional states. Melancholy and excessive sadness thicken the blood with undigested and heterogeneous substances,[29] while anger excites the bile,[30] and "merriment and amusements contribute to the digestion of the foods, the dispensation of the chyles [emulsified fats], and the promotion of the circulations."[31] All human emotions produce corresponding chemical effects on the viscosity, gravity or levity of the blood. Levity in the blood corresponds to levity and joy in the mind.

Every species of human affection, whether corporeal, sensual, rational, or spiritual, operates from and has the center of its gravity in some distinct realm of the multi-leveled scaffolding of the mind, brain and body. All Love has a real center of gravity, for Love is the determination of a real force into some subject or form, where it finds its equilibrium and correspondence. The significance of this to neuroscience, physics, and religion is that quantum gravity involves the application of gravitational order to discrete qualities of force and energy, ranging from the physical to the psycho-spiritual.

There is a real, eternal consequence to all these vicissitudes of gravity. As we saw in Chapter Eleven, our life choices and their attendant loves form the center of gravity for our soul's action. The soul, by means of our Ruling Love, fashions the specific utility and self-organization of our spiritual being. This non-physical bio-complexity allows us to fully function as human beings in the spiritual world. Making life choices as though they had no eternal consequences is truly the "taking-away-of-responsibility"; it is a form of hypnosis and sleep. True religion and God's Holy Word address this diminished cognitive function in their quantum language. Recall the elevated interpretation of the biblical phrase "Let there be Light" in Genesis. It represents not that God created electromagnetic phenomena in the physical universe, but that God started the process of waking up a person's desire to begin spiritual evolution from within.

► Miraculous Hypnotic Cures

Another theory of hypnotism proposes that deep within us there exists a higher form of consciousness. Consider this theory from the perspective of the miraculous cures that have been attributed to hypnotism. Swedenborg was a witness to such things as a youth whenever his father, a Lutheran Bishop, used the power of suggestion and passionate reading of Scripture to cure people of hysteria and other mental ailments.

Besides the subconscious or inner human essence, there is the *unconscious* mind that controls our involuntary or instinctive functions. Swedenborg and modern neuroscience agree that the seat of the unconscious mind is in the cerebellum. When we sleep, the cerebellum becomes more active and restores the imbalances that the cerebrum, with its various worries, concerns and worldly stresses, places on the bodily system during the day. The cerebellum restores the common equilibrium of the body according to cosmic order, not the faulty order of human prudence.

While the cerebrum draws its information from the outside world, the cerebellum perceives what is going on inside us, both our myriad organic processes and our states of mind. What the cerebellum senses comes not into the consciousness of our mind, but into the consciousness of our soul.[32] While the cerebrum senses and is affected by the outside world, the cerebellum senses the passions and affections of the cerebrum and induces the appropriate and corresponding physical change of state upon the organs. For instance, if the eye sees something beautiful or ugly, it is the nerves from the cerebellum that dilate or contract the pupils.

As we saw in Chapter Nine, the cerebellum is fashioned under the Divine order of the soul and gets its active information immediately from heaven. It is instinctive. Its involuntary function heals injuries in the body and as the body's master chemist, it cannot be matched by any human chemist. The cerebellum has a full mastery of science.

The cerebrum responds to external influences. Since the cerebellum accommodates the organs to the qualities of the affections in the cerebrum, its super-physical powers can be manipulated by external influences. A hypnotist *acts as that external influence* and manipulates (unknowingly) the cerebellum's ability to modify the organic structures to be in agreement and correspondence with the interior state of the mind and its affections. As mentioned above, a hypnotist renders a person's false consciousness passive and gains direct access to the subconscious mind. The hypnotist then makes the cerebrum's subconscious mind believe that healing is taking place in a diseased part of the body. The cerebellum detects this new state, and its para-physical powers comply. The subconscious mind, free of the habitual mind, blends with the involuntary wisdom of the body. The power of the subconscious mind to *visualize* the cure "as accomplished" ushers in the action of the soul, through the cerebellum, to make the appropriate modifications upon the physical body. Miraculous cures can then result. Hypnosis has been used as anesthesia in surgery with surprisingly positive results precisely because it disconnects some neural pathways and connects others. Preachers and holy men with powerful personalities produce ecstatic states on believers, even cures for disease, by suspending their habitual minds and gaining access to their subconscious and unconscious levels.

▶ **Astral Projection**

Swedenborg cited another form of ecstatic state bordering on the miraculous caused by a peculiar blood flow. What he described sounds very much like what we would call an out-of-body experience.

> "In northern regions certain persons skilled in the art of magic are credited with being able to fall spontaneously into a kind of ecstasy in which they are deprived of the external senses and of all motion, and with being engaged meanwhile in the operation of the soul alone, in order that after resuscitation they may be able to reveal thefts and declare desired secrets."[32]

Swedenborg was referring to the Lapps of northern Scandinavia. He had learned of their arcane abilities during his earlier years in college. Recall that as a child, he also induced his own unusual mental states through intensified contemplation which stopped normal breathing, producing a more internal respiration. During his later anatomical studies, he seemed to discover the precise blood flow that accompanied the rarified mental state of an out-of-body experience:

> "In persons subject to ecstasy the circulation of the blood seems to have stopped; for the pulse of the arteries is

felt nowhere except the cervical artery, where it is very feeble; the respiration is also gone. Thus there is nothing which shall elevate the muscles and limbs, either on the part of the blood or on the part of the lungs—for both concur in all actions, especially the voluntary. The blood does not run through its aorta, that is, through the common trunk of all the arteries in the body; but that which enters the right auricle of the heart is carried off through the foramen ovale (which in these subjects should be open) or else through the coronary vessels, into the left heart, and from here it is taken up through the ascending trunk of the aorta and through the vertebral artery, and perhaps also the carotid, toward the cerebellum and cerebrum; it afterwards returns by the vertebral vein, which issues from the cranium, and by the spinal veins, which run through the posterior region of the long spinal shaft, and from there it flows off into the vena azygos and for a short space into the inferior vena cava, and thence back again into the right auricle of the heart." [33]

In this particular circulation, most of the blood was going to the vertebral artery and the *cerebellum*. Swedenborg went on to state that just enough blood entered into the cerebrum to keep it alive. The cerebrum also drew in some sustenance from the air through the pores of the skin, but it remained relatively inactive in this trance state. This was a real state of suspended animation of the cerebrum, and its modified blood flow may indeed provide the mechanism behind human after-death or near-death (NDE) experiences.

Swedenborg's own spiritual experiences were unique because during the last 28 years of his life he could enter these states without losing connection to the five physical senses and his waking consciousness. His entire systematic approach to theology came from things revealed to him in this unique cognitive state. The men and women of the Most Ancient Church also had this cognitive ability, which allowed them to learn profound things about God and heaven. Recall that Swedenborg had learned that Sacred Scripture was written by men who were put into a similar state of the "spirit" in which they received visions and revelations directly from God.

"As man's spirit means his mind, therefore, 'being in the spirit' (a phrase sometimes used in the Word) means a state of mind separate from the body; and because in that state the prophets saw such things as exist in the spiritual world it is called "a vision of God." The prophets were then in a state like that of spirits and angels themselves in that world. In that state man's spirit like his mind in regard to sight, may be transferred from place to place, the body remaining meanwhile in its own place. This is the state in which I have now been for twenty-six years, with the difference, that I am in the spirit and in the body at the same time, and only at times out of the body. That Ezekiel, Zechariah, Daniel, and John when he wrote the Apocalypse, were in that state is evident from the following passages."[34]

The spirit took me up, and brought me in a vision by the Spirit of God into Chaldea, to them of the captivity. So the vision that I had seen went up from me. (Ezek. 11:24)

I was in the Spirit on the Lord's day, and heard behind me a great voice, as a trumpet. (Rev. 1:10)

Swedenborg actually observed disconnection and intensification of cognitive functions of those in the spiritual world. He witnessed spirits having their lower faculties lulled to sleep while a more interior sense and mental faculty was aroused to a higher degree of wakefulness.[35] He was also shown how those in hell constantly try to excite and keep others in the lower mind of worldly inclinations.

▶ Anton Mesmer

Anton Mesmer (1733-1815) was a Viennese physician who first brought hypnotism to the attention of western science. He first made the observation that his patients would bleed faster when he moved towards them. He also discovered that touching his patients produced a positive effect on their health. He suspected that he was radiating some kind of *magnetic* or subtle force that augmented those same forces in the ill patient, the normal flow of which had become disrupted.

He discovered that this subtle force and the person's *imagination* were closely linked. He experimented with magnets and found that they enhanced a patient's susceptibility to *suggestion*. Through the use of both magnets and suggestion, Mesmer obtained great success in curing his patients. Later, he added wacky outfits and various props to the procedure.

While Mesmer's subtle cosmic fluid was later resoundingly dismissed by a special scientific committee that included Benjamin Franklin, no one doubted the verity of his considerable cures. What the committee did not grasp was that Mesmer's use of magnets, wacky outfits and props were very effective means for breaking through his patients' habitual minds and gaining access to their subconscious minds.

Swedenborg had made the similar discovery that the cognitive function of imagination operated under a vortical or magnetic force when Mesmer was about ten years old. Swedenborg's Doctrine of Forms identified a magnetic-structured neural substrate in the hierarchical scheme of brain organization. There is no evidence that either man knew of the other's work, but it is certainly possible, because Swedenborg was making headlines throughout Europe when Mesmer was a young man.

Neuroscience may be moving in Swedenborg's and Mesmer's direction. Consider the microtubules discussed in Chapter Five (p. 93). These may represent a distinct neural network that is mathematically identical to kind of magnetic system called "spin glass." Brain researchers are looking at microtubules as a possible distinct nervous system within the neuron.

Research in *Transcranial Magnetic Stimulation* (TMS) may further indicate that Swedenborg and Mesmer were on to something. Scientists have found that focusing magnetic fields on specific areas of the human brain can cause phenomena in volunteers ranging from a loss of eyesight or inability to move their arms and legs, to experiencing euphoric happiness, including mystical states of consciousness and even spiritual encounters with God.[36]

Swedenborg, like Mesmer, believed that the organic activity of both the human body and mind ejected an effluvial sphere around people. The quality of each sphere is a reflection of one's state of mind, specifically to one's affections and loves.

"From every man there goes, nay, rather pours forth and encompasses him, a spiritual sphere from the affections of his love; and this imparts itself to the natural sphere which is from the body, and they conjoin themselves."[37]

It is through these spheres that we can sense the invisible "chemistry" between ourselves and others. When the spheres of two people are powerfully sympathetic and make contact, both individuals can come under the special spell of love. Similarly, a hypnotist creates a *sympathetic connection* with some passion or intense expectation in the patient and blends with his or her sphere.

> "Sympathies and antipathies are no other than exhalations of affections from the mind, which attract another according to similitudes, and cause aversion according to dissimilitudes."[38]

Swedenborg made a remarkable statement concerning the physics of telepathy and the continued connection between close friends, after one has passed on—through sympathetic connection of their spheres. The mental wave fronts in the nerves send out magnetic eddies that affect others over great distances as long as the hearts are similarly attuned. These non-local connections can remain even after the death of one of the individuals!

> "Sympathy is so great and, as it were, magnetic, that there is frequently a communication of many persons at a distance of a thousand miles. Such sympathies, however, are deemed by some as idle tales; and yet experience confirms the truth. Nor would I wish to mention the fact, that the shades of some men have become visible after the fates of the body and the last rites; and they could never have been made manifest (granted but not admitted) unless the animal spirits had been mutually conjoined and not separated from their mutual fellowship."[39]

In Swedenborg's cognitive model, all levels of human consciousness are related to affection, to the quality and bent of our various loves. All perception and recognition involves *harmony*, so some dynamic of Love focuses our attention and organizes our thinking. This hierarchical approach to Love allowed Swedenborg to fashion his multi-leveled cognitive model within a broad person-level framework—from corporeal affection at the bottom to Spiritual Love at the top. Each hierarchically stratified level is a self-organizing system that adapts information to its own disposition. The center of gravity or *Ruling Love* of one's life is always predominant. It subordinates and coordinates all the other types of Love. We become what we love, and the inner bio-complexity of our spirit is fashioned around our subjective love and its utility in reaching some end, good or bad.

Religion attempts to bring this multi-leveled cognitive architecture back into Divine order by elevating Spiritual Love, rather than self-love, to the top. In this way, it organizes the innermost fabric of our being around a higher and nobler principle of utility, which is always to serve the common good, not the individual good. God's entire focus is on our individual ends.

Everything in this book, from physics to evolution, from neuroscience to theology, has dealt with some dynamic of *Love*. If Love is indeed the ultimate science, then it must lead us to a comprehensive model of reality through which scientists can evaluate its explanatory and predictive powers.

Such a schematic exists.

SUMMARY

▶ Hypnotism provides evidence that the human intellect and consciousness consist of distinct cognitive functions.

▶ All forms of trance involve the disconnection of a distinct cognitive function from the others and its intensification.

▶ Humans consist of three basic species of "blood" or essential essences, the flow of which provide the mechanism behind a hypnotic state.

▶ All blood flows accommodate some passion of the mind.

▶ The Lord came into the world to de-hypnotize the human race.

PREDICTION

New scientific and theological revelations will require humanity to evolve spiritually.

(Love equals a "non-material" unit of action)

Chapter Fourteen

LOVE:
THE ULTIMATE
SCIENCE

"Nature is in an obvious sense 'unified.' The universe
we find ourselves in is interconnected, in that everything
interacts with everything else."

— Lee Smolin, Theoretical Physicist
The Trouble With Physics

Doing science is a "rough and tumble" business. As soon as a new theory is offered, the first thing scientists will do is to prove it wrong—aggressively. This procedure doesn't work with the topic of God. The claim that God exists is inherently non-falsifiable. That is, God's existence cannot be conclusively proved or disproved by the methods available to scientists. A scientist can believe in God from his or her faith, but the "possibility" of an intelligent designer doesn't stand up to scientific muster. The reason is that science is determined to explain the universe *solely* through natural laws. If spiritual forces and laws operate on the most fundamental level of reality, then they play no role in a scientific model. Natural science deals with physical models, not truth.

I chose the quote above from Lee Smolin because scientists *do* believe that the universe is unified (which is quite an achievement for mindless matter). They continually seek a final unified theory—a *Theory of Everything*—which embraces all laws and forces in the universe. Smolin points out that such a final theory should offer us more than a formula for making predictions in experiments; it should "give us a picture of what reality is."[1] But this would require that scientists address *foundational* issues. What agency and what first principle would lawfully lead to a unified world? What first principle would create a universe of laws perfectly fine-tuned so that life, much less intelligent life, could emerge?

I argue that it is Love.

As we saw in Chapter One, the Big Bang theory, which posits that the universe started as an essential singularity, rules out that the universe had a physical cause. Some pioneering thinkers in quantum theory seek the origins of the physics ruling the universe in a non-physical principle from the non-temporal and non-spatial realm of "pregeometry." In other words, the universal dynamics of unity somehow existed before there was either space or time.

Love, since it is a state of mental activity, is non-physical. And we don't need scientific proof to tell us that the essence of Love is to unite. Love offers us a wonderful, non-physical, first principle and formative creative force for providing agency in nature and the emergence of a unified world. Whether we start from a Big Bang or a quantum vacuum, nature shows us an unremitting disposition towards self-organization. Things find distinctiveness through unification. Existence is coexistence. Current physics is not capable of explaining this lavish ontological situation of reciprocal union especially from nature's deepest activity. According to theologian/physicist John Polkinghorne, physics lacks an explanation of how "a determinative property of large scale systems could emerge from the indeterminate quantum substrate of their constituents."[2]

Quantum physics hints strongly that *disposition* and *tendency* are a fundamental substance. Substance is the capacity for generating form. Swedenborg argues that Love, which is a disposition to generate forms and outcomes of measurement, is the real underlying non-temporal and non-spatial substance of the world. Not only do determinate results emerge from the microworld of infinite propensity, they also lead to measurements of order and unity through *utility*. Complexity emerges in the universe through mutual cooperation as a physical analog of Love. The "measurement problem" in quantum physics is not unrelated to theology or God's acts in the world.

If Love is indeed the non-physical origin for unified order and structure in the world then Love must lead to a model that both explains and predicts. It must provide us with a full expression of complexity theory. It must show us convincingly how God thinks.

Can the dynamics of *Love* lead us to a grand unified theory of everything? Yes. However, a true Theory of Everything can come only from *revelation*. Such a theory and model will come from neither the efforts of the human rational mind, nor from rigorous scholarship. This knowledge is wisdom itself and is the sacred knowledge of the soul. It will embrace the holonomic patterning principles by which dynamical processes close on themselves and form cooperative unities and systems. This universal model represents angelic perception and the science by which "...the soul at once sees the intrinsic nature of all things set before it."[3] It will enable us to see all process, on all scales, as running through cycles in *comprehensive wholes*.[4] This knowledge of complexity will demonstrate that coherent structure is the outcome of the non-reversible flow of time, for according to Swedenborg, "time runs through these cycles."[5] A true Theory of Everything will not be obtained by scientific reductionism, for it must include theology. It will expand the theory of complexity and whole-part process into the laws and dynamics of spiritual salvation.

▶ Angelic Perception of a Universal Science

Angels directly perceive that The Lord created everything in the universe out of the Divine Love and Divine Wisdom that emanate from the Spiritual Sun as active information containing universal patterning principles. This spiritual influx is an endeavor and striving to take on measured form through both physical process and uses. Spiritual Love cannot be transformed into any form that does not serve some use or is disconnected from Divine Wisdom. As complexity descends from God's Love into the physical universe, utility evolves.

The influx of active information from God is *Love taking form in Wisdom*—indeed, all form is information, and complex forms are wisely arranged. Love is a creative causal agent that adapts information to its own disposition. Since Love can flow only into a corresponding form, it is *the only force in the universe that requires top-down symmetry, order, the constants of law, and a highly fine-tuned universe*. The laws of physics, which describe the rational structure of the universe, can have emerged only from a spiritual first principle. This implies purpose rather than randomness.

Since Love is a conscious, *living force*, it also provides us with the perfect creative agency to fine-tune nature such that the universe anticipates the evolution of life, including human-kind. This fine-tuning is called the *Cosmological Anthropic Principle*. Whether you believe in a Divine Being or not, it is a scientific fact that the universe is fine-tuned. As we will see, the universe, based on the harmonizing principle of Love, is perfectly fine-tuned. Science has yet to "realize" the full extent of this perfection.

Science and theology can both agree that the manifest universe is a process. The most important thing for us to do in order to have a proper meeting of God and science is to come to some agreement about what we mean when we speak of a universe that is in dynamical process.

Process is differentiation. Differentiation requires movement, that is, either change of state or physical change of position. But differentiation is incoherent unless it is unified and coordinated. The kinetic universe coheres into stabilized forms having both *order* and *orientation*. In order to be coordinated, however, process has to be subordinated to some unifying first principle. Coherent process occupies time and space in a way that can generate a fixed form. The process itself, or continuous change, gains some degree of permanence in the form it assumes. This subordination of process requires control, which has a physical equivalence in the coordination of both *change of direction* and *timing*. As we learned in Chapter Three (p. 49), a circle gives us the simplest geometrical expression of directionality and timing in which differentiation and process become a unified whole. This creates constancy through change.

> "Thus in every series there is established a kind of circle, in virtue of which the first thing can have reference to the last, and the last to the first."[6] (Figure 14.1)

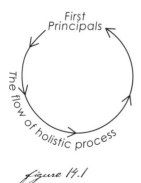

figure 14.1

To build a model for an angelic interpretation of a holistic Theory of Everything within the scheme of dynamical process, we use the circle at left as our starting point. A circle offers us the perfect geometrical model for illustrating constancy of form through the dynamics of continual change. There is no other way that God can act in the physical world or for there to be a nexus between the infinite and the finite unless geometry allowed process to continually move in repeating cycles and therefore respond to an infinite principle and influence. And there is no simpler way for the past, present, and future, that is, the operation of Divine Omniscience and Providence, to have continuity and control based on a rational, geometrical principle.

Because creation is contiguous and proceeds through levels, our scientific model must next express this circular process proceeding through distinct steps—as a series of singular events that are conjoined into a common whole in which the whole consists of a perfect *harmony* of its parts. As Swedenborg states, "In a word, the conclusion and end has respect to all the premises, that the effect may be like to its causes."[7]

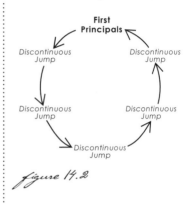

figure 14.2

In the act of creation, the Infinite mind of God simultaneously comprehended a progression of concepts and intentions that were His means to accomplishing the goal of a unified universe. This precision required a *mathematical intuition of ends* as opposed to a Divine "snap of the fingers." Since God acts nonstop, the universe must be fine-tuned so that there is a perpetual connection of all means from the first end to the last. Otherwise, the cosmic chain of events will be disrupted. This holistic approach also requires that all processes proceed by the same rules and on all scales. Divine Love therefore always acts as a force representative of the universe. Love, as causal agency, always follows a Divine template and a mathematically precise patterning principle in its progression towards a goal. Within all progression is a recurring image of creation. This universal patterning principle applies to the cycle of action depicted in Figure 14.2. So as a model depicting a universal science, it needs further mathematical modification. It needs to show how discontinuous steps are harmonically unified.

Proving God comes down to showing that mathematics, proportions, ratios—indeed the laws of the universe—put the unifying operation of Love on display. Ironically, a mathematical philosophy of universals is under our very noses, but its significance is epistemologically overlooked. The sacred ratios and proportions of such a universal mathematics is to be gleaned from octaves.

▶ The Mathematical Challenges of a Universal Science

Any theistic explanation of complexity and nature's propensity for self-organization requires top-down causation. The organizing principle, which subordinates and coordinates all things in universal nature so that everything is held in common connection, must come from "above." The principle of universal harmony, if we attribute it to God's Divine operation, must come from a superior region beyond spacetime. The principle of harmony is derived from the non-physical dynamics of *Spiritual Love.*

Swedenborg writes, "Harmony alone conjoins the entities of nature and sustains the world..."[8] and, "All concords emerge from the octave."[9] Harmonic laws involve real physics and precise mathematics. While energy spectra can be correlated with octaves, and energy creates matter (which also contains frequency values), science does not apply the laws of octaves to tangible material entities like matter. This is an oversight, especially if one wants to maintain the belief that nature is unified. Therefore, any framework, formalism or interpretation of universal law using octaves deserves serious attention in creating a scientifically plausible theory.

Swedenborg believed that universal patterning principles within greatest and least things were difficult to discover because the rich variety in the world often obscured these principles from the mind.[10] Scientific specialization has also led to the exploration of details and to reductionism, both to the detriment of the big picture.

Recall from earlier chapters that Swedenborg viewed order, complexity and unity as a gravitiational organization of the cosmos. All unified process must preserve a common equilibrium. It is a scientific fact that once something is set into motion its trajectory will naturally seek an equilibrium. However, organic process orchestrates a phenomenally large series of events, each seeking its own distinct equilibrium end-state while contributing to a common gravitatinal order that apprehends a single concept. Matter and energy cannot orchestrate anything. The mathematical intuition of ends by which Love disposes things into a unified concept is a harmonious ordering of distinct equilibrium end-states. Love is responsible for the harmonic ordering of all complexity.

To augment the model (Figure 14.2) representing a unified cycle of action proceeding from beginning to end, we need to express it as an octave. This requires us to divide it up into various "nodes" depicting a mathematical order of events in proper chronological progression and gravitational sequence to express harmonic process. Arranging these successive steps around our circle to convey unified process and meet the mathematical rigor demanded by scientists produces a simple schematic representing various octaves.

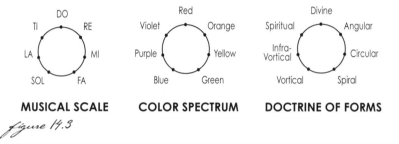

MUSICAL SCALE   **COLOR SPECTRUM**   **DOCTRINE OF FORMS**

figure 14.3

This produces the beginnings of a universal template that can potentially unify science and theology and even show us the mathematical precision with which a God of Love thinks. In religion, the number seven is often referred to as a holy number, but only Swedenborg provides a rational reason why this is so. In *Apocalypse Revealed*,[11] he explained that the number seven represents completeness—an entire period or cycle, from beginning to end. He also stated that the number eight signifies the beginning of a new series.[12] The seven-day creation story in Genesis is a narrative through which Divine ingenuity cleverly expresses a real law of physics!

Holiness involves unified completeness. In a dynamical universe, completeness is Divinely arranged according to circuits. As mentioned above, every series establishes a kind of circle

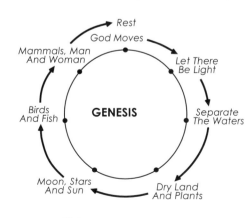

figure 14.4

so that the beginning of a process has constant reference to the last, and the last to the first. Swedenborg offered a very simple example of process forming a comprehensive whole through repetitive cycles involving an octave. The unfolding life of a tree represents a circular process whose directionality moves from seed, root, stem, branches, leaves, flowers, fruit and back to seed (Figure 14.5).[13] In this cycle, distinct steps find common equilibrium in a single idea—the tree. First principles return to first principles—from seed to seed. Octaves make this possible.

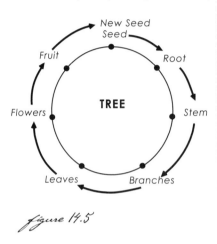

Swedenborg called this arrangement the *Circle of Life*. It consisted of a series of things that had their own distinct existence, yet were subordinated and coordinated into a comprehensive and unified whole through successive and simultaneous order. Another expression he used for this unified harmony of process was *the unanimous conspiring of efficients*.[14] He even added creative prose to the mathematical precision of harmonious process, seemingly to soften the shock that *Love* was to be viewed as a hard and exact science.

> "But the agreement of all these things cannot possibly exist without a spiritual principle of union, or Love, in the veriest rays of Life; from that principle alone beauty derives its harmony, its bloom and reality; and life, its day-dawn and vernal freshness; wherefore love itself shining forth from grace of form, by its hidden and insown virtue, elicits mutual love, and as an index reveals the vein of beauty."[15]

Distinct steps of succession close our circle by means of the octave, a real aspect of physics. But we still need to show how this autocatalytic unit can be kept in a state of thermodynamic integrity. In his work, *Worship and Love of God*[16] Swedenborg suggested that this perpetual orbit depended upon outside inputs. Somewhere in the cycle of action, then, mathematical contingencies must allow for openness to outside influences.

In an earlier chapter, we saw that God creates through contiguity and discrete steps. Discontinuity in nature necessitates that the dynamical universe and everything in it be held together through harmonic ratios and *mutual support*. Since creation is discrete and not continuous with God, ontological gaps appear at precise mathematical points along the path of every process in the scheme of top-down causation. These causal gaps serve the Divine purpose of *exalting* Love in the universe. The physical world was purposely designed to maintain its integrity through reciprocation and mutual *need*. These lawful gaps allow for "openness" and novelty in the universe. Everything in the universe depends on something else. All complexity and self-organization emerges from Divine strategies to fill these gaps with *outside* inputs. The universe is not only in dynamical process; it is bio-centric. The evolution of organic complexity was designed specifically for increased interfacing and mutual dependence with things "outside itself." This is why existence is relationship.

▶ Hungry Octaves

Organic life and its evolution is more than a random outcome of serendipitous events, because existence and neediness arise simultaneously. Even the simplest of life forms must address a myriad of needs at every moment during their organic cycles. When life emerges, it does so with mathematical precision and an intuition of needs. All living process has to be organized around the inescapable need to be fed.

All organic complexity is mathematically organized around nature's ontological gaps, because these are the precise points of openness where inputs are needed to keep the chain of causal links in biological operations and their common gyre in perpetual renewal. The inner organization of a life form takes shape around openings, or apertures, placed precisely at these discrete gaps so that the outer world and its substances can flow in and be processed for sustenance by the life form. (For instance, the organs of digestion are organized from the opening of the mouth.) All the organs within a biological structure depend on each other and are arranged for this *unanimous* and *single purpose*.

George Gurdjieff's insights offer important details that help flesh out a mathematical model of reality based on the bio-centric ideas mentioned above. In the early part of the Twentieth

Century, Gurdjieff introduced a symbolic patterning device that he said held the secret to understanding the intelligent structure of the universe and its unified processes. It portrays an organic model of reciprocal creation and maintenance based on a *doctrine of Love*. He called it the Enneagram (nine-diagram).

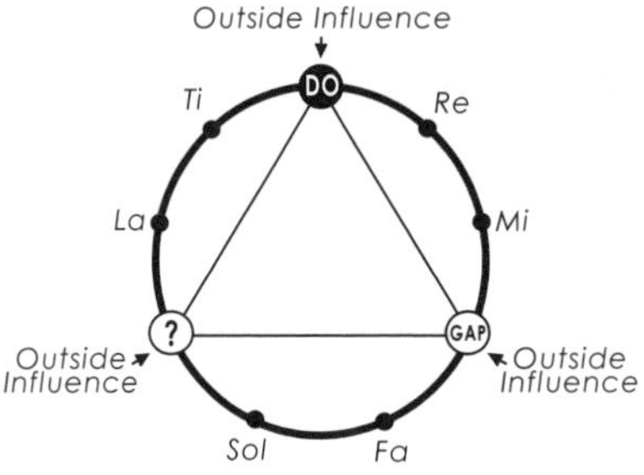

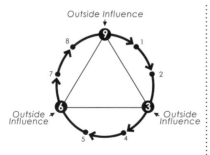

figure 14.6

The Enneagram consists of an octave placed around a circle. It places great importance on the mathematical fact that there are missing semitones between the intervals *mi-fa* and *ti-do* (which can be seen on any piano keyboard). These missing semitones have a cosmic ontological significance, because if reality consists of octaves, these "gaps" are precisely where process needs outside help to overcome nature's inherent discontinuity. God has provided strategies to bridge these ontological gaps by wisely making them the mathematically precise points where all process is open to outside influences and can be sustained and "fed."

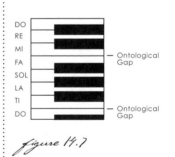

figure 14.7

The Enneagram contains the triangle 3, 6 and 9. Points 3 and 9 represent the missing semitones of an octave. These two points represent the *ontological gaps* where all holistic process is challenged by nature's built-in discontinuity. Points 3 and 9 indicate where external inputs or "food" can lawfully enter into the teleological framework of the system and keep the process *nourished*. Again, the external needs of all organic processes are anticipated with mathematical precision and cannot be the chance outcomes that current evolutionary science suggests.

figure 14.8

What is intriguing about this schematic of holistic process is that it portrays both constancy and contingency, embracing both strict law and openness to novelty and hazard. Point 6 does not correspond to a missing semitone of the octave. You won't find it on a keyboard. Point 6 is special, and as we will shortly see, has its analog in human free will. It is "chaotic" and its quality is determined according to what is happening at points 3 and 9. This is the point where the process can run into mischance or self-correct and promote novelty in the form of continued evolution. This schematic also demonstrates that process is not linear but consists of *triads*, or *trines*. Swedenborg concurred that three discrete modes must come together "in everything, that it may be anything."[17] This dynamic is an expression of God's threefold nature of *Father, Son, and Holy Spirit* being expressed in physical process as *active* force, *passive* force and *equilibrating* force.

The fundamental challenge for any model attempting to depict God's action in creation is to address the issue of an open system with chance and accident versus a closed system with causal predictability. The universe contains both constants and novelty. How does God act in a purposeful way that allows for order, novelty and the seemingly accidental to occur? Theologically, this issue extends to balancing God's will and human free will. The formalism and interpretation of reality of the Enneagram includes both *constancy* and *chance* in the Divine scheme. While a progression through octaves mathematically ensures that a goal will correspond to its initiating cause, proper conditions must be met at points 3, 6 and 9. Otherwise, there will be an unpredictable event. So the *hexad* part of the Enneagram (at the six gravitational nodes 1,2,4,5,7,8) represents constancy of law, and the triangle part represents the system's mathematical openness to new possibilities which might be good or bad. We will see further examples of this later in the chapter.

The Enneagram consists of another mathematical dynamic that ensures intrinsic unity. This unity is mathematically obtained by dividing seven into one, which gives us the following infinitely repeating series:

$$1 \div 7 = 142857142857142857$$

When we graphically apply this new series of numbers (1-4-2-8-5-7) that represent a true mathematical expression of unity within the outer cycle and chronological progression of the Enneagram, we get the direction of flow of influences that represent the inner order that is required for continuity in holistic process. This *inner* dynamic ensures mathematical unity and unanimous connection of causes throughout the whole *outer* process. The *outer* sequence of events or cycle would fall apart if it were not for this inner mathematical aptitude for *unity of process*. Outer sequential events can only be coordinated and subordinated into a *system* if they are inwardly connected in a way that describes mathematically a simultaneous intuition of the full concept. The combined outer and inner sequence of the Enneagram offers us a more graphic way of envisioning Swedenborg's concept of God's Divine scheme in creation as a *mathematical intuition of ends*.

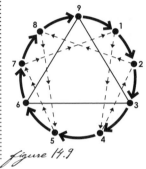

figure 14.9

It may be argued that Swedenborg had no such nine-part schematic in mind when he sought to create a mathematical philosophy of universals. But, in fact, they are one and the same. The Enneagram captures his idea that unified process must preserve a common equilibrium. Second, it expresses the idea that first principles return to first principles. Third, both Swedenborg and Gurdjieff referred to their systems as *hieroglyphic* keys to secret universal knowledge. The Enneagram gives a mathematical portrayal of the *science of correspondences*, within *perpetual* causality, starting from an initial state moving towards a corresponding final goal. It also consists of three distinct series, which also correspond to each other.

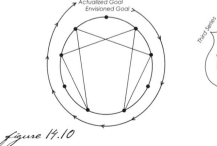

figure 14.10

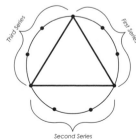

▶ Is God's Holy Word a Scientific Document?

Gurdjieff did not create the Enneagram. He discovered this ancient repository of cosmic knowledge in Central Asia. Some of the most sophisticated cosmological ideas in history have come from ancient Asia. Swedenborg said that the science of correspondences was cultivated in many ancient kingdoms of Asia.[18] Swedenborg was also aware that the number "nine" symbolized "all things conjoined into one complex" to produce utility and common good. The *triangle* and *hexad* in the Enneagram may well correspond to the marriage of truth to good (*three* and *six*) to promote some use from the Holy Word. The number "nine" actually signifies Conjunctive Design from a Holy principle. Furthermore, the Enneagram is a biocentric, rather than a static expression of correspondences, just as Swedenborg maintained the Holy Scripture was (see Chapter Ten, p. 239). Holiness and holistic process are the same things.

In his anatomical works, Swedenborg's posited that organic process consists of subordinated and coordinated trans-flux equilibrium states. Simply put, all unified progression seeking a goal moves from one discrete gravitational or equilibrium end state to another, forming unities, or a common equilibrium through harmony. So God's *living conscious action* of creating organized systems (mutually autocatalytic complexes) always comprises process moving through seven distinct centers of gravity in the mathematically precise pursuit of unity. Love as causal agent always brings a series of events and their connection to a similar harmonious state in which the particular details conspire to generate coincidence in the general, larger complexes they equip and comprise.

Again, this is what Swedenborg meant when he said that creation was a *mathematical intuition of ends* and proceeded out of a universal force that carried an idea *representative of the universe* in its endeavor. The Enneagram can be understood to express what modern physicists call the *Lagrangian*, which represents the most compact mathematical expression of the complete laws of nature, on every scale. However, the Enneagram is an organic Lagrangian.

Both Swedenborg's and Gurdjieff's view of a *mathematical philosophy of universals* represents a degree of fine-tuning in the universe natural scientists are unable to bear. After all, such a model of reality is based on Infinite Intelligence. Even more challenging to current scientific paradigms, Swedenborg offers strong hints that the Lord's Holy Word and its structure is the source and template for the fine-tuning of all ordered process. This knowledge came from an angelic perception of universal laws.

> "Of the Lord's Divine mercy I too have been permitted in the same way to see the Lord's Holy Word in its beauty in the internal sense, and this many times; not as it is while the words are being explained as to the internal sense in detail, but with all things both in general and particular brought together into a single series and connection..."[19]

Theological doctrine, without further revelation, offers us no scientific insight into how the Lord God created everything from the Word (John 1:1-3). In *Views from the Real World*, Gurdjieff stated that the law of trines and octaves represent "God the Word."[20] Could the Enneagram actually turn out to be a template for process and Divine order derived from the sacred architecture of Scripture? Is it a universal patterning principle for how God thinks?

Is the Enneagram the Word of God?

In the book *In Search of the Miraculous*,[21] Russian mathematician, P. D. Ouspensky, who also studied the Enneagram, described a 1922 performance by dancers whose movements followed the inner lines of a large Enneagram drawn on the floor of the room. Gurdjieff had taught his followers to move about the Enneagram in a special way to help them gain a more visceral understanding of the cosmic processes it portrayed. But Ouspensky also recalled seeing the Enneagram depicted in a most unique way two years earlier. Something unexpected had been added to the schematic of the Enneagram.

Ouspensky observed that the four biblical creatures of Ezekiel's vision and of the Apocalypse were rendered inside this particular drawing of the Enneagram, including the image of a dove. He later learned that these creatures signified various human cognitive centers and that the Enneagram portrayed a *universal philosophical language*. What he did not know was that by adding these symbolic features, Gurdjieff had raised the stakes even higher. The Enneagram not only had cosmological importance, but its biocentric dynamics represented the Lord God's living Holy Word! Figure 14.11 is a depiction of the Enneagram based on Ezekiel's strange vision of the creatures and *wheels within wheels*.

figure 14.11

According to Swedenborg, these four creatures were cherubim and represented important aspects of the Holy Word; the Divine Living Truth from which the Lord God speaks and protects the holy things of love and faith. Gurdjieff alluded to this same symbolism in his epic work, *Beelzebub's Tales to His Grandson*, when he describes God as existing with his cherubim and seraphim before the act of creation. The words "cherubim" and "seraphim" are code for God's Holy Word, which existed before creation.

Ezekiel describes his vision thusly:

As for the likeness of their faces, they had the face of a man, and they four had the face of a lion on the right side; and they four had the face of an ox on the left side; and they four also the face of an eagle. (Ezekiel 1:10)

And John:

And the first beast was like a lion, and the second beast like a calf, and the third beast had a face as a man, and the fourth beast was like a flying eagle. (Rev. 4:7)

Thanks to Swedenborg, we can interpret the spiritual meaning of these various symbols. An eagle signifies Divine circumspection and providence, because when it is in flight it has a high vantage point. The lion represents the power of Truth derived from Love, because lions have great power and heart. The face of a man represents Divine truth intelligently understood, because man has rationality and can receive the Lord's teachings. The ox represents the love of natural or civil good and truth, because its value to man is worldly. So like Gurdjieff, Swedenborg also identified these creatures as representing distinct cognitive and emotional functions or centers, for the Holy Word, in its complex, portrays a Grand Human Design. These living creatures also had wings and moved in unity with the wheels. Wings represent an abundance of science and another power of truth—things that lift the mind. Since the wings in Ezekiel's vision make contact with each other, they represent the unification of knowledge—the Theory of Everything—including the unity of natural science and theological doctrine. Wheels represent the power of truth advancing through Divine foresight. That the creatures moved with the wheels indicates that the unity of knowledge and goodness is from the Holy Word. All power to act and move forward in time and space is the power of truth emerging from God's good. God's living influx is always determined into comprehensive wholes for utility and is infinitely purposeful.

In both John's version in Revelation and Ezekiel's vision in the Old Testament, the creatures were "full of eyes" and refer to God's heavenly throne. In a nutshell, the Enneagram represents God's Infinite Wisdom from which all law and causality proceed. Put into scientific language, this is a schematic of active information creating complexity by God's Providence! The fact that Ezekiel observed "wheels within wheels" is highly suggestive that this patterning scheme operated on all scales, as shown in Figure 14.11.

Furthermore, from Swedenborg we learn that a "dove" represents spiritual regeneration and purification. One of the most important lessons Gurdjieff conveyed to his adherents was that the Enneagram explained the process of spiritual re-creation as an extension of biological process into the psychical and mental plane. Figure 14.12 is Gurdjieff's Enneagram representing the gravitational order of the basic human organization. It illustrates how the flow of process moves through various organs which serve as centers of gravity by which the entire system finds a common equilibrium through a sequence of mathematically precise states of stability.

The Enneagram not only shows the proper inputs needed at points 9, 3, and 6 to keep the system "alive," it also shows the points where new possibilities can lawfully enter the system. It is at point 6 that God places His hope for humankind. Gurdjieff called point 6 a "bearer of new direction." In spiritual evolution, when one adopts a higher principle of Love, a new octave will begin to self-organize at point 6 through the *metabolism* of new ideas from a deeper mental perception of *sensory data*. This new spiritual octave will be discussed shortly. Before addressing spiritual evolution we need to see if Swedenborg agreed with the basic organic scheme depicted below.

In Swedenborg's anatomical work, *The Animal Kingdom*, he described the progression of organic process and "digestion" in a similar sequence to that of both the inner and outer movements depicted on the Enneagram. Please be attentive to the fact that this discussion requires that we simultaneously keep track of both inner and outer activities—the action around the circle as well as within it. Refer to Figure 14.12 as you follow my explanation.

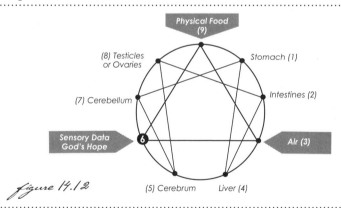

figure 14.12

Terrestrial food *comes from without* and enters through the mouth at point 9 and moves towards the stomach (step 1) where it is turned into *chyme*. Right from the start, the Enneagram illustrates Swedenborg's idea of how Love disposes action because what enters the mouth is driven from appetite, which emerges from some affection and evaluation. Swedenborg knew that the stomach received its nerves from the cerebellum via the *par vagum*, which corresponds to the inner process of step 7 moving towards step 1. From the stomach, the chyme moves into the small intestines (step 1 moving to step 2) where it is turned into *chyle*. Swedenborg observed that the small intestines received bile from the liver to assist this operation (step 4 towards step 2). Step 2 refers to step 8 because sustenance has reproduction as its ultimate goal. Next, the small intestines send the chyle, now referred to as *serum*, to the liver by means of veins and the *vena portae*. Each step represents a change to a different center of gravity.

According to Gurdjieff, the process now meets its first causal "gap" and will come to a stop in the liver unless there is additional help from the outside. This outside help is provided by the atmosphere entering into the lungs as a second kind of "food," bringing in the oxygen that the system uses to create arterial blood. Swedenborg drew the same conclusion that the liver needed external help:

> "Were it not for the hepatic artery, the defecation of the chyle and the lustration of the blood could never be accomplished by the liver."[22]

The liver would clog up if the oxygen-rich arterial blood coming from the *hepatic artery* did not mingle with the impure blood of the vena portae and act as a correcting solvent. If the process stopped here in the liver, blood would not enter the vena cava toward the heart or the brain, and the "game" would be over.

The Enneagram next depicts arterial blood moving to the cerebrum (step 4 to step 5). Swedenborg observed the same phenomenon or arterial blood traveling up the internal carotid arteries to feed the brain. Modern science concurs that capillaries from the *internal carotids* supply nutrients, glucose, and aliment for the neurons by means of astrocytes.

It is less clear how to follow this process anatomically as it progresses from the cerebrum to the cerebellum, as indicated by the Enneagram (step 5 to step 7). Swedenborg suggested that the cerebellum knows everything that the cerebrum is doing. He stated that these two brains are joined in the occiput and the medulla oblongata and that they mingle their nerves in the various plexuses and ganglia of the body. This meshing of nerves may allow the cerebellum to interpret our state of mind in the cerebrum and then send a signal to the stomach (step 7 to step 1) to absorb specific aliments from the food in order to make blood chemistry correspond to our various inclinations and states. A common equilibrium could not be obtained unless the bodily system was designed to respond to a person's affections and subjective loves. From step 7 to 8, we can more easily pick up the anatomical trail of the Enneagram, since the nerves to the human organs of reproduction are involuntary and originate in the cerebellum.

In humans, point 6 on the Enneagram represents a special Divine strategy for outer influences to enter into the system as a third kind of "food." Impressions and data from the physical world reach the brain and its neurons here by way of the five senses. In lower forms of animal life, these influences are handled instinctively. But in humans who possess free will, the process becomes chaotic and uncertain exactly at this critical ontological gap. This is where humans can either continue evolving or thwarting their God-given potential. Gurdjieff suggested that in the normal conditions of human life, impressions and data are not properly assimilated or metabolized. As a result, humans often find themselves taking part in ignoble behavior in spite of what they have learned.

What allows the mind to properly and fully metabolize information? Swedenborg stated that it is the human endeavor and affection to become *wise*. This requires that we seek knowledge not simply to promote our self-interests, but to lead us to the cognition of the *truth* that will lead us to the cognition of *good*. This distinction offers us a new way of understanding the true purpose of religion: to help humans to metabolize ideas further and feed their souls. Religion is God's strategy to help us start to create a new octave at point 6 in the Enneagram. God provides us with proper values for living wisely. These values are real spiritual aliment but require a new conscious effort on our part to assimilate it.

The process by which we seek wisdom and become spiritual is actually the birth of a new octave and structural order within us. Swedenborg describes this metabolic progression in the human intellect by steps that can be perfectly arranged around the Enneagram. As mentioned above, this new octave representing our spiritual evolution starts to form at point 6 on Gurdjieff's diagram (Figure 14.9). When we adopt a higher spiritual value into our lives, love arranges the human mind into a new series of gravitational nodes that allow us to metabolize information in a way that produces *wisdom*. This treatment of the human intellect as an octave can be found in Swedenborg's work, *Worship and Love of God*,[23] where he describes it as a *circle of operations*.

By placing Swedenborg's depiction of the distinct cognitive functions of the human intellect around the Enneagram as

HUMAN INTELLECT

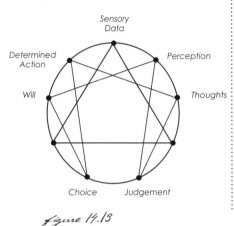

figure 14.13

gravitational nodes we will see further evidence that this ancient symbol is a schematic of his mathematical philosophy of universals which oversees all coherent process. Again, we will need to track both inner and outer movements to see whether the correspondence principle between the human anatomical octave and the octave of the human intellect holds up.

▶ The Enneagram and Mathematical Philosophy of Universals

First: In Figure 14.14, we see that the process of human cognition starts with external *influences* reaching the organs of sensation from the circumfluent world. This is represented by the data entering at the top of the triangle, just like physical food entered at the top of Figure 14.12.

Second: *Perception* then seizes upon information that will be instrumental to our goals and Ruling Love (step 1). Since perception is influenced by what we *will* or *intend*, it fits perfectly with step 7, which has a mutual reference to and influence over step 1. But perception must also be inwardly oriented to *judgment*, which has reference to *thought* and *reasoning* (step 1 moving toward step 4 then to step 2). In *The Five Senses*, Swedenborg actually states that sensory data is "digested."[24]

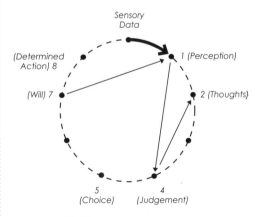

figure 14.14

Third: We take what we perceive as valuable and interesting and submit it to a deeper operation of the mind, where we determine what we will explore by rolling the perception around and inwardly looking at it from many angles. This rational operation is called true *thought*, which is not associative

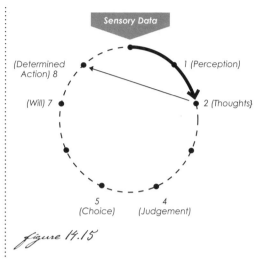

figure 14.15

but reflective. Thought and reasoning (step 2) have reference to some final or *determined action* (step 8).

Fourth: *Judgement* is next. Judgement is the perfection of thought. We form judgements from the causal objects of thought. Judgement collects the ideas and thoughts that focus on the matter proposed and disposes them into a framework and series similar in form to an analytical equation—a "chain entwined by intermediate reasons."[25] This means that thought requires the support of additional ideas already in the memory to move into the function of judgement. Swedenborg insisted that analytical and algebraic calculus proceeds in a similar manner as the rational operation of the mind. His mathematical philosophy of universals, or science of universals, refers to both the arrangement of process into a unified series and the mathematics involved between the distinct steps. In the elevated and distinct mental operation of judgement, the intellect

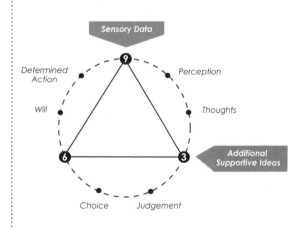

figure 14.16

✳ *Organic Mathematics*

All equation consists of analogy and ratio, and all analogy and ratio relate to form. *Forms consist of ideas and information. The human mind generates thought in the brain and neuron by modifying their structures to take on similar ratios, analogies and harmonies. Swedenborg likened this to musical rules by which a "sharp" mode draws the "grave" mode into coincidence.*

Within the multi-leveled architecture of the brain and neuron (discussed in Chapters Five and Six) the various cognitive substrates act one inside the other. The deeper levels of structure not only communicate with the outer, more compound structures through contiguity, but from a principle of Love, can detect new analogies and ratios from the modifications ("tortional" gyres) of the more general and compound structures. This heirarchical layering gives biological structure the ability to modify its states in a way that produces more distinct ideas from general ones and create new equations that advance the mind's inquiry into truth.

If all this seems obscure, you are not alone. The steps between thought and judgement are not only distinct, they involve structures with different curvatures operating within different kinds of "spaces" and contain infinite new powers of flux and mutability. This is how Swedenborg believed that deeper qualities could be extracted from general qualities and then scientifically expressed through quantities. Current science lacks a mathematical knowledge of qualities. Swedenborg could tap into higher mind and seems to have left us with ideas worthy of a more advanced civilization.

begins to form new harmonic unities* from the general sense of thought to produce more distinct ideas. This analytical process is an inquiriy into truths, arranged into a new series or equation from new analogies and ratios, which are further inquiries into the "good" of these truths.

Fifth: The intellectual functioning of the mind moves into its discretionary mode in the process. The mental calculations up to this point have been "bare" cognitions as to what is deemed true and good. One's intelligence must now be applied to the heart. The analytical process of forming equations (Figure 14.17) is now replaced by an *inverse* synthetic process (Figure 14.18) so that the conclusions of judgment can become ends which are to be pursued. This next operation in the series is called *choice*. An action cannot be determined until some *choice* has first been made (step 8 to step 5).

Sixth: *Choice* must actually be willed by the heart (5 to 7). Again, our will and our *intentions* influence our perception and our focus (7 to 1) as depicted by the inner and outer movements of Figure 14.19. The will contains all the things we have already chosen, reasoned about, and intend to act upon (7 to 8). This process is perfected in the degree to which we adopt the spiritual principle of Love into our lives and goals. Our improved life-choices lead to a new attuning of our lives.

Seventh: If we have *perceived, reasoned, judged, chosen,* and *willed* correctly, that is, according to the Lord God's spiritual dictates, our *determined action* becomes *wisdom* (7 to 8). To the degree that our will coincides with God's will, we can be said to be wise. We are wise to the degree that spiritual goodness and utility is promoted at each step in the series and produces a loving act for the common good.

Eighth: The cycle completes itself as first principles return to first principles through *action* or *deed. Love of the end in view* becomes the living, organizing agent throughout the whole series of operations. The closing of the circular process of the human intellect is the love of some goal or end that finds its proper equilibrium and corresponding form in some concrete activity and utility. Again, this means that Love, as causal agent, has profound mathematical aptitude, since it harmonizes all mediating causes into comprehensive wholes according to its disposition.

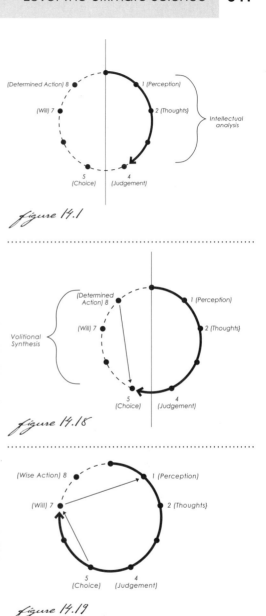

figure 14.1

figure 14.18

figure 14.19

▶ The Mathematical Precision of Religion

The placement of Swedenborg's dynamic progression of the human intellect on the Enneagram seems to be a good fit. We now need to determine if the spiritual evolution of the human heart and mind matches up with the ontological gaps in the process identified by Gurdjieff. This is where new influences must enter the process. According to Swedenborg, religion is God's mathematically precise strategy to fill these causal gaps to allow the human intellect to ascend to wisdom.

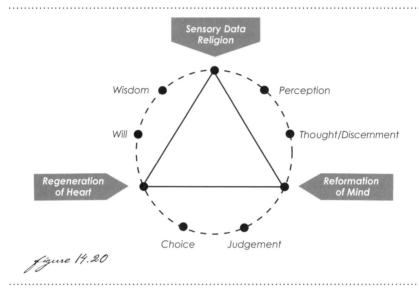

figure 14.20

Religion and its tenets enter the human intellect at point 9. If a person is not exposed to some faith or belief system spiritual evolution is thwarted from the very start. At point 3, human *reasoning* will not advance to spiritual *judgment* unless further supporting data are available and one wishes to be *reformed* by this data. Point 6 represents the uncertainty and vacillation of the human will. It is at point 6 that God works to influence the human heart and regenerate it. But does Scripture support this scheme?

Recall from Chapter Eleven that when Swedenborg's inner mind was opened to the spiritual world, he learned that the Creation story in Genesis symbolically depicts the process in which the human intellect received Divine help for metabolizing information into wisdom. Figure 14.21 shows us a side-by-side comparison between the literal expression of the Creation story and the same story interpreted symbolically—or psycho-spiritually. When God enters our "inner" Enneagram, the process of spiritual evolution begins. God awakens us to make a new evaluation of things. Swedenborg called this spiritual operation the *reformation* of the human intellect and *regeneration* of the human heart.

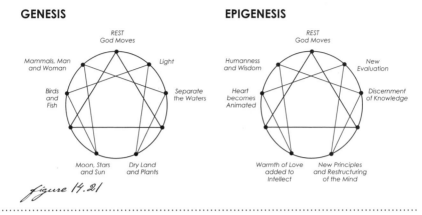

GENESIS

EPIGENESIS

figure 14.21

Spiritual evolution involves the same laws of constancy and contingency that are depicted in the Enneagram. The process of salvation can be thwarted or promoted at each point of the triad where influences enter into the system through mathematically precise gaps. The first step in the progression takes us out of a state of mental obscurity to a new *evaluation*, or "seeing the light." If we reject God from the beginning, when sensory inputs enter the system, religion and its teachings do not enter into our psyche and we remain as we are. If religious tenets do enter the mind, they must then pass through the second gap in the system. Even good ideas play no part in our judgment unless spiritual doctrines have become the foundation or "dry land" of our thinking and produces new thoughts, symbolized by plants. This gap often requires additional study and learning to increase the fertility of the mind and its receptivity to God's lessons. The third gap to cross in spiritual evolution moves our judgments and preferences over to the *will*, where they become true spiritual acts of Love and Wisdom, symbolized by the appearance of warm-blooded animals. Recall that at this interval, point 6 on the Enneagram, things are chaotic; our initial understanding and goal can be deflected at this point. This is where our hearts are really tested.

The ordering of experience in the human intellect and spiritual salvation follow the same rules of self-organization and complexity as everything else: "As above, so below." The Enneagram depicts both a *mathematical intuition of ends* and a graphic of the way order and complexity give us a universal patterning principle *representative of the universe*. The Enneagram is a schematic of the universal science of God's Love, which disposes all things into comprehensive wholes.

Wisdom separates science from religion in the acquisition of knowledge. This is why science does not embrace values. However, Swedenborg believed that science serves an important purpose in God's scheme because the search for knowledge is an important first step that can lead to a search for wisdom.

> "To understand and to be wise are two altogether distinct things, for we may understand and still not be wise; but one leads us to another, namely, science to the cognition of truth (*veri*), and truth (*veritas*) to the cognition of good, and it is the good which is to be sought."[26]

In Swedenborg's worldview, doing "bad" science is not simply straying from the scientific method; it is straying from the inclusion of moral values into our models of reality. Knowledge becomes dead and disjointed if it is not ultimately oriented to the heart. This disjointedness is evidenced by the fact that physicists have embraced contradictory theories (such as relativity and quantum mechanics) and have failed to either produce a unifying equation or identify the first causal principle, which is Love.

> "For what are truths without an ultimate regard for goodness! Or what is the understanding of the mind, or intelligence, unless to know how to choose the Good, prefer the Better, and finally, to will the best."[27]

> "Wherefore to know much and also to understand, and not to be wise, is to rave in the midst of the sciences, or like Tantalus to be surrounded by water, but never to drink a drop; for it is wisdom which completes and crowns intelligence and effects that intelligence may understand."[28]

When Swedenborg wrote these words in *Worship and Love of God*, he was a mere dozen pages away from leaving his career as a scientist and becoming a theologian. The book itself ended abruptly and was left unfinished. He also dropped his goal of formulating a mathematical philosophy of universals. This occurred at the very moment he received his new commission from the Lord and began to focus on the hidden inner meanings of Scripture through the science of correspondences.

While a boon to religion, this was a blow to science. Swedenborg believed that his mathematical philosophy of universals would allow him to submit the ideas of the mind to algebraic calculation and to reduce them "by rules to an equation in the same manner as the calculus of infinities is done."[29] He indicated that he had come close to making the attempt to provide us with an example or two of the mathematics. But it never happened.

Swedenborg admitted that the mathematical philosophy of universals represented the connate science of the soul, *which angels enjoy*, and that he would therefore have to tap into the cognitive functions of higher mind to articulate it. Why did he not make the attempt? The reason is that the development of a universal mathesis and philosophy would lead directly to the science of correspondences.

Because physical process, mental process, and Divine process all correspond, hidden spiritual truths can be intuited without the advanced math that is required to raise ideas and concepts to superior degrees of abstraction. Complex mathematics is unnecessary for the average person to be introduced to the deeper spiritual language contained within the biblical stories. When it comes to revelation, the Lord is pragmatic. In Swedenborg's words, correspondences "more quickly and surely lead us into hidden truths,"[30] whereas one mathematical mistake would send us into many fallacies.

However, in the attempt to unify science and religion, we must consider the powerful notion that a mathematical philosophy of universals really exists and, according to Swedenborg, represents the connate knowledge of the soul. With this mastery of science "the soul at once sees the intrinsic nature of all things set before it."[31] The angelic perception of comprehensive wholes involves a mathematical intuition of octaves, that is, the ability to see things in their proper series. The soul can see at once the seven deflections, nodes, and gravity centers by which all completing processes arrange their order towards a *unanimous conspiring of efficients* or common equilibrium.

A true "Theory of Everything" will show that all things proceed by similar rules; this is the essential feature of holonomic design and a necessary part of causal linkage in top-down process. The Enneagram portrays the science of love and coherence. It has real explanatory and predictive powers. However, since this template portrays the mind of God, it must be protected from every Tom, Dick or Harry who would abuse it. Its information is sacred and cannot become the property of those seeking a naturalistic or materialistic philosophy of reality. As men and women continue to make the proper efforts at spiritual growth, this knowledge will someday become the science of those who will represent the Holy City, The New Jerusalem.

God does not grant premature access to this science of all sciences. Cherubim symbolize the guards that protect Holy things. Cherubim were the sentries in Genesis who guarded the Tree of Life and sat atop the Ark of the Covenant in Exodus. The four creatures depicted in the Enneagram indicated that its knowledge fell under Divine protection. The ultimate science will require a new approach to interpreting Scripture that allows universal and bio-friendly laws to be extracted from its multi-leveled architecture and order. These revelations will be made known to men and women according to the Lord's good pleasure, not academia. This is knowledge that is only obtainable by a spiritual heart.

▶ A Final, Bio-Friendly Theory of Reality

Everything is a series within a series. Swedenborg stated that real human acumen and real science depended on how well a person could see things in their proper series.

> "Consequently, the science of natural things depends on a distinct notion of series and degrees, and of their subordination and coordination. The better a person knows how to arrange into order things which are to be determined into action, so that there may exist a series of effects flowing from their genuine causes, the more perfect is his genius."[32]

Gurdjieff expressed a similar notion. He claimed that a person's understanding of a particular topic could be determined by how well the person could place the specifics of their knowledge around the Enneagram. Anyone who could not place knowledge accurately around the Enneagram did not have real understanding of process. Swedenborg claimed that in our current cognitive state *we see everything but understand nothing.*

Science is nothing if it is not explanatory or predictive. Therefore, if the Enneagram is a universal science based on God's wisdom, it should explain and predict how the Holy Word contains in its structure the patterning principle by which all things proceed by similar rules of coherence and unity. One of the more interesting challenges given to students of Gurdjieff was to place the Lord's Prayer around the Enneagram. Swedenborg might have tackled this challenge from ideas he expressed in a work entitled *A Philosopher's Notebook*.[33]

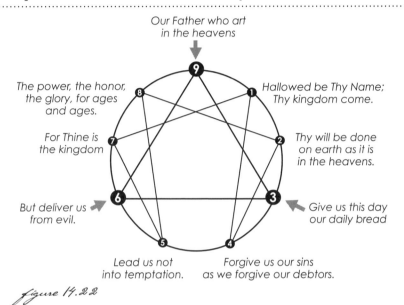

figure 14.22

Swedenborg divided the Lord's Prayer into a chain consisting of *nine* causal links. The way we can gauge the accuracy of this fit is to see how well the Lord's Prayer can help us identify three influences coming from *outside* the process to fill gaps in the process and how well each link in the outer cycle can offer support for the *inner* connections in the series 1-4-2-8-5-7.

The Creator starts the process of salvation at the apex of the triangle as an *outside* influence. Our recognition of God's Holiness and His saving activity (the coming kingdom) puts us at point 1 in the outer cycle. Inwardly, point 1 has reference to point 4, because God's activity is matched by the love and respect we have for others. We move to point 2 when we recognize that in order for God's kingdom to take effect on earth we have to act according to God's will. But to act according to God's will, we need the proper *outside* inputs to fill a gap in the process. Swedenborg's breakdown of the Lord's Prayer places "our daily bread" at point 3 and "feeds" the system with spiritual knowledge and instruction from the outside source of God's revealed tenets. From receiving this doctrinal knowledge we learn to treat others with love and compassion and successfully move to step 4. Point 4 is inwardly connected to point 2 because by loving others we are doing God's will. Point 2 in the outer cycle also has reference to point 8 because doing God's will leads to our alignment with God and our eternal acknowledgment of His power, honor and glory. Point 8 is connected inwardly to point 5 because by giving God dominion His holy power can intervene on our states of temptation and self-love for the purpose of gaining access to the heavenly kingdom (moving from point 5 to point 7). But point 6 is the chaotic interval or "place of uncertainty" in the process where our spiritual challenges and temptations at point 5 can still prevent us from automatically transitioning to the heavenly kingdom (point 7). We need additional external help at this stage. We cannot obtain heaven by ourselves. Sincere humility of the heart is required. "Deliver us from evil" is the point where we ask for God's direct help and mediation to help us safely bridge the gap between temptation and heavenly victory.

The Enneagram predicts that *all* biblical narratives follow this pattern and on all scales. The deeper story of Noah's Ark gives this schematic:

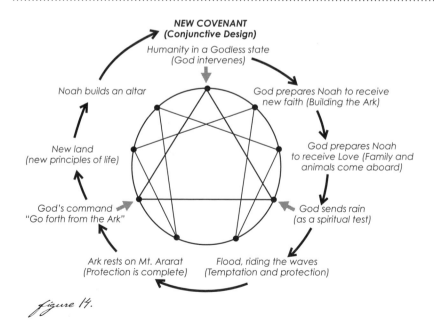

figure 14.

Religion and science are one when it comes to addressing whole-part dynamic process. Religion extends the laws of nature, order, process and complexity into higher, non-physical realms. It is the Divine strategy and lawful contingency by which men and women can overcome personal discontinuity, uncertainty, and hazard in their inner world through Divine order and ideal causality. The Enneagram provides rational and mathematical evidence of God's Infinite Wisdom in reconciling causal determinism with openness to chance in both the cosmic theater and within the human experience. Good and bad things can happen at the ontological gaps within all universal process. Religion fills these gaps by providing proper external inputs at precise points where they are mathematically necessary to promote humankind's continued evolution into a non-physical realm as angelic beings and perfect God's plan for Conjunctive Design.

This cosmic scheme of unified process has its origins in the fountainhead of all law: the Lord God's Infinite Wisdom in governing the universe through the causal agency of Love. A complete model depicting the universal arrangement of knowledge and process into comprehensive wholes can become the property of men and women only through revelation from the Lord. This science has sacred origins.

Swedenborg stated that the Lord is currently establishing a New Church on earth, the New Jerusalem, which will be spearheaded by the reintroduction of the *science of correspondences* to the world. If he was right, a final theory of reality may already be extant in the Enneagram, but simply unrecognized. The Enneagram may well represent the ultimate cultivation of the science of correspondences. It demonstrates the science of correspondences, not simply in static hierarchical terms, but in the context of dynamic, bio-centric processes and their profound interrelations.

Nature does not have this intelligence or foresight. Matter does not possess a *mathematical intuition of ends* in which unified complexity emerges from the *subordination and coordination of things through successive and simultaneous order*. God does. The Enneagram contains an image of God's Love. It is a schematic that shows that existence is subsistence and that being is becoming.

Concerning whole-part process, Swedenborg said "agreement of all these things cannot possibly exist without a spiritual

principle of union, or Love."[34] All conjunction is through reciprocation. Divine Love, as the ultimate formative substance, not
only produces all form, it also produces *forms of uses*, which
are its *ends*. This Divine Order governs both universally, and
in things singularly, through effigy and similitude. In this way
"Nature everywhere observes her rules."[35]

Perfection of life is the perfection of Love and Wisdom,
since both creation and evolution progress into higher-level
structures as physical forms become more perfect recipients of
God's love and wisdom. God cannot flow into, dwell within,
or sustain any form that does not support some Divine quality. This results in a universe consisting of holistic utility and
serviceability.* The perfection of conjunction between Creator
and creation is what drives complexity. Ultimately, complexity
more perfectly expresses first principles by
producing a being capable of conscious Love.

> ✱ *Swedenborg would
> not view the second law of
> thermodynamics simply
> as systems seeking out
> their most stable state,
> but that systems find their
> stability in utility and
> forms of use.*

"That the world subsists by the same
series by which it exists; and, in respect
to its subsistence and existence, has
perpetual relation to its first principle.
That the more perfectly the world exists
and subsists, the more easily may it
refer itself to its first principle..."[36]

This idea of first principles returning to first principles
through Conjunctive Design can be shown by a circular
diagram depicting God's order of creation from involution
(Spiritual Love to inert matter) and back again through evolution (inert matter to Spiritual Love) through human spiritual
development.

The more perfectly that distinct things become unified,
the more perfectly the emerging complexity puts first principles on display. The human race has been given the capacity
to perfectly reflect first principles through the evolution of
mind and spiritual complexity. The Divine goal to be reached
through human evolution is *Spiritual Love*, and Spiritual Love
is the first unifying principle of the universe. "Togetherness-in-
distinctiveness" finds its ultimate living expression in all the
unique ways humans can show love of the neighbor by being
of service to each other.

As Figure 14.24 shows, the directionality of evolutionary
complexity is oriented towards God's Infinite Love and Wisdom
through the creation of angels from the human race. This

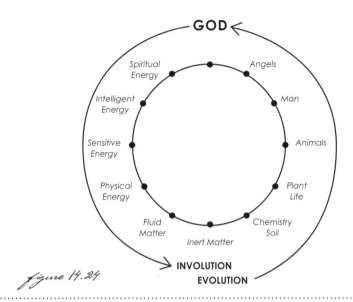

figure 14.24

increased receptivity of forms to Divine influence is why there is a one-way-ness to time, that is, the directionality of God's perpetual endeavor to create a heavenly kingdom.

> *"This is the work of science and wisdom.* It is the part of science to arrange the means in an orderly manner. It is the part of wisdom to direct those means to what is best. This could not be done without Divine influx."[37]

The Enneagram portrays the "work of science and wisdom" Swedenborg alluded to.

The Enneagram as a model for a universal science, offers additional interpretations that can depict involution and evolution in different ways from the illustration above. It arranges things not simply according to their order and progression in time, but also according to where they fit into the Divine organic scheme of cosmic utility and usefulness through unity. The universe is interactive. Every time a new life form appears, so do new relationships. Evolution not only allows God to operate more intimately with creation, but it allows everything in creation to become more organically connected through *need* and *dependency*. Plants need the soil, animals depend on plants, mankind depends on minerals, plants, animals, and information, and angels in heaven depend on humans acquiring the right kind of *psychical* food, that is, doctrinal information concerning God. The Enneagram, if completely understood, presents us with a bio-centric schematic of the way all need, dependency, and serviceability are coordinated and subordinated in the universe by the Lord God into Divine order.

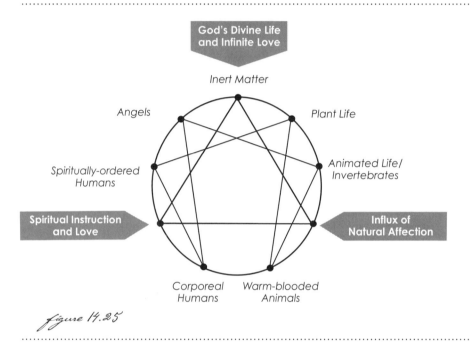

figure 14.25

The Enneagram in Figure 14.25 falls short of showing all these relationships—relationships that would depict different ratios of octaves and triads and the fact that enneagrams on different scales are entangled with each other in a grand organic web of undivided wholeness. But note that human potential occupies point 6 on the ascending process of the Enneagram. This is where the process is chaotic. The human race occupies the spot in the grand cosmic scheme where things are open to novelty, accident and self-correction. This explains the human challenge of free will. Humans can evolve further at this point or fail. Religion is God's strategy to bridge this ontological gap in the Divine scheme of Conjunctive Design.

Scientists know the world is fine-tuned and unified, but not in the mathematically precise way we have been discussing. The next great upheaval in science will address this precision. Evolution is spiritual and perfects holistic interconnectedness, interrelationship, and interdependence by filling nature's ontological gaps with new orders of complexity and utility. Complexity emerges through these gaps with mathematical precision as novel organic forms and new directions of process. Complexity serves God's purpose through the exaltation of Love.

As a scientific model, the Enneagram has both explanatory and predictive qualities. The full power of its knowledge will have to be paid for, not with money, but with spiritual equity.

▶ Eternal Happiness

The universe is goal driven. God's ultimate goal is to create a heaven from the human race. God has given the human race a mind that perpetually strives to seek happiness beyond the Neo-Darwinian synthesis of physical reproductive fitness and gene preservation. Seeking happiness impels the human mind to acquire knowledge that best promotes our desire or love—the knowledge we feel will lead us to the most happiness. This knowledge becomes our *truth*, and what we love becomes our *good*.

Doing what we believe to be "good" is the goal of our acquisition of knowledge. But there is an ontological gap between knowledge and wisdom, beween self-serving good and blessedness. Religion seeks to bridge this gap.

The purpose of religion is to help us to choose what is "best." Doing what is best is wisdom. Doing "good" from a spiritual principle conjoins us to God and allows us to escape the tyranny of time and entropy. When we are conjoined to God, we create an inner organic complexity capable of receiving eternal happiness.

Wisdom is the arrangement of knowledge into higher-level organic structures where the spirit gains the apparatus for more noble utility. Conjunctive Design allows for utility and higher-level structure to lawfully ascend into the non-local and non-temporal realms of the spiritual world. But the mind has to be arranged according to spiritual values or it will be discordant within the heavenly sphere of eternal life and expelled.

"Knowledge has no other purpose than that you may become spiritual through it."[38]

The ontological gap between knowledge and wisdom in human experience is evidenced by the fact that Love is not treated as a hard or an exact science. Love, affection, and emotion not only focuses our attention, they weave the very fabric of our inner being into a real spiritual body. Nothing could be more exact when it comes to the mathematical ordering of process into forms of uses, in this world or the next. Love, as the creative causal agent, seeks form through a mathematical intuition of ends, which is the power of Truth advancing from Good (Ezekiel's wheels).

If Love did not have this *a priori* intelligence, human cognition could not be subordinated and coordinated into distinct functions of *perception, thought, judgment, choice, will, determination* and *action* (an Enneagram). Nor could they act in concord with and conspire to promote the aspirations of the spirit. There are harmonies in all unities.

Love creates geometrical space and perpetually modulates the arrow of time and flow of dynamical process towards the Divine goal. Again, inert matter has no such intelligence or foresight for creating unity through increased distinctiveness.

Because God is simultaneously the Alpha and Omega, all process flows in a circular series in which beginnings and last things have perpetual reference to each other. Swedenborg said that there is nothing in the universe that is not a subordinated and coordinated series in itself or a part of some greater series. This harmonious unity of all things in nature coincides with love, wholeness and "transcendental goodness."

Love, first seeks form through Wisdom, then final completion through *Use*.

Love is the ultimate formative substance. It can take no other form or stay true to itself except through uniting things in use. The created universe, when viewed from the perspective of uses, is an image of God's Infinite Love. When men and women adopt Spiritual Love in their lives, dynamical process completes the grand circle of creation, and first principles return to first principles. This transcendent circle is the ultimate quantum of action. It moves from God's Love in heaven and returns by men and women reciprocating.

Humans play a critical role in maintaining the continuity of universal process through the perfection of reciprocal conjunction with God. This occurs through the exaltation of Love by means of enlightened usefulness and serviceability. Eternal life and happiness in heaven is eternal usefulness. In heaven, spiritual usefulness is immediately turned into correspondingly beautiful psycho-landscapes that reflect all the qualities of our love, wisdom and innermost happiness. Heaven is something we become.

Love is more than a striving, disposition or propensity. It is the ultimate formative substance in the universe. It has intrinsic mathematical aptitude because its ultimate embodiment is God's Divine Wisdom. Love is a vital force representative of the universe as unified process. Love is the ultimate reality as well as the ultimate science.

The ultimate science will glorify God, not man.

Appendix

These notable individuals were all influenced by the writings and ideas of Emanuel Swedenborg (1688-1772).*

In Europe:

Honore de Balzac
Charles Pierre Baudelaire
Ernst Benz
William Blake
Robert Browning &
   Elizabeth Barrett Browning
Thomas Carlyle
Samuel Taylor Coleridge
Herni Corbin
Fyodor Dostoevsky
Johann Wolfgang Goethe
Victor Hugo
Carl Jung
Immanuel Kant
Soren Kierkegaard
Joseph Sheridan Le Fanu
Coventry Patmore
George MacDonald
Czeslaw Milosz
Oscar Milosz
Elizabeth Stewart Phelps
John Ruskin
George Sand
Friedrich Schelling
August Strindberg
Alfred Lord Tennyson
James John Garth Wilkinson
William Butler Yeats

In Japan:

T. K. Suzuki

In The United States:

Bronson Alcott
Henry Ward Beecher
Daniel Hudson Burnham
Andrew Carnegie
John Chapman
   (Johnny Appleseed)
Lydia Maria Child
Ralph Waldo Emerson
Benjamin Franklin
Robert Frost
Margaret Fuller
Edward Everett Hale
Walter Marshall Horton
Henry James, Sr.
   (father of William James)
Helen Keller
Rev. Dr. Martin Luther King, Jr.
Abraham Lincoln
Edwin Markham
Raymond Moody
Dr. Mehmet Oz
Edgar Allen Poe
Robert Ripley (*Believe It or Not*)
Harriet Beecher Stowe
Henry Thoreau
Wilson Van Dusen
Walt Whitman

In Latin America:

Jorge Luis Borges

✳   *Emerson called Emanuel Swedenborg the "Aristotle of the North."*
   The well-known Buddhist scholar T. K. Suzuki called him "the Buddha
   of the North," in his book of the same title. The famous nineteenth
   century preacher, Henry Ward Beecher said, "Nobody was educated
   who had not read Swedenborg." And, as fuel for speculation, it is
   known that George Washington had received books by Swedenborg
   and that Thomas Jefferson's secretary was a Swedenborgian.

GLOSSARY

Active information. Science is beginning to view the fundamental universe as consisting of flows or fluxes of information. All information represents the measure and form that action takes, making it qualitatively and quantitatively knowable. Swedenborg proposed that active information had a knowable source. See influx.

Arrow of time. A phrase used to address the asymmetry or "one-wayness" of time. Swedenborg described time as universal process and whose directionality was cyclic. In such a paradigm it would be impossible to travel backwards in time because it would unravel God's influx and the sanctity and power of goodness advancing through uses.

Big Bang theory. That the universe began from an infinitely compressed state with a zero radius. This state of infinite compression of the prenatal universe is called an essential singularity. The universe violently exploded out from this original zero-dimensional state.

Biocentric universe. The idea that the universe is ordered and arranged in an organic pattern.

Bounded infinity. An infinite number of points can occupy the circumference of a circle, which has a finite radius. The circle therefore represents how an infinite principle can reside in a finite or bounded form. A spiral gives us a form superior to a circle and therefore represents a new analysis of infinity.

Causal nexus. The nexus by which an Infinite God can act in a finite world is through correspondences and similitude. God can only create things which are finite (otherwise God would be recreating Himself). And God can only create finite things so far as they can receive and reflect the various qualities of Divine nature. This allows God to be conjoined to creation and be intimately involved with its governance. God does not create a ratio from the finite but from the infinite, which determines how the infinite resides in and forms relationships with finite things. See top-down causality.

Circle of life. Swedenborg's view that all coherent and holistic process occurring on all scales is cyclic. The process of creation and evolution represents a transcendental circle in which the spiritual dynamics of God's love descends into the world of inert matter then ascends through the evolution of intelligence and increased consciousness in humans who can manifest spiritual love and form an eternal relationship with God. The interconnectedness and unity of all things in the universe is the result of first principles returning to first principles. See Conjunctive Design.

Collapse of the wavefunction. It is assumed that the distribution of probability waves of a quantum event collapse or somehow "jump" into an actual outcome from the crapshoot of quantum indeterminacy. Oddly, this discontinuous jump is not expressed by the Schrodinger equation, which only expresses a smooth variation of the wavefunction through time and the probability of a quantum entity showing up at a location.

Conjunctive Design (CD). Swedenborg's model of purposeful creation by God that goes beyond the current theories of Theistic Evolution or Intelligent Design. Evolution is spiritual and perfects the Creator's relationship with the created. The final product of evolution is a heaven from the human race.

Cosmic background radiation. The thermal after-affects of the Big Bang explosion. The existence of this background radiation throughout the universe is seen as the strongest evidence for supporting the Big Bang theory of creation other than the fact that the universe is still expanding.

Cosmological anthropic principle. The idea that the universe and its laws are "fine-tuned" for the possibility of life, including intelligent life. The laws of the universe are bio-friendly. Some scientists see this as evidence of a purposeful designer while others hold that this is just a chance outcome.

Curvature of space. In Einstein's relativity theory space is essentially flat but the presence of energy or matter causes space to bend and warp, causing gravity.

Dark energy. A theory that an invisible force is overcoming the force of gravity and allowing the universe not only to continue to expand but to accelerate its expansion.

Dark matter. A theory that suggests there is a new kind of matter, not made of atoms, that exists in the universe and gives it enough mass for gravity to keep galaxies of stars together.

Diesm. The theological notion that after God created the world it was left to run own its own without any further Divine commitment. This model makes it impossible to determine first principles and agency in the universe. However, it spares God from being responsible for evil in the world—moral or natural. See theism.

Discrete cognitive function. The operations of sensing, imagining, and reasoning are not continuous but discrete cognitive functions. Each operates on a different level or substrate of organic structure in the brain.

Doctrine of Forms. Swedenborg's unique theory of the discrete structure of space and time. Not only can the structure of spacetime be quantified into distinct units but also into units with different qualities. These different types of spaces give us the hierarchical sandwiching of multi-dimensional space. Intelligence is intrinsic to this model of reality and its extra dimensions include mental realms and the spiritual realm of God. See principle of action.

Dynamical process. Modern science is increasingly embracing the view that the universe is process. The kinetic universe is ordered and has orientation through the cyclic process of first principles returning to first principles. If we embrace the idea that the universe had a beginning and that time and space is an outcome of creation, then the laws and dynamics that govern all process in creating our unified universe must have their origins in a pregeometric world. Science cannot answer foundational questions with mere physical theories. The biggest challenge of modern science for the 21st century will be to grasp how action and process can take place outside of time and space.

Enneagram. An ancient symbol discovered by George I. Gurdjieff in Central Asia. The author believes that it is a schematic representing an idea similar to Swedenborg's Circle of Life, his mathematical philosophy of universals and the intelligent structure of reality based on the Word of God.

Epigenesis. In Swedenborg's terms, epigenesis is the spiritual re-creation of a person. This spiritual process is symbolically represented in the seven-day creation story of Genesis in Scripture.

Epiphenomenon. A term used to suggest that consciousness is not fundamental to the universe but is an outcome of some prior dynamic or interaction of matter especially subatomic particles.

Eschatology. The theological doctrine of the "end times."

Exaltation of Love. The theological purpose for evolution and increased complexity in the world leading to the emergence of the human race. What a person loves affects and organizes the non-physical biostructure of one's spirit. Religion is a guide to help men and women choose nobler loves.

Force representative of the universe. Swedenborg's phrase for a universal force that creates all things and processes on all scales through the same holonomic and holistic laws of order. This is a holographic theory, which states that each created thing contains a reflection of everything else.

Field-entity. In Swedenborg's model, particals of matter are not simply disturbances of some larger field but are miniature kinetic fields themselves.

Glorification. This was the process by which the Lord united his human essence with his Divine essence (Jehovah) while living in the world. This process of glorification can be revealed to men and women who grasp the higher symbolic meanings of the Holy Word. The glorification was the means by which the Word was made flesh, saving the human race from extinction.

God's Holy Word. A multi-dimensional document that was communicated to humankind from heaven through Divinely inspired visions. Beyond the literal meaning of the words in Scripture, higher levels of meaning can be distilled. These levels correspond to the multi-dimensional architecture and top-down causal processes of the universe. God's Holy Word contains the patterning principles and template for creating the physical universe and all things in it. Scripture unifies science and theology when these higher levels are taken into account.

Grand Human (Maximus Homo). The spiritual world and heaven represent communities and societies of humans that, when looked at from the point of view of mutual love and utility, are arranged in the same organic order as we see with the unanimous cooperation of all the organs and tissues of the human anatomy. This heavenly structure is an image of God's love and wisdom. God is the origin of humanness.

Gravity transition. Swedenborg's unique concept that gravity is not only variable but can "jump" into discrete realms and manifest entirely new qualities. Both the process of creation and spiritual re-creation involves forces jumping into discontinuous equilibrium end states representing all the different kinds of organized structure in the universe, including the bio-complexity of angels. The hierarchical universe consists of distinct gravitational orders. Swedenborg would view the modern idea of the "collapse of the wavefunction" as a gravity transition.

Heaven. The Lord's kingdom of mutual love. Heaven is not a place for loving people to go. It is a non-physical state we enter through spiritual choices. Therefore its landscape, trees and climate are all reflections of the qualities of one's heart and mind, which consist of spiritual substances. See Grand Human.

Heavenly cybernetics. God's continuous efforts to make adjustments to the human psyche and religious doctrine in order to keep the door to salvation open.

Hell. The spiritual environment that human souls find themselves in who prefer self-love and ego-centeredness over loving God and the neighbor. In such a realm where everyone loves only themselves, they create their own living hell. Evil is its own punishment.

Hierarchical universe. The idea that reality consists of a layering of discrete ontological dimensions and orders. The higher orders are prior to and form the lower orders—from heaven to earth.

Holotropic states. A term coined by Stanislav Grof to represent non-ordinary and profound states of human consciousness. This is a state that mystics and prophets tap into. Swedenborg claimed to have been given this ability from the Lord, which he experienced for 29 years.

Hypnotism. A unique result from the multi-leveled cognitive function of the human mind and brain. Some cognitive functions can be intensified while others are put to sleep, leading to various forms of consciousness and trance. Each state of mind comes with its own special blood flow. Irresponsibility is a form of hypnosis.

Implicate orders. A theory put forth by David Bohm that there are higher levels of order that are entwined in and give law to lower visible (explicate) orders. This idea fits well with Swedenborg's concept of discrete degrees.

Inflationary model. A theory that attempts to explain why the universe seems so equally spread out from a Big Bang beginning. It proposes that for a short period of time the early universe expanded faster than the speed of light.

Influx. God's influence and spiritual energy flowing into recipient forms. All forms in the universe are recipient forms and terminals representing the different metrics and vicissitudes of God's Divine nature. See causal nexus.

Interior sensories. Subtle organs of sensing in the deepest biostructures of the neuron which produce the higher cognitive functions of imagination and reasoning. See discrete cognitive function.

Jacob's Ladder. Symbolic code for God's theological and multi-dimensional model of causal processes in the universe.

Love. The ultimate and foundational substance of reality. All law, force, energy and form are derived from Love's disposition to become a concrete reality. Love is the *living* causal agent and first principle of the universe. See non-physical extension and spiritual substance.

Mathematical philosophy of universals. This is the profound and holistic science of the soul. The soul has a full mastery of all scientific law. A theory of everything represents the connate knowledge of the soul which simultaneously perceives the intrinsic nature and order of all things set before it. See mathematical intuition of ends.

Mathematical intuition of ends. The conscious and intrinsic intelligence of God's universe to unfold its powers in a way that is purposeful and mathematically precise—from non-material first principles to the ultimate effects in the visible universe, including the arrival of the human race. See fine-tuned universe.

Measurement problem. Scientists have yet to understand how the quantum world of potentials and possibilities lead to observable outcomes with different specific results during experimentation made at different times.

Metaphysical digestion. The human brain and mind metabolizes information in a way that corresponds to the operations of the stomach and intestines. This process allows humans to internalize the world and create a spiritual reality and subtle body that can survive the death of the physical body.

Morphogenetic field. A concept from the early 20th century that is similar to Swedenborg's depiction of the Spiritual Sun whose substance is pure Love and represents the ultimate formative field of the universe from which all things emerge and exist.

Multi-dimensional exegesis. A phrase used to express Swedenborg's idea that Holy Scripture can be interpreted on several levels. The higher levels contain more expanded meanings revealing deeper revelations about the Lord God and His grand scheme for the human race. This layering of meaning also provides the patterning principle for the multi-dimensional order of the universe and its top-down causality.

Multi-dimensional space. The concept of Multi-dimensional space was formulated to help reconcile the differences between relativity theory and quantum theory. It assumes that the universe consists of more than Einstein's notion of three spacial dimensions and one time dimension (also called hyperspace). These extra dimensions are created by twisting and bending the idea of continuous space into complicated micro-pretzels. While these extra dimensions cannot be scientifically proven as ontologically real, they provide scientists with more mathematical flexibility to unify the laws of physics. Swedenborg's view of multi-dimensional space does not bend or warp the idea of continuous space into new configurations, but rather describes further dimensions as discree units of qualitatively different kinds of spaces. Therefore, Swedenborg suggests there can be physical spaces, mental spaces and Divine spaces.

Multi-leveled invariance theory. Swedenborg's notion that there are constants of law on different levels of reality, from the physical to the mental and even to spiritual dynamics. Each of these levels describe actions that are distinct and invariant, relative to each other.

Neural substrate. The fundamental framework of neural structure. Swedenborg suggests that the neuron itself has a deeper substrate such as its own superior or simple cortex. Modern neuroscience is beginning to suspect that there are deeper substructures in the neuron to account for the higher cognitive functions of the human intellect.

Noncommutativity. The idea that ab - ba (changing the order of the letters "a" and "b") equals more than zero. Philosopher and inventor Arthur Young interpreted noncommutativity in quantum physics as suggesting there is more to physical process than meets the eye, reflecting a deeper, non-material dynamic. Professor of Philosophy Michael Heller believes that fundamental physics is based on a noncommutative realm of "pregeometry"—a real dynamical world removed from the physical ideas of time and space.

Non-locality. The phenomenon that shows that all things are instantaneously connected on the smallest scales of reality. Science views this as physical signals being transferred from one place to another faster than the speed of light. Swedenborg understood this concept from the spiritual view that everything in the universe has emerged from one non-physical substance—Love. Love provides us with the ultimate field-theory by which everything is simultaneously connected, non-locally.

Non-physical extension. The metrics or measurement of Love. Love, the primal, non-physical substance of the universe, forms boundaries and topologies in the spiritual world according to the limitations and dimensions of the hearts and minds of its various inhabitants. A person's state of mind has a non-physical parameter. See psycho-topological equivalent.

Ontological or causal gaps. This does not refer to the idea of "God of the gaps," which states that any hole in our knowledge can simply be filled in by some miracle of God. Rather, this book portrays causal gaps as the mathematical consequence of creation proceeding from God through discontinuous steps rather than a smooth continuous process (which is also partially addressed by modern quantum physics). These gaps serve God's plan as the mathematically precise points in which the universe is open to contingency and novelty.

Pregeometry. A non-physical form that contains all possible forms and kinetic outcomes which can exist in time and space. In Swedenborg's Doctrine of Forms, pregeometry includes singularities, spiritual forms and the Divine form or God.

Principle of action. In classical physics the principle of action expresses the ubiquitous law of the universe that everything in nature seeks the most economic route. Interpretations of quantum physics challenge this Newtonian principle, stating that the laws ruling the microworld of atomic and subatomic events are indeterminate and represent happenstance. Swedenborg's model of the universe preserves the law of conservation, that is, the law of action to remove resistance even on smaller scales. In fact, the principle of least action applies to physical action, mental action and Divine action.

Proleptic process. God's eternal goal existing in every single sequential event as if it were "already done." This produces a holistic trajectory. In God's omniscience and providence, the future is always present.

Psycho-topological equivalent. Everything in the physical universe has its equivalent in a spiritual reality. All physical things and processes can portray a deeper level of ontological reality when abstracted from the ideas of time and space. The stories of Scripture can all be interpreted by similar abstraction, which yields deeper revelations concerning faith. See Science of correspondences.

Psycho-topological turf. The inner landscape and ecosystem of the human heart and mind that forms from our values and life choices made on earth. After death this landscape provides the eternal abode and spiritual environment of the individual. We become who we are and live where we belong. Heaven (and hell) is not the same for everyone.

Punctuated equilibrium. A theory that challenges the Darwinian view of evolution as a gradual and continuous process. Instead, it suggests that nature shows that evolution advances in spurts of increased activity.

Purpose of Creation. God's ultimate purpose for the creation and evolution of life in the physical world is to create a heaven from the human race. See Conjunctive Design.

Quantum entanglement. Nothing exists alone. Everything is fundamentally interconnected simultaneously from a non-local reality or universal field. See non-locality.

Quantum gravity. The idea that gravity can be expressed in terms of quantum physics. However, since gravity is spacetime curvature and spacetime structure is continuous rather than discrete, scientists are puzzled as to how curved space can exist as a particle-like "graviton." This is why relativity theory and quantum theory cannot be unified. This predicament also suggests to some physicists that there is a deeper unifying principle that is yet to be discovered (which the author believes Swedenborg had already solved).

Quantum language. Just as phenomena become expanded and non-local within the deepest levels of nature, the words of Scripture also contain deeper narratives with expanded meanings entirely abstracted from things spatial and temporal. Metaphor and symbolism are examples of quantum language.

Religion. God's scientific strategy to help the human race attain eternal life and happiness. Spiritual salvation is an extension of order and law into the non-physical realm of the heart and mind.

Rope-a-Dope. A boxing fighting style commonly associated with Muhammad Ali (who coined the term). The rope-a-dope is performed by a boxer assuming a protected stance, lying against the ropes, and allowing his opponent to hit him, in the hope that the opponent will become tired and make mistakes which the boxer can exploit in a counterattack. In competitive situations other than boxing, rope-a-dope is used to describe strategies in which one party purposely puts itself in what appears to be a losing position, attempting thereby to become the eventual victor. The Lord applied a similar tactic by taking on a human body and allowing the hells to be drawn into a fight they couldn't win.

Science of Correspondences. Everything in the physical world represents and corresponds to something in the spiritual world. For instance, water corresponds to spiritual truth because water cleanses the physical body just as truth cleanses the soul. All the narratives of God's Holy Word contain this correspondence principle within the literal meaning of its words. The knowledge of correspondences represents the secret knowledge and perennial philosophy of all who have sincerely sought God's guidance in spiritual enlightenment.

Scientism. Science based only on the evidence of the senses. Also known as naturalism and physicalism. Since Swedenborg believed that natural law was derived from spiritual law, science could be unified with theological and metaphysical approaches to truth. See Doctrine of Forms.

Second Coming of the Lord. It is not a physical event. It represents new enlightenment from God and the transformation of one's life, based on spiritual principles. This new enlightenment is obtained from gaining access to higher levels of meaning that are contained and can be distilled from the literal words of Holy Scripture. The Lord has chosen to reveal these heavenly secrets to the world through the theological writings of Emanuel Swedenborg.

Singularity. A state in which spacetime is crumpled into a zero-dimensional "point" of infinite curvature and gravity before the birth of the physical universe. It is a boundary between physical and pre-physical realities. According to Swedenborg's spiritually-based model, a singularity is God's Divine action focused at a point and serves as a medium between spiritual and physical realms. Through the singularity, or first natural point, the birth of geometry and birth of nature were the same event.

Spacetime continuum. Einstein's idea that the structure of space is continuous. This is in contradiction to quantum theory, which suggests that the structure of reality is discrete. This contradiction is why general relativity theory, which addresses gravity and the macroworld, resists being unified with quantum physics and phenomena of the microworld. Physicists are uneasy that their model of the universe consists of two major contradictory laws. This is a case where the mysteries of science are no less unreasonable than the mysteries of faith.

Spacetime fabric. See spacetime geometry.

Spacetime geometry. The form (or forms) in which spacetime takes its texture and topology, such as curvature. See Doctrine of Forms.

Spacetime warps. The plasticity of time and space. The geometry of reality can be stretched or contracted according to the speed of objects traveling through it. See curvature of space. See Doctrine of Forms.

Spiritual body. A subtle body that is lawfully formed within one's physical body from the choices one makes in life. Ideas and feelings are real substances and embody who we really are on the spiritual plane. See spiritual substance.

Spiritual substance. Love is the essence of spiritual substance and Truth represents the various forms and measurements that Love takes in the spiritual world.

Spiritual world. A non-spatial, non-physical realm whose topological features, landscapes and cities perfectly correspond to the inner qualities and measure of one's heart and mind. All activity and periods of time in the spiritual world reflect an individual's changing state of mind. See Grand Human and heaven.

String and membrane theory. The theory that the universe contains more than four dimensions of spacetime. By adding extra dimensions scientists gain additional mathematical freedom to unify the laws of the universe. However, these additional dimensions are created out of bending the spacetime continuum in different directions. This procedure, while complicated, never takes one beyond the physical realm. Swedenborg's Doctrine of Forms treats higher dimensions as distinct kinds of "active" space with different dynamical qualities. Therefore, his model of reality includes physical spaces, mental spaces and Divine spaces. Each of these discrete spaces represents the boundaries and parameters of physical, mental or heavenly principles of action. Swedenborg's view of hyperspace is also simpler because unlike string theory, his model does not rely on energetic filaments vibrating on a folded pretzel of spacetime. Instead, he describes both the dynamics and the higher dimensional structure as one and the same thing. The geometry of higher dimensional space is the boundary condition of a principle of action freed up from the physical constraints of classical laws.

Superior cortex. Swedenborg's idea that the neuron has deeper substrates, including its own nervous system and cortex, which are responsible for higher cognitive function in the human intellect, such as imagination and reasoning. The cerebrum is to the neuron as the neuron is to the superior cortex.

Superposition. The idea in quantum theory that a subatomic particle can exist in a mixed state of multiple locations—simultaneously.

Swedenborg, Emanuel. An 18th century scientist/theologian whose scientific theories and systematic theology were more than the products of his own era. He offers unique new insights for the post-modern world, including ideas to unify science and theology.

Teleology. Purposeful creation (which is a no-no in science).

Theodicy. The issue of evil in a world created by a God of Infinite Love, Wisdom, and Potency.

Theological patterning principle. The patterns and orderly arrangement of the universe and its process is a mirror-image of God's Divine character.

Theory of Everything (TOE). In contemporary physics, the ultimate mathematical equation of reality will consist of unifying the four fundamental forces of the physical universe—gravity, electro-magnetic force, weak nuclear force and strong nuclear force. Swedenborg not only challenges the idea that these forces are fundamental but he includes psychology and theology in his grand matematical model of reality. See mathematical philosophy of universals.

Top-down causality. The order of process moving downward in discrete or discontinuous steps from first principles, through causes, and finally taking form in some measurable physical result. God uses this order to create the physical world from spiritual first principles.

Trans-flux equilibrium states. All coherent and organized process consists of forces flowing between various equilibrium end states or gravitational nodes in order to create a common equilibrium representing a unified idea within organized structure. See gravity transition.

Unanimous conspiring of efficients. A phrase used by Swedenborg to express the dynamic process and order by which powers and forces proceed through a coordinated succession of events. This process proceeds from causes into effects, in a constant series from greatest things to least and from least things to the greatest, in a chain of events called the Circle of Life. A rational universe with understandable laws requires that everything in nature can refer itself to a common source and a universal idea. See force representative of the universe.

Uncertainty principle. The dynamics of the microworld cannot be measured with complete accuracy. If you measure the speed of a quantum entity you will lose the certainty of being able to describe where it is and vice versa.

Virtual Particle. An invention by physicists to talk about processes which cannot be observed or described in the usual terms of their energy, momentum and mass. Their action does not create a real or definite measurement outcome.The author believes that Swedenborg's non-dimensional natural points are virtual particles which can only create real outcomes through special unified action.

Wave function. A mathematical quantity representing a quantum state or state of motion in wave mechanics. It is used in the dynamics of solving the Schrodinger equation.

Wave/particle duality. In quantum physics, entities act like both waves and particles. Scientists have no other way to explain the weirdness of the quantum world which suggests that a particle can exist in multiple locations simultaneously, in a state of pure potentiality. Scientists cannot even agree on whether the wave/particle is a real material entity, or a wholly conceptual, non-material entity. See superposition principle.

NOTES

Chapter One
IN THE BEGINNING (WE HAVE A PROBLEM)

1   Davies, Paul. *God & The New Physics*. New York: Simon & Schuster, Inc., 1984, p. 55.

2   Russell, Robert R. Berkeley, CA: classroom presentation at the Graduate Theology Union, 2006.

3   Wolf, Fred Alan. *The Spiritual Universe: One Physicist's Vision of Spirit, Soul, Matter and Self*. Portsmouth, NH: Moment Point Press, 1999, p. 99.

Chapter Two
EMANUEL SWEDENBORG: WHO WAS HE?

1   Sigstedt, Cyriel Odhner. *The Swedenborg Epic: The Life and Works of Emanuel Swedenborg*. New York: Bookman Associates, 1952, p. 340.

2   *Ibid.*, p. 341

3   *Ibid.*, p. 342

4   Swedenborg, Emanuel. *The Animal Kingdom*. Vol.1. Part 1. Translated by James John Garth Wilkinson. London: Swedenborg Scientific Association, 1960, p. 10

5   *The Swedenborg Epic*, p. 185

6   *Ibid.*, p. 211

Chapter Three
TIME, SPACE AND MATTER FROM THE INFINITE

1   *God & The New Physics*, p. 56.

2   *Ibid.*, p. 59.

3   Rucker, Rudy. *Infinity and the Mind: The Science and Philosophy of the Infinite*. New York: Bantam Books, 1983, p. 7

4   *Ibid.*, p. 7-8.

5   *Ibid.*, p. 46

6   Polkinghorne, John. *Science and Creation*. Philadelphia: Templeton Foundation, 2006, p. 98. Origin quote from Rucker, Rudy, *Infinity and the Mind*, pp. 173-4.

7   Bohm, David and F. David Peat. *Science, Order, and Creativity*. London: Routledge, 2000, p. 200.

8   Davies, Paul. *The Mind of God: The Scientific Basis for a Rational World*. New York: Simon & Schuster Inc., 1993, pp. 229-230.

9   Swedenborg, Emanuel. *The Infinite and the Final Cause of Creation.* Translated by James John Garth Wilkinson. Introduction by Lewis Field Hite. London: The Swedenborg Society, 1908, p. xiii.

10 Swedenborg, Emanuel. *The Principia.* Part 1 Chap ll. Translated by the Rev. Augustus Clissold. Bryn Athyn, PA: Reprinted by the Swedenborg Scientific Association, 1976, p. 57.

11 IanThompson.org. Paper on Metaphysics entitled *Propensities–* "*Derivative Dispositions and Multiple Generative Levels.*" I have also had email conversations with theoretical nuclear physicist Ian J. Thompson on the concept of "dressed mass."

12 Smolin, Lee. *The Trouble With Physics: The Rise of String Theory and the Fall of a Science, and What Comes Next.* New York: Mariner Books, 2007, p. 239.

13 *Science, Order, and Creativity*, p. 287.

14 Heller, Michael. "*Generalizations: From Quantum Mechanics to God.*" Russell, Robert John, Phillip Clayton, Kirk Wegter-McNelly, John Polkinghorne, eds. *Quantum Mechanics: Scientific Perspectives on Divine Action.* Vatican City State: Vatican Observatory Publications, 2001, pp. 196-197.

15 *The Trouble With Physics*, p. 241.

16 Greene, Brian. *The Fabric of the Cosmos: Space, Time, and the Texture of Reality.* New York: Vintage Books, 2005, p. 491.

17 Swedenborg, Emanuel. *The True Christian Religion.* Translated by John C. Ager. New York: The Swedenborg Foundation, 1984, p. 87, n. 60.

Chapter Four

LOVE AND QUANTUM GRAVITY

1   Swedenborg, Emanuel. *Divine Love & Wisdom.* Translated by John C. Ager. New York: Swedenborg Foundation, 1971, n. 48.

2   Penrose, Roger. *The Emperor's New Mind: Concerning Computers, Minds, and the Laws of Physics.* Great Britain: Oxford University Press, 1999, p. 475.

3   Young, Arthur M. *The Reflexive Universe: Evolution of Consciousness.* Mill Valley, CA: Robert Briggs Associates, 1984, p. 40.

4   *Principia*, Part 1, Ch. 7, n. 14.

5   Russell, Peter. *Waking Up in Time: Finding Inner Peace in Times of Accelerating Change.* Mt. Shasta, CA: Origin Press, 1998, p. 169.

6   Swedenborg, Emanuel. *The Five Senses.* Translated by Enoch S, Price. Bryn Athyn, PA: Swedenborg Scientific Association, Reprinted in 1976, n. 454.

7   *Quantum Mechanics: Scientific Perspectives on Divine Action*, p. 204.

8   *The Emperor's New Mind: Concerning Computers, Minds and The Laws of Physics*, pp. 480-481.

Chapter Five

THE DOCTRINE OF FORMS

1   Smolin, Lee. *Three Roads to Quantum Gravity*. New York: Basic Books, 2001, p. 217.

2   *The Fabric of the Cosmos: Space, Time, and the Texture of Reality*, pp. 487-488.

3   Swedenborg, Emanuel. *Worship and Love of God*. Translated by Alfred H. Strom and Frank Sewall. Boston: Massachusetts New-Church Union, 1956, p. 220, n. 93 [3].

4   Swedenborg, Emanuel. *The Economy of the Animal Kingdom: The Fibre*. Translated by Alfred Acton. Bryn Athyn, PA: Swedenborg Scientifis Association, 1976, p. 182, n. 264 [a].

5   Berlinski, David. *A Tour of the Calculus*. New York: Vintage Books, 1995, p. 177.

6   Berlinski, David. *Newton's Gift*. New York: Touchstone, 2002, p. 48.

7   Barbour, Ian G. *When Science Meets Religion*. Harper San Francisco, 2000, p. 104.

8   *Science, Order, and Creativity*, p. 122.

9   *The Economy of the Animal Kingdom: The Fibre*, n. 271.

10 Satinover, Jeffrey. *The Quantum Brain: The Search For Freedom And The Next Generation of Man*. New York: John Wiley & Sons, 2001, pp. 56, 158, 171.

11 Swedenborg, Emanuel. *The Cerebrum*. Bryn Athyn, PA: Swedenborg Scientific Association, 1976, n. 409.

12 *Quantum Mechanics: Scientific Perspectives on Divine Action*, p. 203.

13 Swedenborg, Emanuel. *The Spiritual Diary*. Translated by George Bush. New York: Lewis C. Bush, 1849, n. 3484.

14 *The Reflexive Universe*, pp. 229-230.

Chapter Six

LOVE, NEUROSCIENCE AND HIERARCHICAL DESIGN

1   March, Arthur and Ira M. Freeman. *The New World of Physics*. New York: Random House, 1962, p. 142.

2   *The Mind of God*, p. 20.

3   *Ibid.*, p. 24.

4   Hagelin, John. *Manual For A Perfect Government: How to Harness the Laws of Nature to Bring Maximum Success to Governmental Administration*. Fairfield, Iowa: Maharishi University of Management Press, 1998, p. 60.

5   *Ibid.*, p. 51

6   *Ibid.*, p. 52

7   This quote was taken from a lecture given by Stanislav Grof at the International Science and Consiousness Conference in Albequerque, New Mexico, 2003.

8   Watts, Frazier. "Cognitive Neuroscience And Religious
     Consciousness." Russell, Robert John, Nancey Murphy,
     Theo C. Meyering, Michael A. Arbib, eds. *Neuroscience And The
     Person: Scientific Perspectives on Divine Action*. Vatican City State:
     Vatican Observatory Publications, 2002, p. 335.

9   *Emanuel Swedenborg: A Continuing Vision*. Larsen, Robin, Stephen
     Larsen, James F. Lawrence, William Ross Woofenden, eds.New York:
     Swedenborg Foundation, 1988, Introduction by George F. Dole.

10   "Emanuel Swedenborg (1688-1772), Natural Scientist, Neurophysiologist,
     Theologian." JAMA, the Journal of the American Medical Association,
     Vol. 206, No. 4, Oct. 21, 1968, pp. 887-8. Reprinted in *The New
     Philosophy, Vol. LXXII, No. 1,* Jan – March, 1969, pp. 20-4.

11   Fodstad, Harald. "The Neuron Theory." *Stereotactic and Functional
     Neurosurgery*. Vol. 77, No.1-4, 2001, pp. 20-24.

12   Gross, Charles G. "Emanuel Swedenborg: A Neuroscientist Before
     His Time." Reprinted in the *The New Philosophy*, January - June 1999,
     p. 439.

13   *Ibid.*, p. 439.

14   Swedenborg, Emanuel. *De Anima*. London: Newberry 1849.[Sewell
     F. trans. *The Soul or Rational Psychology*, New York: New Church
     1887] Preface.

15   Eiseley, Loren. Class Reader, p. 89.

16   Swedenborg, Emanuel. *Rational Psychology*. Translated by Norbert
     H. Rogers and Alfred Acton. Philadelphia, Pa: Swedenborg Scientific
     Association, 1950, n. 142.

17   *Ibid.*, n. 142.

18   *Ibid.*, n. 145.

19   *Worship and Love of God*, p. 264, footnote [b].

20   *Rational Psychology*, n. 132 (10).

21   *Ibid.*, n. 131 (9).

Chapter Seven

DID NOAH'S ARK TRAVEL ACROSS POSSIBILITY WAVES?

1   Swedenborg, Emanuel. *The Economy of the Animal Kingdom: An
     Introduction To Rational Psychology*. Vol. ll. Translated by the Rev.
     Augustus Clissold. New York: The New Church Press. Reproduced
     by Swedenborg Scientific Association, 1955, n. 626.

2   Swedenborg, Emanuel. *Psychological Transactions: A Hieroglyphic
     Key To Natural And Spiritual Arcana By Way Of Representations
     And Correspondences*. Translated by Alfred Acton. Bryn Athyn: PA:
     Swedenborg Scientific Association, 1984, p. 158, n. 2.

3   *The Animal Kingdom*, p. 451, n. 293, note [u].

4   Swedenborg, Emanuel. *Divine Love & Wisdom*. Translated by John C.
     Ager. New York: The Swedenborg Foundation, 1971, n. 56.

5   *Psychological Transactions*, p. 175, n. 37.

6   Gangadean, Ashok K. *Between Worlds: The Emergence of Global
     Reason*. Peter Lang, 1998, Introduction p. xxiii.

Chapter Eight

WELCOME TO THE SPIRITUAL WORLD

1   *The Economy of the Animal Kingdom.* Vol. 1. Introduction, p. 14, n. 26.

2   *Ibid.,* p. 8.

3   *Ibid.,* p. 11.

4   Trobridge, George. *Swedenborg, Life and Teaching.* New York: Swedenborg Foundation, 1971, p. 96.

5   Swedenborg, Emanuel. *Heaven And Its Wonders And Hell: From Things Heard And Seen.* Translated by John C. Ager. New York: Swedenborg Foundation, 1978, p. 297-8, n. 479.

6   *Ibid.,* p. 299, n. 479 [5].

7   Swedenborg, Emanuel. *The True Christian Religion.* Translated by Wm. C. Dick. Bloomsbury Way, London: The Swedenborg Society, 1950, p. 36, n. 29.

8   *The Five Senses,* p. 263, n. 610 [11].

9   Ellis, George F. R. "Quantum Theory And The Macroscopic World." Russell, Robert John, Philip Clayton, Kirk Wegter-McNelly, John Polkinghorne, eds. *Quantum Mechanics: Scientific Perspectives on Divine Action.* Vatican City State: Vatican Observatory Publications, 2001, p. 277 [5.3].

10   Swedenborg, Emanuel. *Angelic Wisdom about Divine Providence.* Translated by William Frederic Wunsch. New York: Swedenborg Foundation, 1975, p. 321, n. 326 [10].

11   Sheldrake, Rupert. *The Presence Of The Past: Morphic Resonance & The Habits Of Nature.* Rochester, New York: Park Street Press, 1995, p. 303.

12   *Heaven And Its Wonders And Hell,* p. 265-6, n. 434.

13   Swedenborg, Emanuel. *The Delights Of Wisdom Concerning Conjugial Love.* Translated by Alfred Acton. Bloomsbury Way, London: The Swedenborg Society, 1970, p. 10, n. 7 [1].

14   *Ibid.,* p. 11, n. 7 [3].

15   *Divine Love & Wisdom,* p. 43, n. 82.

16   *The True Christian Religion,* p. 857, n. 798 [8].

17   *Ibid.,* p. 857, n. 798 [9].

18   *Ibid.,* p. 128, n. 103 [3] & p. 872, n. 827

19   *Heaven And Its Wonders And Hell,* pp. 270-273, n. 448-450.

Chapter Nine

THE STARTLING SPIRITUAL HISTORY
OF THE HUMAN RACE

1   Swedenborg, Emanuel. *Arcana Coelestia.* Revised by John Faulkner Potts. New York: Swedenborg Foundation, 1984, p. 266, n. 607.

2   Swedenborg, Emanuel. *The Spiritual Diary*. Translated by J. H. Smithson. Newbury, London: 1846, p. 172, n. 530

3   Stringer, Christopher and Clive Gamble. *In Search of the Neanderthals: Solving the Puzzle of Human Origins*. New York: Thames And Hudson, 1993, p. 26.

4   *Arcana Coelestia*, Vol. 2, p. 2, n. 1118.

5   Swedenborg, Emanuel. *The Earths In The Universe*. Boston, Mass: B. A. Whittemore, 1928, p. 32, n. 54.

6   *Ibid.*, p. 33, n. 54.

7   *The Spiritual Diary*, p. 21, n. 3488.

8   *In Search of the Neanderthals*, p. 83.

9   Swedenborg, Emanuel. *The Apocalypse Explained*. Revised by John Whitehead. New York: Swedenborg Foundation, 1976, p. 67, n. 61.

10   *Arcana Coelestia*, Vol. 4, p. 237, n. 3257.

11   *Ibid.*, Vol. 6, p. 92, n. 4326.

12   *Ibid.*, Vol. 1, p. 266, n. 607 [3].

13   *Ibid.*, Vol. 2, pp. 3-4, n. 1120.

14   *Ibid.*, Vol. 1, p. 266, n. 608.

15   *Ibid.*, Vol. 6, p. 93, n. 4326.

16   *Ibid.*, Vol. 1, p. 282, n. 640.

17   *Ibid.*, Vol. 1, p. 267, n. 608.

18   *Ibid.*, Vol. 1, p. 267, n. 609.

19   *Ibid.*, Vol. 5, p. 238, n. 3884.

20   *Ibid.*, Vol. 1, p. 246, n. 644.

21   *Ibid.*, Vol. 1, p. 446, n. 920.

22   *Ibid.*, Vol. 1, p. 447, n. 920.

23   Odhner, Carl Theophilus. *The Golden Age: The Story Of The Most Ancient Church*. Bryn Athyn, PA: The Academy Book Room, 1975, Ch. 9, p. 122.

24   *The Apocalypse Explained*, p. 479, n. 728.

25   *In Search of the Neanderthals*, p. 160.

26   Sandstrom, Erik E. "Adam, Noah and the Stone Age." *The New Philosophy*. Vol. LXXIII, No. 2, April-June, 1975, p. 221.

27   *The True Christian Religion*, Vol.2, p. 875, n. 833[2].

28   *Ibid.*, Vol. 1, p. 316, n. 265.

29   *Ibid.*, Vol. 1, p. 317, n. 265 [3].

30   *The Spiritual Diary*, n. 3315.

31   "Adam, Noah and the Stone Age." *The New Philosophy*. 1975, p. 221.

32   Swedenborg, Emanuel. *Miscellaneous Theological Works: The New Jerusalem and its Heavenly Doctrine*. Translated by John Whitehead, New York: Swedenborg Foundation, 1982, p. 180, n. 276.

33   *The True Christian Religion*, Vol.1, p. 131, n. 105.

34   *Ibid.*, n. 194.

35   *Ibid.*, n. 189.

Chapter Ten

THE SECOND COMING

1   *The True Christian Religion*, Vol. 2. Translated by William C. Dick. London: The Swedenborg Society, 1950, pp. 820-1, n. 760.

2   *Ibid.*, Vol. 1, pp. 173-4, n. 132.

3   *Ibid.*, Vol. 2, p. 826, n. 766.

4   *Ibid.*, Vol. 2, pp. 883-5, n. 846.

5   *Ibid.*, Vol. 2, p. 886, n. 847.

6   *The Delights of Wisdom Concerning Conjugial Love*, n. 63

7   *The True Christian Religion*, Vol. 2, pp. 887-8, n. 851.

8   Russell, Robert John. "Divine Action And Quantum Mechanics: A Fresh Assessment." Russell, Robert John, Philip Clayton, Kirk Wegter-McNelly, John Polkinghorne, eds. *Quantum Mechanics: Scientific Perspectives on Divine Action*. Vatican City State: Vatican Observatory Publications, 2001, p. 323.

9   *The True Christian Religion*, Vol.1, p. 176, n. 134 [2].

10 *Ibid.*, Vol. 1, p. 177, n. 134 [4].

11 *Arcana Coelestia*, Vol. 3, p. 334, n. 2607 [2].

Chapter Eleven

THE SCIENCE OF SALVATION

1   *The True Christian Religion*, Vol. 2, p. 564, n. 498 [3].

2   Peters, Ted and Martinez Hewlett. *Evolution From Creation To New Creation*. Nashville: Abingdon Press, 2003, pp 147, 173. Quoted from Philip Hefner, *The Human Factor: Evolution, Culture, and Religion*. Minneapolis: Fortress Press, 1995, p. 45.

3   *The True Christian Religion*, p. 367, n. 329.

4   *Ibid.*, Vol. 2, p. 600, n. 530.

5   *The Delights of Wisdom Concerning Conjugial Love*, p. 108, n. 115 [3].

6   Swedenborg, Emanuel. *Angelic Wisdom About Divine Providence*. Translated by George F. Dole. Westchester, PA: Swedenborg Foundation, 2003, p. 125, n.120.

7   *Ibid.*, p. 124, n. 118.

Chapter Twelve

THE THEODICY ISSUE

1   *Divine Love & Wisdom*, p. 85, n. 170.

2   *Arcana Coelestia*, Vol. 1, pp. 89-90, n. 205.

3   *Angelic Wisdom About Divine Providence*, p. 266, n. 277a[1].

4   *Ibid.*, p. 266, n. 276[2].

5   *Ibid.*, p. 240, n. 251[3].

6   Swedenborg, Emanuel. *Heaven & Hell.* Translated by George Dole. West Chester, Pennsylvania: Swedenborg Foundation, 1979, pp. 446-7, n. 537[2].

7   *Ibid.*, p. 447, n. 536.

8   *Arcana Coelestia*, Vol. 1, p. 321, n. 715.

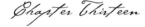

HYPNOSIS: THE COSMIC MANIPULATION OF LOVE

1   Swedenborg, Emanuel. *The Apocalypse Revealed.* Translated by John Whitehead. New York: Swedenborg Foundation, 1984, p. 152, n. 158.

2   *The True Christian Religion*, Vol. 2, p. 651, n. 606.

3   Gurdjieff, George I. *Herald of Coming Good: First Appeal To Contemporary Humanity.* New York: Samueel Weiser, 1974, p. 19.

4   Gurdjieff, George I. *Beelzebub's Tales to His Grandson: An Objectively Impartial Critiscism of the Life of Man.* New York: E. P. Dutton, 1978, Second Book, p. 155.

5   *The Economy of the Animal Kingdom*, Vol. 1, p. 410, n. 449.

6   *Ibid.*, Vol. 1, p. 411, n. 451.

7   *Ibid.*, Vol. 1, pp. 399-400, n. 432.

8   *Ibid.*, Vol. 1, p. 404, n. 440.

9   *Divine Love & Wisdom*, p. 413, n. 415.

10   The True Christian Religion, p. 643, n. 593.

11   Swedenborg, Emanuel. *The Animal Kingdom: Generation.* Translated by Alfred Acton. Reproduced by Swedenborg Scientific Association, 1955, pp. 146-8, n. 153.

12   *Ibid.*, p. 136, n. 145.

13   Swedenborg, Emanuel. *The Infinite And The Final Cause Of Creation.* Translated bt James John Garth Wilkinson. Bloomsbury, London: The Swedenborg Society, 1908, Chap. 2, Part 4, n. 11, p. 213.

14   Swedenborg, Emanuel. *On Tremulation.* Translated by C. Th. Odhner. Reprinted by Swedenborg Scientific Association, Bryn Athyn, Pennsylvania, 1976, p. 40.

15   *The Infinite And The Final Cause Of Creation*, Chap. 2, Part 4, p. 207, n. 9.

16   *On Tremulation*, p. 40.

17   *Ibid.*, p. 51.

18   Swedenborg, Emanuel. *The Cerebrum.* Translated by Alfred Acton. Philadelphia, PA: Swedenborg Scientific Association, 1938, p. 260, n. 417.

19   *The Economy Of The Animal Kingdom: The Fibre*, p. 334, n. 525.

20   *The Infinite And The Final Cause Of Creation*, Chap. 2, Part 4, n. 9, p. 207.

21 *On Tremulation*, p. 28.

22 *The Economy of the Animal Kingdom: The Fibre.* n. 429, 430, 431, 432.

23 *On Tremulation*, p. 6.

24 *Ibid.*, p. 7.

25 *The Economy Of The Animal Kingdom: The Fibre*, p. 285, n. 427.

26 *Beelzebub's Tales to His Grandson, Second Book*, p. 150.

27 *The Five Senses*, n. 16.

28 *Ibid.*, n. 24.

29 *The Economy Of The Animal Kingdom: The Fibre*, n. 454.

30 *Ibid.*, n. 486.

31 *Ibid.*, n. 494.

32 *The Five Senses*, n. 63.

33 *The Economy Of The Animal Kingdom: The Fibre*, n. 537.

34 *Ibid.*, n. 538.

35 Swedenborg, Emanuel. *The True Christian Religion.* Translated by John C. Ager. New York: Swedenborg Foundation, 1984, n. 157.

36 *Arcana Coelestia*, Vol. 2, n. 1982.

37 Balfour, Grant. "Transcranial Magnetic Stimulation: An Introduction." Created for Erowid.org. May 7, 2002.

38 Swedenborg, Emanuel. *Conjugial Love.* Translation revised by Samuel M. Warren. New York: Swedenborg Foundation, 1980, p. 184, n. 171.

39 *The True Christian Religion*, p. 460, n. 365.

40 *Rational Psychology*, p. 295, n. 518.

Chapter Fourteen

LOVE: THE ULTIMATE SCIENCE

1 *The Trouble With Physics*, p. 7.

2 Polkinghorne, John. "Physical Process, Qunatum Events, And Divine Agency." Russell,Robert John, Philip Clayton, Kirk Wegter-McNelly, John Polkinghorne, eds. *Quantum Mechanics: Scientific Perspectives on Divine Action.* Vatican City State: Vatican Observatory Foundation, 2001, p. 183.

3 *Rational Psychology*, p. 319, n. 563.

4 Swedenborg Emanuel. *The True Christian Religion.* Translated by Wm. C. Dick. London: The Swedenborg Society, 1950, n. 756.

5 *Ibid.*, n. 756.

6 *The Economy Of The Animal Kingdom.* Vol.2. "An Introduction To Rational Psychology," p. 25, n. 608.

7 *Psychological Transactions: And Other Posthumous Tracts*, p. 24.

8 *Ibid.*, p. 174, n. 34.

9   Swedenborg, Emanuel. *A Philosopher's Note Book: "Musical Harmony, Sounds."* Translated by Alfred Acton. Philadelphia, PA: Swedenborg Scientific Association, 1931, p. 470.

10 *Divine Love & Wisdom*, n. 228.

11 *Apocalypse Revealed*, n. 10.

12 *Arcana Coelestia*, n. 8400.

13 *Divine Love & Wisdom*, n. 351.

14 *The Animal Kingdom*, Vol. 1, p. 276, n. 203

15 *Worship And Love Of God*, Part Second, p. 234, n. 100.

16 *Ibid.*, p. 227.

17 *Divine Love & Wisdom*, n. 167.

18 *The True Christian Religion*, n. 202.

19 *Arcana Coelestia*, n. 1772.

20 Gurdjieff, George I. *Views from the Real World.* New York: E. P. Dutton, 1976, pp. 195-6.

21 Ouspensky, P. D. *In Search Of The Miraculous.* Harcourt, Inc., 2001, pp. 294-5.

22 *The Animal Kingdom*, Vol. 1, p. 284, n. 208[r].

23 *Worship And Love Of God*, Part Third, n. 122 [footnote b].

24 *The Five Senses*, n. 628.

25 *Worship And Love Of God*, Part Third, n. 122 [footnote b].

26 *The Five Senses*, n. 625-31.

27 *Worship And Love Of God*, Part Third, n. 122 [footnote b].

28 *Ibid.*

29 *Ibid.*

30 *Rational Psychology*, n. 566.

31 *Ibid.*, n. 567.

32 *Ibid.*, n. 563.

33 *The Economy of the Animal Kingdom.* Vol. 2, "An Introduction To Rational Psychology," p. 12, n. 587.

34 *A Philosopher's Note Book.* "The Lord's Prayer or Paternoster." p. 466.

35 *Worship And Love Of God, Part Second*, n. 100.

36 *The Five Senses*, n. 50.

37 *The Principia*, Vol. 2, Part 3, Ch. 2, p. 244, n.2.

38 *The Five Senses*, n. 636.

39 *Arcana Coelestia*, Vol. 1, n. 1100.

Notes on Images and Quotations

Many of the images and charts in this book have been specially created for this project. Others are public domain and still others are used with permission. Attributions for quotations used are also listed here. We thank all contributors.

CHAPTER 2

- Quote from *Believe It or Not - Emanuel Swedenborg*, by Robert L. Ripley September 16, 1934. © Ripley Entertainment, Inc. Used by permission.
- Portrait of Emanuel Swedenborg from the Swedish journal *Svenska Familj-Journalen* (1864-1887).
- Sketch of Swedenborg's Flying Machine from 1714.

CHAPTER 3

- *Ouroboros* by Bryan Derkesen.

CHAPTER 4

- Weighing of the heart scene, with en:Ammit sitting, from the *Egyptian Book of the Dead of Hunefer*. British Museum.

CHAPTER 5

- From the 53rd plate from Ernst Haeckel's *Kunstformen der Natur* (1904), depicting organisms classified as Prosobranchia.
- Two-dimensional hypersurface of the quintic Calabi-Yau three-fold.
- Atomic particles in Bubble Chamber. © CERN.

CHAPTER 6

- "Emanuel Swedenborg (1688-1772) Natural Scientist, Neurophysiologist, Theologian" JAMA, the Journal of the American Medical Association, Vol. 206 No. 4, Oct. 21, 1968. Reprinted by permission.
- "Emanuel Swedenborg: A Neuroscientist Before His Time" by Charles G. Gross, *The Neuroscientist*, Vol. 3, No. 2, March 1997, pp. 142-147. Copyright © 1997 by Sage Publications. Reprinted by permission of SAGE Publications.
- Cave Paintings – Lascaux, France 15,000 – 10,000 B. C.
- *Ishango Bone*, from the Royal Belgian Institute of Natural Sciences, Brussels, Belgium.
- *Venus with a horn*, from Laussel in the Dordogne. Musee des Antiquites Nationales, St. Germain-en-Laye, France.

CHAPTER 10

- Quote from *Pogo*, by Walt Kelly.
  Used by permission of Okefenokee Glee & Perloo, Inc.

CHAPTER 12

- *Lucifer, King of Hell* by Paul Gustave Doré, from Dante's *Divine Inferno*.

CHAPTER 13

- Detail of a page from a *Biblia Pauperum*, 15th century.

Helpful Links

If you would like to learn more about the scientific and spiritual discoveries of Emmanuel Swedenborg, please contact:

The Swedenborg Foundation
swedenborg.com

Swedenborg Scientific Association
thenewphilosophyonline.org
swedenborg-philosophy.org

The Swedenborg Society, London
swedenborg.org.uk

The Online Swedenborgian Library
Swedenborg's Theological, Scientific and Philosophical works
swedenborg.org/library_list

Samara Center for Practical Spirituality
samaracenter.org

TheGodGuy Blog
thegodguy.wordpress.com

Staircase Press
staircasepress.com

Books, articles, sermons and blogs
based on the Writings of Emanuel Swedenborg
swedenborgstudy.com

EDWARD F. SYLVIA, M.T.S.

Philosopher/Theologian Edward F. Sylvia attended the School of Visual Arts in New York and went on to a career writing for some of the world's largest advertising agencies in New York, St. Louis and Chicago.

He now advertises for God. He received his Master of Theological Studies at the Pacific School of Religion in Berkeley, CA and a Certificate of Swedenborgian Studies from the Swedenborgian House of Studies. He is a member of the Center for Theology and the Natural Sciences (CTNS) and the Swedenborg Scientific Association (SSA).

He lives with his wife Susan in a solar house on a ten acre "homestead" just outside St. Louis, in southern Illinois. Edward is committed to natural gardening and has many other projects in the works, including books, articles, lectures and documentaries.

Author of *Sermon From the Compost Pile: Seven Steps Toward Creating An Inner Garden*, his book, *Proving God*, fulfills a continuing vision that God's fingerprints of love can be found everywhere in the manifest universe.

He has been a student of the ideas of both Emanuel Swedenborg and George I. Gurdjieff for over thirty years.

INDEX

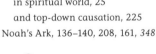

Special Thanks

....

To the team who helped create this book.

Robert Goodman – Editor
silvercat.com

Susan Sylvia – Designer
staircasepressdesign.com

Christine Frank & Associates – Index
christinefrank.com

Stephen Koke, Oliver & Rachel Ohdner
Sarah McMaster – Proofreading